Methods in Molecular Biology

Series Editor
John M. Walker
School of Life Sciences
University of Hertfordshire
Hatfield, Hertfordshire, AL10 9AB, UK

For further volumes:
http://www.springer.com/series/7651

Artificial Riboswitches

Methods and Protocols

Edited by

Atsushi Ogawa

Ehime University, Matsuyama, Ehime, Japan

Editor
Atsushi Ogawa
Ehime University
Matsuyama
Ehime, Japan

ISSN 1064-3745 ISSN 1940-6029 (electronic)
ISBN 978-1-62703-754-9 ISBN 978-1-62703-755-6 (eBook)
DOI 10.1007/978-1-62703-755-6
Springer New York Heidelberg Dordrecht London

Library of Congress Control Number: 2013958315

Printed on acid-free paper

Humana Press is a brand of Springer
Springer is part of Springer Science+Business Media (www.springer.com)

Preface

Ribonucleic acids or RNAs are versatile. mRNAs serve as mediators of hereditary information from DNAs to proteins, tRNAs carry amino acids into the ribosome on mRNA, and rRNAs in the ribosome promote peptide bond formation between the amino acids. In addition, many diverse noncoding RNAs directly regulate gene expression in cis and/or trans. There are also many other types of RNAs known in nature, including RNA enzymes (ribozymes). It is considered that in an ancient RNA world (where proteins and DNAs did not still exist), a greater variety of functional RNAs would have flourished, though almost all of them are now defunct.

RNA functions are dependent on their sequences, which are composed of four kinds of nucleotides: A, G, U, and C. The sequences determine the higher order structures, which in turn make possible their unique functions. In other words, we could create new functional RNAs by appropriately lining up the nucleotides. Although RNAs are not as diverse as proteins, which have 20 amino acids as building block, RNAs have a large advantage in that they are easy to be amplified with the following three steps: reverse transcription, polymerase chain reaction (PCR), and transcription. We can therefore obtain functional RNAs from a randomized RNA pool by repeating selection–amplification cycles (in vitro selection). In fact, many types of functional RNAs have been selected via in vitro selection. The representative examples are aptamers that tightly bind to their specific ligand molecules and ribozymes that catalyze specific biochemical reactions. In vitro selection also enables us to obtain a ligand-responsive ribozyme called aptazyme, which is generally a conjugate between two foregoing functional RNAs, an aptamer and a ribozyme.

A riboswitch is also a functional RNA: a ligand-dependent and cis-acting gene regulator in mRNA. This regulatory RNA is composed of an aptamer domain, which binds to the specific ligand as described above, and an expression platform, which regulates transcription termination, translation initiation, self-cleavage, or splicing of its own mRNA. The interaction between the aptamer domain and the ligand induces conformational changes of the expression platform to up- or down-regulate gene expression. Natural riboswitches acting in response to endogenous metabolites have been identified mainly in bacteria and rarely in eukaryotes over the last decade. In parallel with these discoveries, a variety of methods have also been reported for artificially constructing arbitrary molecule-dependent riboswitches with the corresponding in vitro-selected aptamers or self-cleaving aptazymes.

This volume is mainly focused on the state-of-the-art methods developed in recent years for creating artificial riboswitches. Approximately half of the total chapters are devoted to screening or rational design methods for obtaining artificial riboswitches that function in either bacterial or eukaryotic translation systems. In these methods, an aptamer or an aptazyme is generally required for a gene regulator to acquire ligand responsiveness. Although an already identified aptamer or aptazyme is available, several chapters cover in vitro selection methods for obtaining a new aptamer or aptazyme for a ligand molecule that the reader might be interested in (a small molecule, protein, or photo-responsive molecule). In this context, one chapter is devoted to a computational method for designing a starting library of RNA sequences for in vitro selection. Protocols for evaluating the activities of the resultant riboswitches are also presented. Some other chapters include protocols for construction of ligand-dependent, trans-acting gene regulators.

Artificial riboswitches and other ligand-responsive gene regulators that are obtainable through the cutting-edge methods described here make it possible to switch protein synthesis ON or OFF with arbitrary ligand molecules, which can be freely chosen when selecting the corresponding aptamers to be implemented in the gene regulators. Therefore, this book can be regarded as a collection of recipes for the gene circuit elements in synthetic biology and metabolic engineering. However, I would recommend this cookbook not only to bioengineers who aim to reprogram cell behaviors and molecular biologists who leverage these regulators for genetic studies but also to all researchers who just want to regulate the expression of a specific gene by an arbitrary molecule in various organisms, to detect a specific molecule with reporter protein expression in vitro or in vivo, or to design ligand-dependent RNA switches by using aptamers or aptazymes.

All chapters are written by experts from all around the world who are active in the front lines of the relevant research areas. I would like to express my gratitude to the contributors, all of whom were willing to write their lab protocols in an easily comprehensible manner, despite their busy research lives. I believe that the readers will be able to easily understand and follow the experimental procedures, thanks to the intelligible explanations and notes, and to construct their own artificial riboswitches. Last but not least, I am also grateful to Prof. John Walker for giving me the opportunity to edit this book.

Matsuyama, Ehime, Japan *Atsushi Ogawa*

Contents

Contributors

CHASE L. BEISEL • *Department of Chemical and Biomolecular Engineering, North Carolina State University, Raleigh, NC, USA*

RYAN J. BLOOM • *Department of Bioengineering, Stanford University, Palo Alto, CA, USA*

RONALD R. BREAKER • *Department of Molecular, Cellular and Developmental Biology, Howard Hughes Medical Institute, Yale University, New Haven, CT, USA; Department of Molecular Biophysics and Biochemistry, Yale University, New Haven, CT, USA*

JAMES M. CAROTHERS • *University of California, Berkeley, CA, USA; U.S. Department of Energy Joint BioEnergy Institute, Emeryville, CA, USA*

MOLLY E. CHAPLEAU • *UES, Inc., Dayton, OH, USA*

JORGE L. CHÁVEZ • *UES, Inc., Dayton, OH, USA*

YAROSLAV G. CHUSHAK • *The Henry M. Jackson Foundation, Bethesda, MD, USA*

NEIL DIXON • *Manchester Institute of Biotechnology and School of Chemistry, The University of Manchester, Manchester, UK*

KEI ENDO • *Department of Reprogramming Science, Center for iPS Cell Research and Application, Kyoto University, Kyoto, Japan*

CASEY C. FOWLER • *Department of Biochemistry and Biomedical Sciences, McMaster University, Hamilton, ON, Canada; Department of Chemistry and Chemical Biology, McMaster University, Hamilton, ON, Canada; Michael G. DeGroote Institute for Infectious Disease Research, McMaster University, Hamilton, ON, Canada*

KAZUHIRO FURUKAWA • *Department of Molecular, Cellular and Developmental Biology, Yale University, New Haven, CT, USA*

ZOHAIB GHAZI • *Department of Biochemistry and Biomedical Sciences, McMaster University, Hamilton, ON, Canada; Department of Chemistry and Chemical Biology, McMaster University, Hamilton, ON, Canada; Michael G. DeGroote Institute for Infectious Disease Research, McMaster University, Hamilton, ON, Canada*

JONATHAN A. GOLER • *University of California, Berkeley, CA, USA; U.S. Department of Energy Joint BioEnergy Institute, Emeryville, CA, USA*

HONGZHOU GU • *Department of Molecular, Cellular and Developmental Biology, Howard Hughes Medical Institute, Yale University, New Haven, CT, USA*

MASAKI HAGIHARA • *Graduate School of Science and Technology, Hirosaki University, Hirosaki, Japan*

SVETLANA V. HARBAUGH • *The Henry M. Jackson Foundation, Bethesda, MD, USA*

JÖRG S. HARTIG • *Department of Chemistry, University of Konstanz, Konstanz, Germany*

GOSUKE HAYASHI • *Department of Chemistry and Biotechnology, Research Center for Advanced Science and Technology, The University of Tokyo, Tokyo, Japan*

JIAN-DONG HUANG • *Department of Biochemistry and Shenzhen Institute of Research and Innovation, The University of Hong Kong, Pokfulam, Hong Kong; The Center for Synthetic Biology Engineering Research, Shenzhen Institutes of Advanced Technology, Shenzhen, People's Republic of China*

KOHEI IKE • *Department of Applied Chemistry and Biotechnology, Graduate School of Engineering, Chiba University, Chiba, Japan*

YE JIN • *Department of Biochemistry, Li Ka Shing Faculty of Medicine, University of Hong Kong, Hong Kong SAR, People's Republic of China*
SHUNNICHI KASHIDA • *Department of Reprogramming Science, Center for iPS Cell Research and Application, Kyoto University, Kyoto, Japan*
JAY D. KEASLING • *University of California, Berkeley, CA, USA; U.S. Department of Energy Joint BioEnergy Institute, Emeryville, CA, USA*
NANCY KELLEY-LOUGHNANE • *Human Effectiveness Directorate, 711 Human Performance Wing, Air Force Research Laboratory, Wright-Patterson AFB, OH, USA*
YINGFU LI • *Department of Biochemistry and Biomedical Sciences, McMaster University, Hamilton, ON, Canada; Department of Chemistry and Chemical Biology, McMaster University, Hamilton, ON, Canada; Michael G. DeGroote Institute for Infectious Disease Research, McMaster University, Hamilton, ON, Canada*
LIBING LIU • *Key Laboratory of Organic Solids, Institute of Chemistry, Chinese Academy of Sciences, Beijing, People's Republic of China*
SHEREF S. MANSY • *CIBIO, University of Trento, Mattarello, Italy*
JENNIFER A. MARTIN • *The Henry M. Jackson Foundation, Bethesda, MD, USA*
LAURA MARTINI • *CIBIO, University of Trento, Mattarello, Italy*
JASON MICKLEFIELD • *Manchester Institute of Biotechnology and School of Chemistry, The University of Manchester, Manchester, UK*
ASAKO MURATA • *The Institute of Scientific and Industrial Research, Osaka University, Ibaraki, Japan*
KAZUHIKO NAKATANI • *The Institute of Scientific and Industrial Research, Osaka University, Ibaraki, Japan*
YOKO NOMURA • *Department of Biomedical Engineering, University of California, Davis, Davis, CA, USA*
ATSUSHI OGAWA • *Proteo-Science Center, Ehime University, Matsuyama, Japan*
SHOJI OHUCHI • *Graduate School of Biostudies, Kyoto University, Kyoto, Japan*
CHARLOTTE REHM • *Department of Chemistry, University of Konstanz, Konstanz, Germany*
CHRISTOPHER J. ROBINSON • *Manchester Institute of Biotechnology and School of Chemistry, The University of Manchester, Manchester, UK*
HIROHIDE SAITO • *Department of Reprogramming Science, Center for iPS Cell Research and Application, Kyoto University, Kyoto, Japan*
SHIN-ICHI SATO • *Institute for Integrated Cell-Material Sciences, Kyoto University, Uji, Japan*
CHRISTINA D. SMOLKE • *Department of Bioengineering, Stanford University, Palo Alto, CA, USA*
MORLEY O. STONE • *Human Effectiveness Directorate, 711 Human Performance Wing, Air Force Research Laboratory, Wright-Patterson AFB, OH, USA*
DAISUKE UMENO • *Department of Applied Chemistry and Biotechnology, Graduate School of Engineering, Chiba University, Chiba, Japan*
HELEN A. VINCENT • *Manchester Institute of Biotechnology and School of Chemistry, The University of Manchester, Manchester, UK*
SHU WANG • *Key Laboratory of Organic Solids, Institute of Chemistry, Chinese Academy of Sciences, Beijing, People's Republic of China*
MING-CHENG WU • *Manchester Institute of Biotechnology and School of Chemistry, The University of Manchester, Manchester, UK*
YOHEI YOKOBAYASHI • *Department of Biomedical Engineering, University of California, Davis, Davis, CA, USA*

Chapter 1

Computational Design of RNA Libraries for In Vitro Selection of Aptamers

Yaroslav G. Chushak, Jennifer A. Martin, Jorge L. Chávez, Nancy Kelley-Loughnane, and Morley O. Stone

Abstract

Selection of aptamers that bind a specific ligand usually begins with a random library of RNA sequences, and many aptamers selected from such random pools have a simple stem–loop structure. We present here a computational approach for designing a starting library of RNA sequences with increased formation of complex structural motifs and enhanced affinity to a desired target molecule. Our approach consists of two steps: (1) generation of RNA sequences based on customized patterning of nucleotides with increased probability of forming a base pair and (2) a high-throughput virtual screening of the generated library to select aptamers with binding affinity to a small-molecule target. We developed a set of criteria that allows one to select a sequence with potential binding affinity from a pool of random sequences and designed a protocol for RNA 3D structure prediction. The proposed approach significantly reduces the RNA sequence search space, thus accelerating the experimental screening and selection of high-affinity aptamers.

Key words Screening library design, Patterned sequences, RNA structure prediction, Virtual screening, RNA–small molecule docking

1 Introduction

Aptamers can be selected in vitro using an iterative process called systematic evolution of ligands by exponential enrichment (SELEX) [1, 2]. The SELEX process consists of multiple cycles of selection and amplification: (1) a random pool of RNA molecules is screened and sequences demonstrating binding to a target molecule are separated from non-binders and (2) retained sequences are amplified by the polymerase chain reaction (PCR) to create a pool of sequences for the next round of enrichment. The entire selection process typically requires up to 15 rounds of selection and can take from a few days to a few months to complete [3]. The initial pool generally contains 10^{13}–10^{15} unique sequences designed with two PCR primer regions flanking a random region of 30–50 nucleotides. Most of the sequences present in this initial pool are not functional

Atsushi Ogawa (ed.), *Artificial Riboswitches: Methods and Protocols*, Methods in Molecular Biology, vol. 1111,
DOI 10.1007/978-1-62703-755-6_1, © Springer Science+Business Media New York 2014

and have no potential to function as binders, as the majority has a simple stem–loop structure [4]. The occurrence of more complex, high-affinity aptamers in the starting random library is very low, and therefore, many aptamers selected from such random pools have a simple stem–loop structure. Researchers have investigated various methods of biasing an initial selection pool to lessen prevalence of random or simple structures and to increase the presence of more complex potential binders. This was experimentally performed by Davis and Szostak [5] who selected aptamers from a mixture of fully random and partially structured RNA libraries created by incorporation of an internal loop which is characteristic of many known RNA aptamers. They identified six sequence families of aptamers. Four of these families came from the designed library, while two families were from the random pool. The aptamers with the highest affinity were selected from the designed library of sequences and demonstrated more complex structures, while aptamers from the random pool had a simple stem–loop structure. Thus, increasing the complexity of structures formed in the initial pool (including internal loops, bulges, junctions, and mismatches) gives a higher probability of selecting high-affinity binders [6].

Recently, there have been several attempts to design enhanced initial pools of sequences for selection of high-affinity aptamers. In one approach Ruff et al. [7] designed libraries with varying degrees of predicted average secondary structure by patterning pyrimidine- and purine-rich positions. For the three targets used for individual selection, the patterned library substantially outperformed the random library in two cases and performed at least as well as random in the third case. Kim et al. [8, 9] used 4×4 "nucleotide transition probability" matrices to design and generate pools with an optimal yield of specific motifs. By using a combination of different matrices it is possible to generate pools with different motif distributions characteristic to the target [9]. In another approach, Luo et al. [10] have developed random filtering and genetic filtering methods to increase the number of five-way junctions or to design a uniform structure distribution in the starting RNA/DNA pools. One of the aptamers selected by SELEX from the designed structural libraries displayed higher affinity for ATP than the previously selected low-complexity aptamer, while other aptamers showed weaker affinity.

We present here a computational approach for designing a starting library of RNA sequences with increased formation of complex structural motifs and increased affinity to a desired target molecule. Our approach consists of two steps (Fig. 1): (1) generation of RNA sequences based on customized patterning of nucleotides with increased probability of forming a base pair and (2) a high-throughput virtual screening of the generated library to select

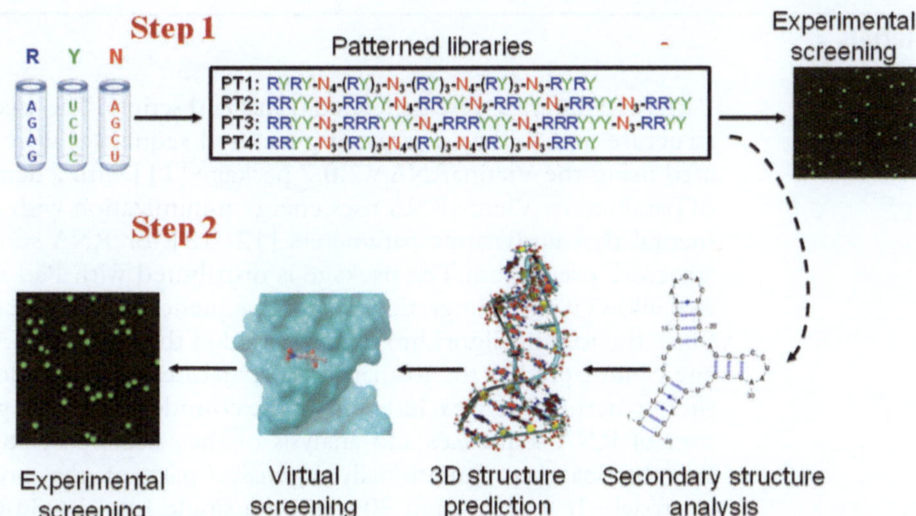

Fig. 1 Generation of initial RNA sequence pool. Step 1: Generate large sequence pool of sequences using patterned libraries with the alternation of purines (R = A, G), pyrimidines (Y = U, C), and random bases (N = A, C, G, U). At this step generated sequences are not target specific and can be used for SELEX-like selection of binding aptamers or can be used in step 2. At step 2 sequences are filtered based on their secondary structure, their 3D structure is generated, and virtual screening is used to identify RNA molecules that bind to a target molecule. Selected RNA molecules that are now specific to the desired target are placed into a pool of sequences for experimental screening and selection of high-affinity aptamers

aptamers with binding affinity to a desired small-molecule target. In step (1), we designed four different patterned libraries (to provide diversity of structures) of RNA sequences that can be used as the combined initial pool for aptamer selection using the SELEX protocol. At this screening level, the selected sequences are not target specific but rather designed to contain complex structures. In the second step, we used computational docking to identify which of the patterned RNA molecules can bind to a specific target ligand. We have developed a set of criteria that allows one to preselect sequences with potential binding affinity from a pool of patterned sequences and developed a protocol for RNA 3D structure prediction and virtual screening. At this point, the selected RNA molecules are specific to the desired target molecule, and they are placed into a pool of sequences for experimental verification and selection of high-affinity aptamers, e.g., using microarray techniques. The developed computational approach allows one to preselect sequences from the initial pool of RNA molecules and effectively leads to a reduction in sequence search space by eliminating sequences with low binding potential before experimental selection and providing starting libraries with higher percentages of potential binders present.

2 Materials

Patterned libraries were generated using Perl scripts. The secondary structure and folding energy of generated sequences were evaluated using the ViennaRNA v.2.0.7 package [11] with a default set of parameters. ViennaRNA uses energy minimization with experimental thermodynamic parameters [12, 13] for RNA secondary structure prediction. The package is distributed with Perl utilities that allows it easy integration with the sequence-generating scripts. Since the folding algorithm implemented in the ViennaRNA package cannot predict the formation of structures with pseudoknots, such structures were excluded from the consideration. The generation of RNA sequences and analysis of their secondary structure are the least computationally intensive part of the proposed approach. It took around 30 min on a single Intel P4 3.06 GHz CPU to generate a patterned library with 10,000 members.

The tertiary structure of generated sequences in the library was predicted using the Rosetta v.3.4 software suite [14]. The Rosetta software suite focuses mainly on the prediction and design of protein structures and protein–protein interactions, but it also addresses aspects of RNA design, protein–ligand, and protein–DNA docking. The prediction of RNA 3D structures consists of two steps [15]: (a) the fragment assembly of RNA molecule based on the simplified low-resolution energy function followed by (b) high-resolution optimization with a full-atom Rosetta energy function.

While most docking methods have been developed for proteins, recent evaluation of AutoDock and DOCK programs has demonstrated their ability to dock compounds to RNA molecules [16]. We used the AutoDock Vina [17] and AutoDockTools (ADT) packages (http://autodock.scripps.edu/downloads) to perform screening of a library of RNA molecules.

Docking tools are usually used to screen a library of small molecules in order to find a ligand that binds to a specific protein or RNA receptor. In our approach, we screened the library of RNA molecules to find receptors with the highest binding affinity to a desired small molecule.

To perform conversion of molecules in different file formats we used OpenBabel package (http://openbabel.org).

3 Methods

The proposed method for generating starting pools of RNA sequences consists of two major steps (Fig. 1). In the first step we used patterned libraries to generate a large pool of RNA sequences.

The using of patterned libraries increases the probability of folding and formation of a more complex secondary structure.

Four different patterned libraries of 40 nt RNA sequences were designed as an initial pool for aptamer selection:

PT1: RYRY-N_4-$(RY)_3$-N_3-$(RY)_3$-N_4-$(RY)_3$-N_3-RYRY
PT2: RRYY-N_3-RRYY-N_4-RRYY-N_2-RRYY-N_4-RRYY-N_3-RRYY
PT3: RRYY-N_3-RRRYYY-N_4-RRRYYY-N_4-RRRYYY-N_3-RRYY
PT4: RRYY-N_3-$(RY)_3$-N_4-$(RY)_3$-N_4-$(RY)_3$-N_3-RRYY

R = (A, G), Y = (U, C), and N = (A, C, G, U).

The first library (PT1) has a pattern similar to one proposed by Ruff et al. [7]. It consists of alternating purines (R = A, G) and pyrimidines (Y = U, C) separated by completely random regions of N_4 and N_3. The second pattern (PT2) was designed to maximize the number of three-way junctions as proposed by Luo et al. [10]. Pattern PT1 has 14 completely random bases and 1.8×10^{16} possible sequences, while pattern PT2 has 16 random bases and 7.2×10^{16} possible sequences. Patterns PT3 and PT4 were designed with the same number of random bases, 14, but PT3 allows placing three consecutive purines or pyrimidines in a row, while PT4 forces alternation of purines and pyrimidines. Pools of RNA sequences generated with the proposed patterns are not target specific and can be used for experimental aptamer selection using SELEX or further analyzed in step 2.

In the second step we use virtual screening to identify RNA molecules that bind to a specific target molecule.

3.1 RNA Secondary Structure Analysis

Analysis of experimentally selected aptamers shows a significant correlation between the dissociation constant and the free energy of secondary structure formation. Based on the literature, free energy of aptamers is significantly lower than the median same-length random sequence value [18]. Figure 2 shows the distribution of folding energies for 10,000 randomly chosen members of patterned libraries in comparison with the random N_{40} library. The mean folding energy for random library $\Delta G_{\mathrm{mean}} = -6.3$ kcal/mol, and the standard deviation SD = 3.2 kcal/mol. The mean energy for pattern PT1 is −8.98 kcal/mol with standard deviation of 3.55 kcal/mol, for pattern PT2 −7.85 kcal/mol and 3.8 kcal/mol, and for pattern PT3 −8.03 kcal/mol and 3.8 kcal/mol, respectively. The cutoff energy for sequence selection was chosen as $\Delta G_{\mathrm{mean}} - \mathrm{SD} = -9.5$ kcal/mol based on values for the random sequences which is higher than the mean energy value for the patterned sequences.

Next, we analyzed the interaction of nucleic acids with ligands for several aptamer–ligand complexes based on their experimental three-dimensional structures [19]. It was found that RNA bases involved in molecular recognition do not form Watson–Crick pairs with other bases. There are two possible reasons for this: first,

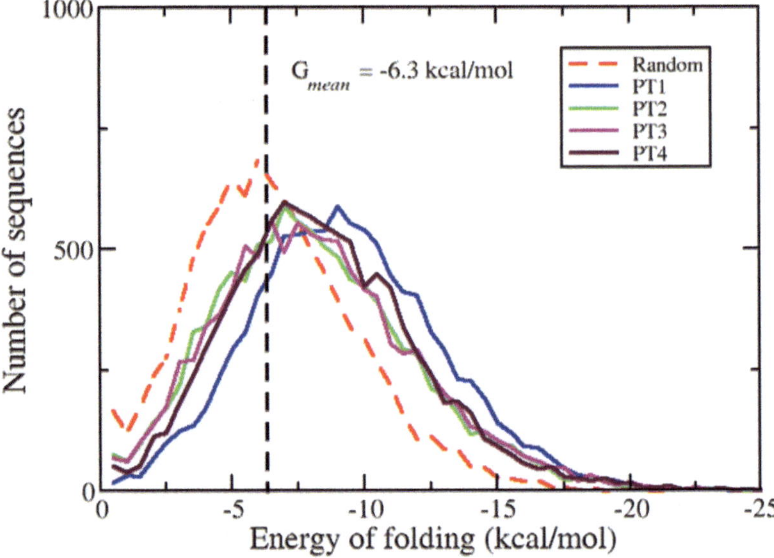

Fig. 2 Distribution of folding energies for 10,000 randomly chosen members of patterned libraries in comparison with the random N_{40} library. The mean folding energy for random library $\Delta G_{\text{mean}} = -6.3$ kcal/mol and the standard deviation SD = 3.2 kcal/mol. The mean energy for pattern PT1 is −8.98 kcal/mol with standard deviation of 3.55 kcal/mol, for pattern PT2 −7.85 kcal/mol and 3.8 kcal/mol, and for pattern PT3 −8.03 kcal/mol and 3.8 kcal/mol, respectively

unpaired RNA bases are more flexible, so they can easily change their conformation to form a binding pocket and accommodate a ligand; secondly, unpaired bases have available donor/acceptor atoms for potential formation of hydrogen bonds with the ligand. Therefore, we set a constraint that the secondary structure of our sequences with 40 bases should have at least 11 unpaired bases. This number seemed optimal, since a higher number of unpaired bases will significantly reduce the presence of sequences with high secondary structure free energy while a lower number will reduce the potential sites for ligand binding.

Based on these findings, we developed a set of criteria that limited the presence of sequences with abundant simple structural motifs and maximized the presence of stable low-energy structures. For the pool of 40-mer sequences these criteria are as follows [4]: (1) in the lowest energy conformation bases 1 and 2 form pairs with bases 40 and 39; (2) the free energy of secondary structure formation is less than -9.5 kcal/mol; (3) there are at least 11 bases that do NOT form Watson–Crick pairs. Only the sequences that passed the selection criteria were forwarded to the next step in the selection process.

The typical steps for the secondary structure generation and analysis are outlined below.

3.1.1 Generation of Sequence Library

To generate random or patterned libraries in Perl script we use function `rand()`. By default it generates a number between 0 and 1. However, we can pass it a specific number to generate random numbers between 0 and that number. Initially, we create tree arrays:

```
my @base4 = ('A', 'U', 'C', 'G');
my @baseR = ('A', 'G');
my @baseY = ('U', 'C');
```

Next, we specify two numbers for `rand()` function:

```
my $range_4 = 4;
my $range_2 = 2;
```

Now, example of the random part of the sequence:

```
my $seq1= '';
for($i=0; $i<4; $i++)
 {
   $ind = int(rand($range_4));
   $seq1 .= $base4[$ind];
 }
```

An example of the patterned part of the sequence RRYY:

```
my $link= '';
$ind = int(rand($range_2));
$link .= $baseR[$ind];
$ind = int(rand($range_2));
$link .= $baseR[$ind];
$ind = int(rand($range_2));
$link .= $baseY[$ind];
$ind = int(rand($range_2));
$link .= $baseY[$ind];
```

Finally, we assemble all part of the sequence into a variable `$sequence`.

3.1.2 Calculation of Folding Energy of the Generated Sequences

Next step is to calculate the secondary structure and folding energy of the generated sequence. We perform that using `RNA::fold` function from the ViennaRNA package:

```
use RNA;
($s_struct, $f_energy) = RNA::fold($sequence);
```

The secondary structure of the generated sequence is presented in bracket-dot notations. In these notations unpaired bases are presented by dots and base pair between bases *i* and *j* is presented by "(" at position *i* and ")" at position *j*, e.g., "(((…((…))..)))".

3.1.3 Analysis of the Secondary Structure of Generated Sequences

Now, perform analysis of the secondary structure. If it satisfies our criteria, we save the sequence and its secondary structure into the output files:

```
open(fasta_FILE, ">Seq1.fasta") || die("Could not
open file:fasta_file !");
```

```
open(temp_FILE, ">Seq1.temp") || die("Could not
open file:temp_file !");
my $cut_en = -9.5;
$seq_len = length $sequence;

#Check energy < $cut_en
if ($f_energy < $cut_en )
{
#Check that nt_1 forms base pair with nt_end
@sp_struct = split(//, $s_struct);
if($sp_struct[0] eq "(" && $sp_struct[$seq_len] eq
")" )
{
# Check that we have more than 10
# but less than 30 unpaired bases
@ARG = split(/\./, $s_struct);
$n_points = @ARG;
if (($n_points > 10) && ($n_points < 25))
{
# Save sequence in FASTA format
print fasta_FILE ">RNA_sequence\n$sequence\n";
# Save sequence and structure into temporary file
print temp_FILE "$sequence\n$s_struct\n";
} } }
```

3.1.4 Generation of Files for 3D Modeling

Saved sequences and their secondary structure are used to generate three-dimensional structures by Rosetta package. RNA 3D structure modeling in Rosetta requires as input sequence in FASTA format (saved in **step 3.1.3**). As optional file it uses ".param" file that specifies Watson–Crick base pairs, contained in `$s_struct`. Parameter file looks like the following:

```
STEM PAIR 1 40 W W A
STEM PAIR 2 39 W W A
STEM PAIR 3 38 W W A
STEM PAIR 8 29 W W A
STEM PAIR 9 28 W W A
```

Our experience shows that predicted structures are more accurate with the use of parameter file. There is no easy way to create parameter file from the `$s_structure` string. One possible solution is to use b2c.c function that is distributed with the ViennaRNA package and is located in folder "Utilities." This function converts bracket-dot notations into mfold *.ct file. We modified `write_ct_file` subroutine of b2c.c function to produce output file in Rosetta parameter format. That modified function is named b2param.c. As input it uses temporary file created in **step 3.1.3**, and it is used inside the Perl script:

```
system ("b2param < Seq1.temp > Seq1.param";
```

3.2 Generation of RNA Three-Dimensional Structures

During the last few years, several groups have proposed different approaches for de novo prediction of RNA tertiary structure (*see* ref. 20 for recent review). We used the Rosetta package [14, 15] developed by the Baker group at the University of Washington to predict three-dimensional structures of selected sequences. In Rosetta modeling algorithm, the RNA sequence is divided into short fragments containing three nucleotides. The assembly of fragment from a library of experimental RNA crystal structures is based on Monte Carlo process with a simplified low-resolution energy function. The generated models are then refined with the all-atom potential to provide high-resolution atomistic structures. This two-step protocol has been named Fragment Assembly of RNA with Full Atom Refinement (FARFAR) [15]. In the previous section we created two files for each of the sequences in the pool: a sequence file in FASTA format (`Seq1.fasta`) and a parameter file (`Seq1.param`) that contains information about the Watson–Crick base pairing. These two files are used as input files for the 3D structure prediction (*see* **Note 1**).

3.2.1 Generation of Three-Dimensional Structures

The command to run RNA 3D structure modeling in Rosetta is as follows:

```
$ROSETTA_HOME/rosetta_source/bin/rna_denovo.
<exe> @flags_rna - database $ROSETTA_HOME/rosetta_
database
```

where <exe> is the extension of the `rna_denovo` command and file `flags_rna` contains a list of options for the RNA modeling:

```
-fasta Seq1.fasta
-params_file Seq1.param
-nstruct 500
-out:file:silent Seq1.out
-minimize_rna
-vary_geometry
-mute core.io.database
```

In the above example we generated 500 minimized 3D structures in the output file `Seq1.out`. For each of these structures the program runs 10,000 (default value) of Monte Carlo cycles (*see* **Note 2**). The detailed description of the parameters can be found on Rosetta website http://www.rosettacommons.org/manuals/archive/rosetta3.4_user_guide/d2/d82/rna_denovo.html.

3.2.2 Sorting Structures Based on Their Score

The generated structures in output file `Seq1.out` are in order they were generated and not in order of their score. Therefore, next step is to sort structures based on their score. The first line in the `Seq1.out` is a sequence, e.g.:

```
SEQUENCE: ggcgucacaccuucgggugaagucgcccaccuucgg
```

Next line is the description of different score fields:

```
SCORE: score fa_atr fa_rep fa_intra_rep lk_nonpolar hack_elec_rna_phos_phos ch_bond rna_torsion rna_sugar_close hbond_sr_bb_sc hbond_lr_bb_sc hbond_sc geom_sol atom_pair_constraint angle_constraint linear_chainbreak rms rms_stem N_WC N_NWC N_BS f_natWC f_natNWC f_natBP description
```

We used the first field "score" and the last one "description" that has a tag for a given model, e.g., `S_000001`.

The sorting of the output results can be done using another Perl script. The part of script that performs sorting looks like the following:

```
$i = 0;
while ($line = <in_FILE>)
{
chomp($line);
@ARG = split(/ +/,$line);
($a1, $a2) = @ARG;
$name = pop(@ARG);
if (substr($a1, 0, 6) eq "SCORE:" && substr($name, 0, 2) eq "S_")
{
$S_name[$i] = $name;
$S_score[$i] = $a2;
++$i;
}
}
@hash{@S_score} = (0..$#S_score);
@S_sort = sort {$a <=> $b} @S_score;
@S_sort_index = @hash{@S_sort};
```

3.2.3 Extracting Best Structures

Next, we need to extract structures with the lowest score (score is negative; therefore, best score is the lowest score) from the `Seq1.out` file into the *.pdb files. It is widely accepted that ligand binding can drastically alter the receptor's conformation [21]. To account for such conformational flexibility, five to ten of the lowest score structures for each sequence can be selected and placed into a library of RNA molecules (*see* **Note 3**). This library will be used for virtual high-throughput screening to select aptamers with binding affinity to a specified small molecule.

To extract structures we use Rosetta `extract_pdb` command:

```
$ROSETTA_HOME/rosetta_source/bin/extract_pdbs.<exe> @flags_extract -database $ROSETTA_HOME/rosetta_database
```

where file `flags_extract` looks like the following:

```
-in:file:silent Seq1.out
-in:file:fullatom
```

```
-out:prefix Seq1
-in:file:tags S_000469 S_000021 S_000004 S_000145 S_000358
```

The last line specifies a list of structures with the lowest scores that need to be extracted. By including five 3D structures we perform ensemble docking and thus account for conformational flexibility of the RNA molecules.

3.3 Virtual Screening of the RNA Molecule Library

Computational docking is a common tool used to identify small-molecule ligands that bind to proteins [22]. While most docking methods have been developed mostly for proteins, recent evaluation of AutoDock and DOCK programs has demonstrated their ability to dock compounds to RNA molecules [16]. We used the AutoDock Vina and ADT to prepare RNA molecules and small-molecule ligand for docking and to perform docking.

In the previous section we generated PDB files with 3D coordinates of RNA molecules. Now, we need to create a PDB file with the coordinates of a small-molecule ligand. One way of doing that is to use SMILES representation of the molecule. Usually, SMILES representation of molecule can be found on Wikipedia on the right side of the screen. Under the "Chemical data" section there is a tag SMILES. Online SMILES translator at http://cactus.nci.nih.gov/translate/ can be used to generate 3D coordinates from SMILES string and to save them into PDB file. Another way to generate coordinates is to use a commercial package to build a molecule and to create a PDB file with the coordinates.

Next step is to prepare RNA and ligand molecules for docking using AutoDock Vina. But, before we proceed to that step, we need to center the RNA molecules around (0,0,0) coordinate. This can be done using Open Babel package. The command is as follows:

```
babel -ipdb Seq1.pdb -opdb Seq1_Ct.pdb -C
```

where the output file `Seq1_Ct.pdb` contains the centered coordinates of the molecule.

To prepare molecules for docking we can use GUI version of ADT or scripts provided by ADT for batch mode. To use ADT, please follow the tutorial at http://autodock.scripps.edu/faqs-help/tutorial/using-autodock-4-with-autodocktools. To prepare files in batch mode we used two commands provided by ADT. To prepare ligand molecule:

```
prepare_ligand4.py -l ligand.pdb
```

The input file for this command is a PDB file with 3D of ligand created above. As output it produces `ligand.pdbqt` file that includes Gasteiger charges, AutoDock atom types, and special keywords that establish the torsional flexibility of ligand molecule (*see* **Note 4**).

The similar command is used to prepare receptor (in our case RNA) molecules:

```
prepare_receptor4.py -r Seq1_Ct.pdb
```

The output file `Seq1_Ct.pdbqt` contains coordinates of receptor with added charges, AutoDock atom types, and merging nonpolar hydrogens. Two files, `Seq1_Ct.pdbqt` and `ligand.pdbqt`, are required for docking software AutoDock Vina. Another required file is the configuration file `config.txt` that contains docking options for Vina. Here is the example of configuration file:

```
receptor = Seq1_Ct.pdbqt
ligand = ligand.pdbqt

center_x = 0
center_y = 0
center_z = 0

size_x = 25
size_y = 25
size_z = 25
```

The first two lines specify the receptor and ligand PDBQT files. The next three lines define the center of the search space. We use (0,0,0) coordinates as we centered out receptor molecule before using `babel` command. The last three lines specify size of the search space in X, Y, and Z directions for the ligand docking. The full usage summary can be obtained using the command "`vina --help`". To run the docking of ligand molecule to RNA molecule specified in file config.dock user need to use the command "`vina --config config.txt --log Seq1.log`". Two output files are created: `Seq1.log` file with the summary of docking results and `ligand_out.pdbqt` file with the predicted ligand binding conformations. This file can be analyzed and visualized using ADT.

To run virtual screening of many RNA molecules in Linux/Mac environment we can use shell scripting. Suppose we are in a directory containing ligand molecule in file `ligand.pdbqt` and RNA molecules in files `Seq1.pdbqt, Seq2.pdbqt`, etc. We want to run docking with the same ligand but different receptor molecules. In the `config.txt` file we remove a line that specifies receptor and that file looks now as follows:

```
ligand = ligand.pdbqt

center_x = 0
center_y = 0
center_z = 0
size_x = 25
size_y = 25
size_z = 25
```

Now, we are using the script in Bash shell to perform virtual screening:

```
 #! /bin/bash

for f in Seq*.pdbqt; do
    r=`basename $f .pdbqt`
```

```
        echo Processing receptor $r
        vina --config config.txt --receptor $f --out
        ${r}_out.pdbqt --log ${r}.log
done
```

For each receptor molecule we have now separate result files `Seq*_out.pdbqt` and `Seq*.log`.

Next step is to analyze the output files and select RNA molecules with the highest binding affinity to the ligand molecule. At this point, the selected RNA molecules are specific to the desired target molecule, and they are placed into a pool of sequences for experimental verification and selection of high-affinity aptamers, e.g., using microarray technology (*see* **Note 5**). The accurate prediction of the three-dimensional structures of RNA molecules and modeling of RNA–small molecule interactions still faces considerable challenges [23, 24] and requires a careful protocol and analysis of the results.

4 Notes

1. Our experience with the prediction of 3D structures using Rosetta package shows that predicted structures are more accurate with the use of parameter file with the secondary structure obtained using ViennaRNA package.
2. Prediction of RNA tertiary structure requires computational sampling of a huge conformational space. Typically it requires generation of hundreds to thousands of models that is computationally very expensive. Therefore, there should be some compromise between the number of structures generated in a reasonable amount of time and the accuracy of predicted structures.
3. RNA molecules are very flexible, and ligand binding can drastically change the receptor conformation. To account for such conformational flexibility, five to ten of the lowest score structures for each sequence should be selected and placed into a library of RNA molecules.
4. By default, `prepare_receptor4.py` and `prepare_ligand4.py` assign Gasteiger charges to both RNA molecule and ligand. While for small molecules Gasteiger charges work well, they are not the best charges for biomolecules, including RNA. User can consider an option to assign charges that were specifically designed for biomolecules such as CHARMM [25] or AMBER [26] to RNA molecules. The PDB file format does not have a field for atomic charges. Therefore, RNA molecules should be saved in SYBYL mol2 file format which is readable by AutoDock. In this case the `prepare_receptor4.py` should be used with option `-C` that means to preserve assigned charges.

5. The biggest challenge in the modeling of RNA–small molecule interaction lies in the highly charged nature of RNA molecules. It results in strong solvation and the association of ionic molecules. It is well documented that divalent ions such as Mg^{2+} significantly affect the structure of RNA molecules and binding to small-molecule ligand. Unfortunately, these effects are not incorporated into current versions of docking programs. Therefore, we propose to take thousands of designed sequences for experimental validation and screening.

Disclaimer

The opinions and assertions contained herein are the private views of the authors and are not to be construed as official or as reflecting the views of the US Air Force or the US Department of Defense. This chapter has been approved for public release with unlimited distribution.

References

1. Ellington AD, Szostak JW (1990) In vitro selection of RNA molecules that bind specific ligands. Nature 346:818–822
2. Tuerk C, Gold L (1990) Systematic evolution of ligands by exponential enrichment: RNA ligands to bacteriophage T4 DNA polymerase. Science 249:505–510
3. Osborne SE, Ellington AD (1997) Nucleic acid selection and the challenge of combinatorial chemistry. Chem Rev 97:349–370
4. Chushak Y, Stone MO (2009) In silico selection of RNA aptamers. Nucleic Acids Res 37:e87
5. Davis JH, Szostak JW (2002) Isolation of high-affinity GTP aptamers from partially structured RNA libraries. Proc Natl Acad Sci 99:11616–11621
6. Carothers JM, Oestreich SC, Davis JH, Szostak JW (2004) Informational complexity and functional activity of RNA structures. J Am Chem Soc 126:5130–5135
7. Ruff KM, Snyder TM, Liu DR (2010) Enhanced functional potential of nucleic acid aptamer libraries patterned to increase secondary structure. J Am Chem Soc 132: 9453–9464
8. Kim N, Gan HH, Schlick T (2007) A computational proposal for designing structured RNA pools for in vitro selection of RNAs. RNA 13:478–492
9. Kim N, Shin JS, Elmetwaly S et al (2007) RagPools: RNA-as-graph-pools—a web server for assisting the design of structured RNA pools for in vitro selection. Bioinformatics 23:2959–2960
10. Luo X, McKeague M, Pitre S et al (2010) Computational approaches toward the design of pools for the in vitro selection of complex aptamers. RNA 16:2252–2262
11. Lorenz R, Bernhart SH, Hoener Zu Siederdissen C et al (2011) ViennaRNA package 2.0. Algorithms Mol Biol 6:26
12. Mathews DH, Sabina J, Zuker M, Turner DH (1999) Expanded sequence dependence of thermodynamic parameters improves prediction of RNA secondary structure. J Mol Biol 288:911–940
13. Mathews DH, Turner DH (2009) NNDB: the nearest neighbor parameter database for predicting stability of nucleic acid secondary structure. Nucleic Acids Res 38:280–282
14. Leaver-Fay A, Tyka M, Lewis SM et al (2011) ROSETTA3: an object-oriented software suite for the simulation and design of macromolecules. Methods Enzymol 487:545–574
15. Das R, Karanicolas J, Baker D (2010) Atomic accuracy in predicting and designing noncanonical RNA structure. Nat Methods 7:291–294
16. Detering C, Varani G (2004) Validation of automated docking programs for docking and database screening against RNA drug targets. J Med Chem 47:4188–4201
17. Trott O, Olson AJ (2010) AutoDock Vina: improving the speed and accuracy of docking with a new scoring function, efficient

optimization and multithreading. J Comput Chem 31:455–461

18. Carothers JM, Oestreich SC, Szostak JW (2006) Aptamers selected for higher-affinity binding are not more specific for the target ligand. J Am Chem Soc 128:7929–7937
19. Patel DJ, Suri AK (2000) Structure, recognition and discrimination in RNA aptamer complexes with cofactors, amino acids, drugs and aminoglycoside antibiotics. J Biotechnol 74:39–60
20. Laing C, Schlich T (2011) Computational approaches to RNA structure prediction, analysis, and design. Curr Opin Struct Biol 21:306–318
21. May A, Sieker F, Zacharias M (2008) How to efficiently include receptor flexibility during computational docking. Curr Comput Aided Drug Des 4:143–153
22. Taylor RD, Jewsbury PJ, Essex JW (2002) A review of protein-small molecule docking methods. J Comput Aided Mol Des 16: 151–166
23. Cruz JA, Blanchet MF, Boniecki M et al (2012) RNA-Puzzles: a CASP-like evaluation of RNA three-dimensional structure prediction. RNA 18:610–625
24. Fulle S, Gohlke H (2010) Molecular recognition of RNA: challenges for modelling interactions and plasticity. J Mol Recognit 23:220–231
25. Brooks BR, Brooks CL III, Mackerell AD Jr et al (2009) CHARMM: the biomolecular simulation program. J Comput Chem 30:1545–1614
26. Case DA, Cheatham TE, Darden T et al (2005) The amber biomolecular simulation programs. J Comput Chem 26:1668–1688

Chapter 2

In Vitro Selection of RNA Aptamers for a Small-Molecule Dye

Asako Murata and Shin-ichi Sato

Abstract

Artificial riboswitches can be generated by functional coupling of an RNA aptamer for synthetic small molecule to an expression platform. RNA aptamers that can bind strongly and selectively to their target are feasible to be used for obtaining more potent artificial riboswitches. In this chapter, we describe tips and notes for in vitro selection of RNA aptamers targeting synthetic small molecules.

Key words RNA aptamer, In vitro selection, Small molecule, Riboswitch, Chemical biology

1 Introduction

Bacteria and some eukaryotes use natural RNA aptamers for metabolite sensing and gene regulation [1–3]. These RNA aptamers specifically bind to their targets and exhibit conformational changes that are transduced into the changes in the secondary structure of a downstream expression platform, resulting in a transcriptional or a translational modulation of gene expression. This RNA-based gene regulatory system, termed riboswitch, was first experimentally validated by RR Breaker in 2002 [4]. To date, more than 20 classes of natural riboswitches have been identified in many species [3].

The discovery of natural riboswitches has encouraged greater interest among researchers who seek to create a new type of gene expression system. In fact, a number of studies have demonstrated the creation of "artificial riboswitches" for the conditional control of gene expression in bacteria by engineering natural riboswitches [5–7]. On the other hand, in vitro-selected RNA aptamers for synthetic small molecules have also been employed to create artificial riboswitches [8–12].

In general, the use of RNA aptamers selected against synthetic small molecules for controlling gene expression might offer several advantages: orthogonality to endogenous system and choice of a

Atsushi Ogawa (ed.), *Artificial Riboswitches: Methods and Protocols*, Methods in Molecular Biology, vol. 1111, DOI 10.1007/978-1-62703-755-6_2, © Springer Science+Business Media New York 2014

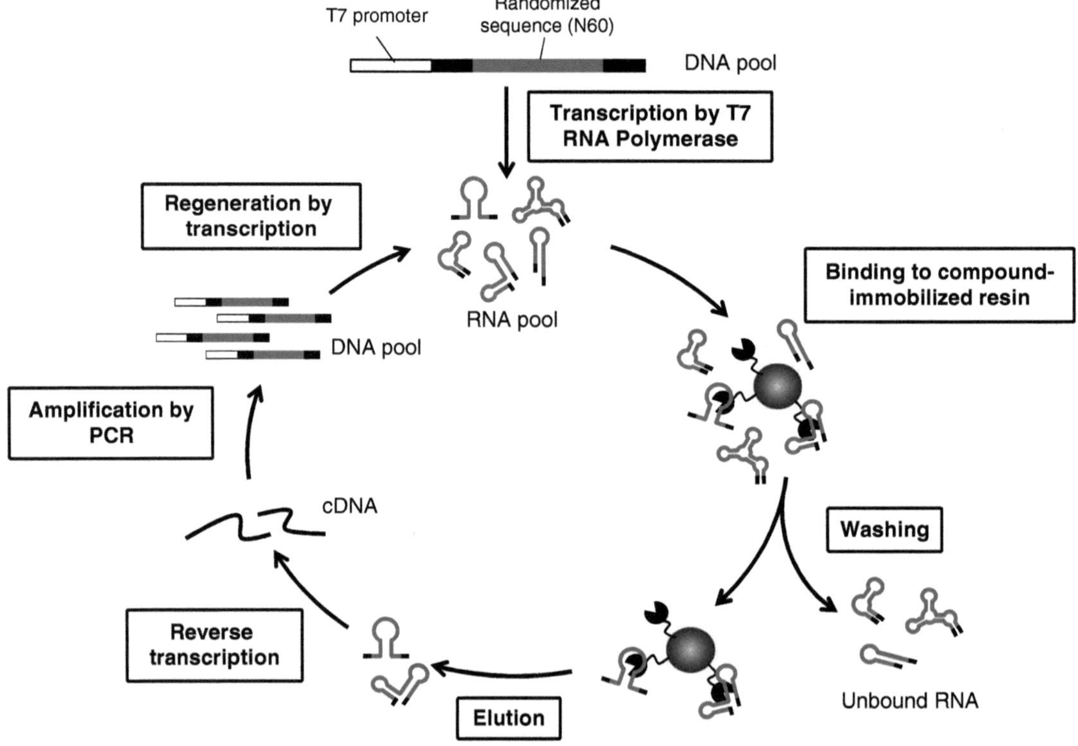

Fig. 1 General scheme of in vitro selection

wide range of target ligands. RNA aptamers can be isolated from a pool of random RNA sequences through a process called in vitro selection or Systematic Evolution of Ligands by Exponential enrichment (SELEX) (Fig. 1) [13]. In vitro selection experiments consist of several steps: (1) preparation of a ligand-immobilized matrix, (2) transcription of RNA from a synthetic pool of double-stranded DNA, (3) isolation of RNA species that can specifically bind to the ligand on the matrix, and (4) regeneration of dsDNA pools through reverse transcription and PCR amplification. Steps 2–4 must be repeated several times to yield aptamers having higher binding affinity to the ligand. Here we focus on steps 1 and 3 that can mostly affect the successful selection of high-affinity aptamers from a random RNA pool and describe notes and tips for these steps.

In the first step of an in vitro selection experiment, an affinity matrix is prepared by immobilizing a target ligand on a resin. Although there is no consensus for good targets for in vitro selection, several considerations may help researchers to choose target ligands. Ligands that contain chemical moieties that are complementary to those of nucleic acids should be good targets. For example, in vitro selections against aromatic compounds, heterocyclic compounds, or nucleotide analogs have often been

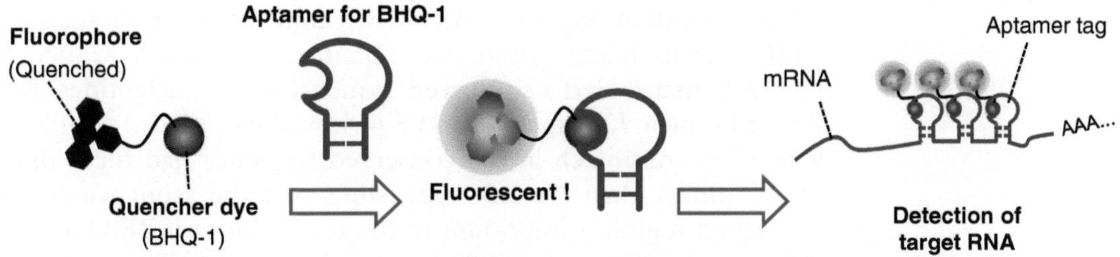

Fig. 2 Application of in vitro-selected aptamers against a small-molecule dye for RNA tagging

successful. In addition, positively charged functional groups (e.g., primary amino groups) of a ligand may facilitate an aptamer selection, as they increase binding affinity to RNA by electrostatic interaction with the phosphate backbone. Next, the ligand of choice can be attached to a resin for making a ligand-immobilized matrix. There are two major methods available for ligand attachment: one is covalent coupling of ligands to a reactive functional group on matrix and another is immobilization of biotinylated ligands on avidin-coated matrix. Typically, the former coupling method gives the matrix with higher concentration of ligand (~20 mM) than that of the latter one (~100 nM), although it may depend on the concentration of functional groups that were originally available and the coupling efficiency. The concentration of the ligand on the matrix can significantly influence the course of a selection experiment. Decreasing the concentration of the ligand on the matrix increases the chance for having aptamers that bind more strongly to the target ligand, encouraging competition for a smaller number of binding sites.

Binding and elution conditions also directly affect the stringency of the selections. Binding buffer containing monovalent and divalent cations (Na^+, K^+, and Mg^{2+}) may be used to stabilize RNA structures, which promotes the formation of specific aptamer structures or interactions with the ligand. Negative selection to remove matrix-binding species using an underivatized resin is critically important to obtain specific aptamers, because it may be difficult to select aptamers if once the matrix binders accumulate and become dominant. To obtain specific RNA aptamers to the target ligand, affinity elution with the specific ligand is highly preferred, though denaturing agents such as 6–8 M urea and 4 M guanidine can elute bound RNA from the matrix by denaturing secondary structures of RNA.

We present our recent work on in vitro selection of RNA aptamers targeting a small-molecule dye (Black hole quencher-1: BHQ-1). Our study has focused on utilizing the aptamers as a tag for RNA detection in living cells in combination with the designed fluorescent probe (Fig. 2) [14]. In vitro selection was performed to isolate aptamers for BHQ-1 from an RNA pool containing a

60-base random sequence. After 13 rounds of selection against a BHQ-1-immobilized resin, we obtained four RNA sequences (A1–A4) that shared a conserved sequence of 17 nucleotides. A1–A3 had similar K_d values around 5 μM, whereas A4 that contained a one-base mismatch in the conserved sequence had three times lower affinity than A1–A3, suggesting that the sequence of the conserved region is important in the recognition of BHQ-1. A1–A4 were predicted to form a stem–loop, with the conserved sequence in the loop region. The prevention of stem–loop formation caused a marked decrease in binding affinity of these aptamers, indicating that the stem–loop is the minimal binding motif. In fact, a short version of A1 containing only the stem–loop retained binding ability to BHQ-1 with lower affinity than parental A1. Using A1 sequence as a tag, a specific mRNA was successfully detected by restoration of fluorescence probe in an *E. coli* lysate.

2 Materials

2.1 Equipment

1. Thermal cycler.
2. Horizontal and vertical electrophoresis system.
3. UV transilluminator.
4. Tabletop centrifuge.
5. Real-Time PCR System (7500 Fast Real-Time PCR System, Applied Biosystems).
6. Microplate reader (MTP-800, CORONA).
7. Fluorophotometer (LS-55, PerkinElmer).

2.2 Reagents and Solutions

1. BHQ-1 carboxylic acid, BHQ-1 amine (Bioresearch Technologies).
2. 4-(4,6-Dimethoxy-1,3,5-triazin-2-yl)-4-methylmorpholinium chloride: DMT-MM (KOKUSAN CHEMICAL Co., Ltd).
3. Triethylamine (Wako Pure Chemical Industries, Ltd).
4. Platinum® *Pfx* DNA polymerase (Invitrogen).
5. 10 mM dNTP mix (Invitrogen).
6. T7 RNA polymerase (TAKARA).
7. 25 mM Nucleotides rNTPs Mixture (TOYOBO).
8. RQ1 DNase I (Promega).
9. Prime Script reverse transcriptase (TAKARA).
10. DNA Ligation Kit (TAKARA).
11. Plasmid isolation kit (QIAGEN).
12. MEGAscript T7 Kit (Ambion).
13. In-Fusion Dry-Down PCR Cloning Kit (Clontech Laboratory Inc.).

14. RTS 100 *E. coli* HY Kit, 5 (Prime Inc.).
15. 3 M sodium acetate solution (pH 5.2).
16. 7.5 M ammonium acetate solution (Sigma-Aldrich).
17. Annealing buffer: 10 mM Tris–HCl, pH 7.5, 100 mM KCl, 50 mM NaCl.
18. Binding buffer: 10 mM Tris–HCl, pH 7.5, 100 mM KCl, 50 mM NaCl, 5 mM $MgCl_2$.
19. 250 mM $MgCl_2$.

2.3 Supplies

1. Microtubes, 1.5 and 2.0 mL.
2. Falcon tubes, 15 and 50 mL.
3. 0.2 mL PCR tubes.
4. EAH sepharose™ 4B resins (GE Healthcare).
5. Poly-Prep Chromatography Columns (BIO-RAD).
6. Micro Bio-Spin Chromatography Columns (BIO-RAD).
7. NAP-5 columns (GE Healthcare).
8. 96-well glass bottom black plates (655896, Greiner Bio-One).

3 Methods

All solutions and buffers should be made with DEPC-treated water. Use RNase/DNase-free disposable plasticware and micropipet tips where possible.

3.1 Preparation of BHQ-1-Immobilized Resin

1. Place 0.5 mL of EAH sepharose™ 4B resin in a Poly-Prep Chromatography Column and allow to drain by gravity.
2. Wash the resin thoroughly with at least three bed volumes of dioxane–H_2O (3:1, v/v).
3. Add BHQ-1 carboxylic acid (0.5 mg, 1 μmol), triethylamine (0.1 mmol), and DMT-MM (29.5 mg, 0.1 μmol) in 0.5 mL of dioxane–H_2O (3:1, v/v) to the drained resin, and shake the suspension at room temperature overnight (*see* **Note 1**).
4. Drain the coupling solution from the resin, and add the freshly prepared acetic acid (30 mg, 0.5 mmol) and DMT-MM (147.4 mg, 0.5 mmol) in 0.5 mL of dioxane–H_2O (3:1, v/v) to block unreacted amino groups on the resin.
5. After overnight incubation, drain the resin and wash thoroughly with dioxane–H_2O (1:1, v/v) followed by methanol–H_2O (1:1, v/v).
6. Suspend the resin with three bed volumes of methanol–H_2O (1:3, v/v) to make a 25 % slurry. Store at −20 °C until use.

Table 1
Reaction mixture for PCR amplification of DNA pools

Template DNA*1	10 μL
10× *Pfx* amplification buffer	50 μL
10 mM dNTP mixture	15 μL
50 mM $MgSO_4$	10 μL
Forward primer*2	5 μL
Reverse primer*3	5 μL
Platinum *Pfx* DNA polymerase	10 μL
d.D.W	395 μL
Total	500 μL

3.2 In Vitro Selection Procedures

In vitro selection was performed to isolate RNA aptamers for BHQ-1 according to the standard procedure (Fig. 1).

3.2.1 Preparation of DNA Pools

1. Assemble the reaction mix in PCR tubes as described in Table 1 (*see* **Note 2**). *1Template DNA: 5′-GAA TTC CGC GTG TGC ACA CC-N60–GTC CGT TGG GAT CCT CAT GG-3′ (for preparation of first DNA pool a reverse-transcribed product was used for the other selection round). *2Forward primer: 5′-GCT AAT ACG ACT CAC TAT AGG GAA TTC CGC GTG TGC ACA CC-3′ (T7 promoter sequence is underlined). *3Reverse primer: 5′-CCA TGA GGA TCC GAA CGG AG-3′.
2. Place the tubes into a thermal cycler and perform the PCR amplification using the following program: 2-min initial denaturation at 94 °C, followed by sequential cycles of denaturation at 94 °C for 30 s, annealing at 55 °C for 30 s, extension at 68 °C for 1 min, and final extension at 68 °C for 7 min (*see* **Note 3**).
3. Precipitate the DNA by adding a half volume of 7.5 M ammonium acetate solution and 1.5 volume of isopropanol. Centrifuge the sample at 10,000 ×*g* for 30 min to pellet the DNA.
4. Remove the supernatant carefully, and rinse the pellet with 70 % ethanol.
5. Air-dry the pellet for 5 min, and dissolve it in 50 μL of RNase-free water.

3.2.2 Preparation of RNA Pools

1. Assemble the transcription reaction at room temperature in the order shown below (Table 2).
2. Incubate the reaction mix at 37 °C overnight.
3. Add 10 μL of DNase I and incubate for 30 min at 37 °C.

Table 2
Reaction mixture for in vitro transcription of reaction

Template DNA	X μL
10× T7 RNA polymerase buffer	50 μL
NTP mixture (25 mM each)	15 μL
50 mM DTT	10 μL
T7 RNA polymerase	5 μL
RNase-free water	420 – X μL
Total	500 μL

4. Precipitate the RNA by adding a half volume of 7.5 M ammonium acetate solution and 1.5 volume of isopropanol. Centrifuge the sample at 10,000 × *g* for 30 min at 4 °C to pellet the RNA.
5. Thoroughly remove the supernatant, and dissolve the pellet in 0.5 mL of RNase-free water.
6. Apply the sample to the pre-equilibrated NAP-5 column, and elute the sample with 1 mL of RNase-free water (*see* **Note 4**).
7. Precipitate the RNA by adding 0.5 mL of 7.5 M ammonium acetate solution and 1.5 mL of isopropanol. Centrifuge the sample at 10,000 × *g* for 30 min at 4 °C to pellet the RNA.
8. Remove the supernatant carefully, and rinse the pellet with 70 % ethanol.
9. Air-dry the pellet for 5 min, and dissolve it in 100 μL of the annealing buffer.
10. Denature the RNA pool for 3 min at 80 °C, and then let it fold by slow cooling to room temperature (*see* **Note 5**).
11. Add 2 μL of $MgCl_2$ solution to a final concentration of 5 mM (*see* **Note 6**).

3.2.3 Binding of RNA Pool to BHQ-Immobilized Resin

1. Place 50 μL of the BHQ-1-immobilized resin (25 % slurry) into an empty column (Micro Bio-Spin Chromatography Columns) (*see* **Note 7**), and wash the resin three times with 150 μL of the binding buffer.
2. Apply the RNA pool solution to the resin and incubate for 30 min on ice. Vortex the sample briefly every 10 min.
3. Drain and wash the resin to remove the unbound RNA species (*see* **Note 6**).
4. Elute bound RNA species three times with 130 μL of the binding buffer saturated with free BHQ-1 amine (*see* **Note 7**).

Table 3
Reverse transcription reaction

a	
Template RNA	8 μL
10 mM dNTP mix	1 μL
10 μM reverse primer*	1 μL
Total	10 μL
[*Reverse primer: 5′-CCATGAGGATCCGAACGG AC-3′]	
	↓ 65 °C, 5 min
b	
Template RNA/primer mixture	10 μL
5× PrimeScript buffer	4 μL
RNase inhibitor	0.5 μL
PrimeScript reverse transcriptase	1 μL
RNase-free water	4.5 μL
Total	20 μL
	↓ 42 °C, 60 min, then 70 °C, 15 min

5. Collect the eluate in a 1.5 mL collection tube, and add 43 μL of 3 M sodium acetate solution and 1 mL of ethanol to precipitate recovered RNA. Centrifuge the sample at 10,000×*g* for 30 min at 4 °C.
6. Remove the supernatant carefully, and rinse the pellet with 70 % ethanol.
7. Air-dry the pellet for 5 min, and dissolve it in 8 μL of RNase-free water.

3.2.4 Reverse Transcription

1. Prepare the following mixture (Table 3a) in a PCR tube. Heat the sample for 5 min at 65 °C and then chill immediately on ice. Assemble the reaction mix as shown in Table 3b, and perform cDNA synthesis reaction for 60 min at 42 °C, followed by heat-inactivation of RTase at 70 °C for 15 min.
2. Use 10 μL of the reverse-transcribed product as a template for the PCR amplification to generate the DNA pool for the next round of selection (*see* **Note 8**).
3. Repeat the selection steps for several rounds to enrich the RNA species with high affinity to BHQ-1.

3.3 Cloning and Sequencing of Each RNA Species in the Enriched Pool

1. After the final round of selection, digest the resulting DNA pool with *Eco*RI and *Bam*HI as well as pUC19.
2. Ligate the digested insert and vector using DNA Ligation Kit.
3. Transform DH5α with the ligation product, and heat shock and spread the cells on LB plate containing 100 μg/mL of ampicillin. Incubate the plate at 37 °C overnight.
4. Culture from a single colony, and isolate the plasmid using standard plasmid isolation procedures.
5. Sequence the insert using primers in the vector backbone.

3.4 Fluorescence Measurements

To determine if the selected aptamers have the ability to restore the fluorescence of the fluorophore–BHQ-1 conjugates, fluorescence titration of the fluorophore–BHQ-1 conjugates with the aptamers can be performed. The synthesis of the fluorophore–BHQ-1 conjugate was described in our previous report [14].

1. Prepare the dsDNA template containing T7 promoter and each BHQ-1 aptamer by PCR amplification using the individual clone as a template. Primers used in the reaction are as follows: the forward primer [5′-GCT AAT ACG ACT CAC TAT AGG GAA TTC CGC GTG TGC ACA CC-3′ (T7 promoter sequence is underlined)] and the reverse primer [5′-CCA TGA GGA TCC GAA CGG AC-3′].
2. Perform a transcription reaction, and purify RNA as described above in Subheading 3.2.2.
3. Prepare a solution of conjugate AL at the concentration of 2 μM in the binding buffer.
4. Prepare RNA solution at the concentrations of 0, 0.6, 2, 6, 20, 60, and 200 μM in the binding buffer.
5. Mix an equal volume of the RNA solution with the conjugate solution, and measure the fluorescence of each sample using either a plate reader or a fluorophotometer. The excitation maximum of each conjugate and the emission maximum in the RNA-bound state are as follows: conjugate AL (Ex/Em = 543 nm/610 nm), conjugate Cy (Ex/Em = 519 nm/562 nm), and conjugate FL (Ex/Em = 497 nm/515 nm).

The K_d values of each aptamer can be obtained by fitting fluorescence titration data using the following equation:

$$F_{obs} = A\left(\frac{\left([conjugate]_T + [aptamer]_T + K_d\right) - \left(\left([conjugate]_T + [aptamer]_T + K_d\right)^2 - 4[conjugate]_T[aptamer]_T\right)^{1/2}}{2[conjugate]_T}\right)$$

Table 4
In transcription/translation reaction mixture

		Final conc.
Template DNA	X μL	0.5 μg
E. coli lysate	4.8 μL	
Reaction mix	4 μL	
Amino acids	4.8 μL	
Methionine	0.4 μL	
Reconstitution buffer	2 μL	
100 μM fluorophore–BHQ-1 conjugate	0.5 μL	2.5 μM
RNase-free water	3.5 − X μL	
Total	20 μL	

where A is the increase in fluorescence at saturating aptamer concentrations ($F_{max} - F_{min}$), K_d is the dissociation constant, and $[\text{conjugate}]_T$ and $[\text{aptamer}]_T$ are the total concentrations of a conjugate and an aptamer, respectively.

3.5 Real-Time Monitoring of Protein Synthesis

Detailed protocol of the preparation of template DNAs is described in our previous report [14].

1. Assemble the reaction according to Table 4.
2. Incubate the reaction mixture at 30 °C, and collect the fluorescence emission data every 3 min at appropriate wavelength by using a real-time PCR system with the filter for FAM and ROX dye.
3. After 1-h incubation, add DNase I to the mixture and incubate for 30 min at 37 °C to digest the template DNA.
4. Analyze the resulting reaction to verify RNA and protein synthesis by RT-PCR and western blotting, respectively.

4 Notes

1. We use DMT-MM as a coupling reagent, but any other water-soluble coupling reagents such as EDC can be used.
2. For the preparation of first DNA pool, use a sufficient amount of template DNA to minimize the loss of individual sequences in the random library.
3. Perform a cycle course PCR amplification to determine the relative quantities of the bound RNA species using one-tenth aliquot of the reaction mixture. For each sample, 4 μL of aliquots

are removed from the reaction after 4, 8, 12, 16, 20, 24, 28, and 32 cycles and separated on an 8 % native polyacrylamide gel. As the population of binding species becomes enriched, a band of the desired product will appear at an earlier PCR cycle than the previous selection round.

4. We use NAP-5 column to remove unincorporated NTPs, but any other gel filtration system can be used.
5. Do not heat the RNA solution in the presence of Mg^{2+}. This is important to avoid possible chemical degradation of RNA with divalent cations.
6. We usually measure the UV absorbance of the RNA pool solution before and after binding to the resin to determine the percentage of bound RNA in each cycle. The percentage of RNA bound to the resin is calculated as 100(AbsT − AbsF)/AbsT, where AbsT is UV absorbance of the RNA solution before binding to the resin, and AbsF is UV absorbance of the flow-through fraction after binding to the resin.
7. Do not use a spin column with a microporous membrane such as PVDF, because BHQ-1 amine may adsorb to the membrane, probably due to the hydrophobicity of BHQ-1.
8. We usually save a half of the reverse-transcribed product for archival purposes. If subsequent amplification or selection reactions fail for any reason, the procedure can be started from the archival pool.

Acknowledgment

We are indebted to Professor M. Uesugi of Kyoto University for helpful suggestions.

References

1. Roth A, Breaker RR (2009) The structural and functional diversity of metabolite-binding riboswitches. Annu Rev Biochem 78:305–334
2. Breaker RR (2011) Prospects for riboswitch discovery and analysis. Mol Cell 16(43): 867–879
3. Serganov A, Nudler E (2013) A decade of riboswitches. Cell 152:17–24
4. Winkler W, Nahvi A, Breaker RR (2002) Thiamine derivatives bind messenger RNAs directly to regulate bacterial gene expression. Nature 419:952–956
5. Nomura Y, Yokobayashi Y (2007) Reengineering a natural riboswitch by dual genetic selection. J Am Chem Soc 129:13814–13815
6. Wieland M, Benz A, Klauser B, Hartig JS (2009) Artificial ribozyme switches containing natural riboswitch aptamer domains. Angew Chem Int Ed 48:2715–2718
7. Dixon N, Duncan JN, Geerlings T, Dunstan MS, McCarthy JE, Leys D, Micklefield J (2010) Reengineering orthogonally selective riboswitches. Proc Natl Acad Sci U S A 107: 2830–2835
8. Werstuck G, Green MR (1998) Controlling gene expression in living cells through small molecule-RNA interactions. Science 282:296–298
9. Suess B, Hanson S, Berens C, Fink B, Schroeder R, Hillen W (2003) Conditional gene expression by controlling translation with

tetracycline-binding aptamers. Nucleic Acids Res 31:1853–1858

10. Suess B, Fink B, Berens C, Stentz R, Hillen W (2004) A theophylline responsive riboswitch based on helix slipping controls gene expression in vivo. Nucleic Acids Res 32:1610–1614
11. Lynch SA, Desai SK, Sajja HK, Gallivan JP (2007) A high-throughput screen for synthetic riboswitches reveals mechanistic insights into their function. Chem Biol 14:173–184
12. Weigand JE, Sanchez M, Gunnesch EB, Zeiher S, Schroeder R, Suess B (2008) Screening for engineered neomycin riboswitches that control translation initiation. RNA 14:89–97
13. Ellington AD, Szostak JW (1990) In vitro selection of RNA molecules that bind specific ligands. Nature 346:818–822
14. Murata A, Sato S, Kawazoe Y, Uesugi M (2011) Small-molecule fluorescent probes for specific RNA targets. Chem Commun 47:4712–4714

Chapter 3

Development of Photoswitchable RNA Aptamer–Ligand Complexes

Gosuke Hayashi and Kazuhiko Nakatani

Abstract

Photoresponsive artificial riboswitch has the potential to offer a de novo method for spatiotemporal control of gene expression in living cells. Because, even today, it is difficult to design a small molecule binding to a specific RNA sequence, generating an artificial riboswitch that possesses highly specific affinity to a ligand of interest basically depends on in vitro selection procedure where a variety of RNA–ligand complexes can be obtained in established methods. Here, we describe the protocol for in vitro aptamer selection against a photoresponsive peptide ligand containing azobenzene moiety that undergoes photoisomerization through light irradiation. Furthermore, we explain a procedure for surface plasmon resonance assay to detect photoswitchable association and dissociation of RNA–ligand complex on gold surface.

Key words In vitro selection, RNA aptamer, Azobenzene, Photoresponsive association and dissociation, Gene regulation

1 Introduction

Riboswitches are functional RNA structures usually positioned at 5′ or 3′ untranslated regions of mRNA, where they interact with specific metabolite and control gene expression. Upon metabolite binding, alteration of mRNA secondary structure occurs followed by translational or transcriptional modulation [1–3]. To date, about 20 classes of riboswitches have been discovered, which recognize amino acids, nucleobases, sugars, and coenzymes [4]. Such *trans*-acting mechanism of riboswitching is attractive for engineers who aim to develop artificial gene regulation tools. Actually, creating artificial riboswitches started more than a decade ago [5, 6], and, in recent years, a broader range of small molecules such as tetracycline, theophylline, and atrazine have been used as their ligands [7–10]. A specific RNA–ligand complex, which is a key component of riboswitch, has been obtained through in vitro RNA aptamer selection procedure from RNA library containing random sequences [11]. Considering that gene expression levels of these synthetic and

Atsushi Ogawa (ed.), *Artificial Riboswitches: Methods and Protocols*, Methods in Molecular Biology, vol. 1111,
DOI 10.1007/978-1-62703-755-6_3, © Springer Science+Business Media New York 2014

natural riboswitches depend on concentration of their ligands, it takes considerable period to turn on/off gene expression. Furthermore, as the small molecular ligands freely diffuse in cell, spatiotemporal modulation of gene expression is extremely difficult. We hypothesized that photoresponsive riboswitches that catch and release their ligands upon light irradiation would provide a powerful technology to rapidly turn on/off gene expression. To this end, we have developed photoresponsive ligand–aptamer complexes, in which aptamer binding to the ligand that undergoes photoisomerization upon photoirradiation can be controlled in a repetitive fashion [12, 13]. Photoresponsive peptide KRAzR (Lys-Arg-azobenzene-Arg) containing azobenzene chromophore was designed and synthesized as a ligand. Here, we report procedures for in vitro RNA aptamer selection against KRAzR and surface plasmon resonance (SPR) analysis, demonstrating that interactions between KRAzR and selected aptamers were successfully controlled by appropriate irradiation to the SPR sensor chip.

2 Materials

All samples and solutions were prepared using ultrapure water unless otherwise specified.

2.1 In Vitro Selection Components

1. Affi-Gel 10® Gel (Bio-Rad).
2. 50 mM KRAzR peptide in water. After solid-phase peptide synthesis, weigh the lyophilized KRAzR powder and add appropriate amount of water. The synthesis method was previously reported [12].
3. Econo column (Bio-Rad) for packing selection gels.
4. Buffer Kit (Life technologies). This kit contains RNase-free solutions including 1 M Tris, pH 7.0, 1 M Tris, pH 8.0, 5 M NaCl, 1 M $MgCl_2$, 0.5 M EDTA, pH 8.0, 2 M KCl, 5 M ammonium acetate, and DEPC-treated H_2O.
5. TaKaRa Ex Taq® (TaKaRa).
6. AmpliScribe™ T7 High Yield Transcription Kit (Epicentre).
7. Binding buffer (5×): 250 mM Tris, pH 8.0, 1.25 M NaCl, 25 mM $MgCl_2$. Mix 23.75 mL of water, 12.5 mL of 1 M Tris, pH 8.0, 12.5 mL of 5 M NaCl, and 1.25 mL of 1 M of $MgCl_2$ to prepare 50 mL of binding buffer (5×) (*see* **Note 1**).
8. Reverse Transcription System (Promega).
9. Gel washing solution: 1 mM HCl.
10. Coupling buffer: 50 mM $NaHCO_3$, pH 8.
11. Glycine.
12. Synthetic oligonucleotide (131 mer, 5′-GCTAATACG ACTCACTATAGG**GAATTC**CGCGTGTGCACACC(N70)

GTCCGTTC**GGATCC**TCATGG-3′) containing 70 random nucleotides, T7 promoter (underlined), and recognition sites for restriction enzymes, *EcoR*I and *BamH*I (bold).

13. Primers for DNA library construction: forward primer, 5′-GCTAATACGA CTCACTATAGGGAATTCCGCGTGTG CACACC-3′, reverse primer, 5′-CCATGA GGATCCGAAC GGAC-3′.
14. 2-Propanol.
15. 3 M Sodium acetate buffer solution, pH 5.2.
16. 70 % Ethanol.
17. NAP-10 Columns (GE Healthcare).
18. TE buffer solution, pH 8.0.
19. Dr. GenTLE (TaKaRa).
20. *BamH*I.
21. *EcoR*I.
22. Mighty Mix (TaKaRa).
23. pUC18 DNA (TaKaRa).
24. *ECOS*™ *Competent E. coli* DH5a (Nippon Gene).
25. Difco™ LB Broth, Miller (BD).
26. Agar.
27. Ampicillin sodium.
28. QIAprep Spin Mniprep Kit (50) (QIAGEN).

2.2 Biacore Assay Components

1. Biacore 2000 (GE Healthcare).
2. Sensor Chip CM5 (GE Healthcare).
3. Running buffer for aptamer binding assay: Same as the binding buffer used in aptamer selection.
4. 10 mM NaOH.
5. Amine coupling kit (GE Healthcare) contains EDC, NHS, and 1 M ethanolamine hydrochloride solution, pH 8.5 for immobilization of KRAzR peptide to CM5 sensor chip.
6. Running buffer for ligand immobilization: HBS-N Buffer (GE Healthcare).
7. MAX-302 Xenon Light Source 300 W (ASAHI SPECTRA).
8. αCLEAN DUSTER ECO (TRUSCO).

3 Methods

Carry out all procedures at room temperature and under room light unless otherwise specified.

3.1 In Vitro Selection Against KRAzR Peptide

3.1.1 Preparation of KRAzR- and Glycine-Agarose Columns

1. Shake the bottle of Affi-Gel 10® Gel to uniformly suspend gel.
2. Take 1 mL of the agarose gel (about 2 mL of the slurry) into Econo column (*see* **Note 2**).
3. Discard the preloaded solution from the bottom of the column.
4. Add 3-column volume of gel washing solution (3 mL) to completely remove the preloaded solution.
5. Equilibrate the gel by adding 2-column volume of coupling buffer (2 mL).
6. Cap the bottom of the column and add 2 mL of 5 mM KRAzR dissolved in the coupling buffer.
7. Leave the column on a rotating shaker for 1 h (*see* **Note 3**).
8. Drain the reaction solution, and then wash the column with 5-column volume of binding buffer (5 mL).
9. To make agarose gel for negative selection (*see* **Note 4**), use 0.5 mL of gel slurry and 1 mL of 10 mM glycine dissolved in coupling buffer instead of 1 mL gel and 5 mM KRAzR solution, respectively. The other procedures are same as that described above.

3.1.2 DNA Library Construction

1. In a 15 mL tube, prepare 8 mL of PCR reaction mixture containing about 70 μg of the 131 mer synthetic oligonucleotide containing a randomized region, 1 μM of forward and reverse primers, 200 μM of each dNTP, and 200 units of Ex Taq.
2. Dispense the reaction mixture into 80 tubes (100 μL each), and perform PCR in four cycles of 98 °C, 10 s; 55 °C, 30 s; and 72 °C, 30 s (*see* **Note 5**).
3. After PCR reaction, mix together the PCR solution in 80 tubes to two 15 mL tubes (4 mL each) and then add 400 μL of 3 M sodium acetate buffer solution and 4 mL of 2-propanol to each tube.
4. Vortex the mixture and leave them on ice for 10 min. Centrifuge the tubes at 13,000 rpm for 15 min at 4 °C, and discard the supernatant.
5. Add 3 mL of 70 % ethanol and centrifuge at 13,000 rpm for 5 min.
6. Discard the supernatant, and dry the pellet for 10 min.
7. Add 250 μL of water to both the tubes and mix them together (*see* **Note 6**).

3.1.3 RNA Pool Construction

1. Prepare 2.7 mL of transcription reaction containing 170 μg of DNA library (PCR product), 10 mM DTT, 7.5 mM each NTP, and appropriate amount of T7 RNA polymerase and RNase inhibitor according to the manufacturer's protocol.

2. Dispense the reaction mixture into 27 tubes (100 μL each) and incubate at 37 °C for 6 h.
3. Add 5 units (5 μL) of RNase-free DNase I to each tube and incubate at 37 °C for 30 min.
4. Directly charge the reaction mixture onto an equilibrated NAP-10 column. For the use of NAP-10 columns including column preparation and sample elution, see the manufacturer's instruction.
5. Add appropriate amount of 3 M sodium acetate buffer solution and 2-propanol to the eluted sample from NAP-10 column for further purification by 2-propanol precipitation. Following procedure is same as that described in the previous section.
6. Dissolve the pellet into 200 μL of TE buffer.
7. Measure UV_{260} to calculate the concentration of the obtained RNA pool (*see* **Note 7**).

3.1.4 Selection

1. Equilibrate KRAzR-agarose for positive selection and glycine-agarose for negative selection with 5 and 2.5 mL of binding buffer, respectively.
2. Fix both the columns to a pole stand and tandemly place glycine-agarose column over KRAzR column (Fig. 1).
3. Prepare 500 μL of RNA pool containing 550 μg of transcribed RNA dissolved in binding buffer, and load total volume of the

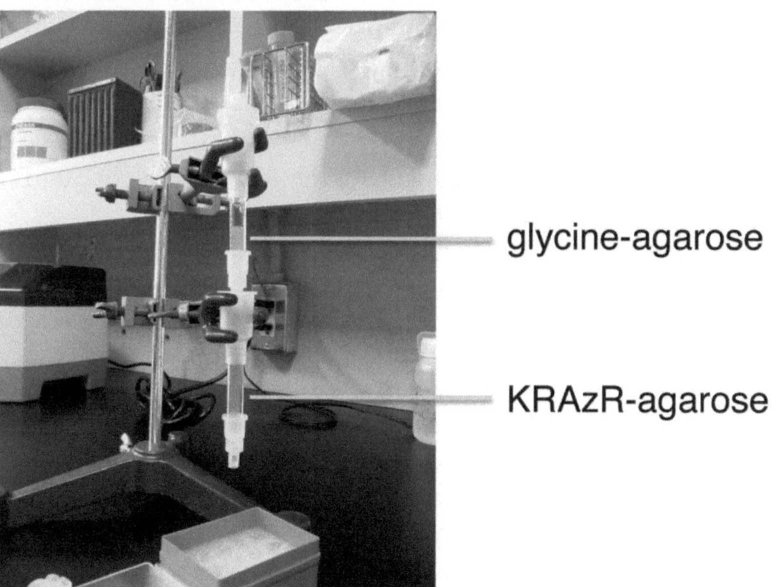

Fig. 1 Apparatus used in our in vitro selection against KRAzR peptide. Glycine-agarose packed column and KRAzR-agarose packed column were tandemly placed to efficiently transfer flow-through of glycine column onto KRAzR column

sample onto glycine column for negative selection. Discard eluted solution during the sample loading.

4. Rinse the glycine column with 1 mL of binding buffer, and load the flow-through directly onto the KRAzR column.
5. Add 4.5 mL of binding buffer to wash out non-binding RNAs. Discard the flow-through.
6. Prepare 3 mL of 6 mM KRAzR solution dissolved in binding buffer for specific elution, and add the total volume onto the column. Discard first 500 μL of the eluate, and then start collecting the flow-through.
7. When the elution is done, add 3 mL of binding buffer and collect all eluate (*see* **Notes 8** and **9**).
8. Mix all fractions together to a 15 mL tube and add appropriate amount of 3 M sodium acetate buffer solution (one-tenth volume of collected solutions), 2-propanol (same volume of collected solutions), and Dr. GenTLE (precipitation carrier) (*see* **Note 10**) for 2-propanol precipitation.
9. Vortex the solution, and leave the tube on ice for 10 min. Centrifuge the tubes at 13,000 rpm for 15 min at 4 °C, and discard the supernatant.
10. Add 3 mL of 70 % ethanol and centrifuge at 13,000 rpm for 5 min. Discard the supernatant, and dry the pellet for 10 min.
11. Dissolve the pellet with 20 μL of TE buffer.

3.1.5 Reverse Transcription (RT)-PCR

1. Incubate the recovered RNA solution at 80 °C for 3 min, and then place it on ice (*see* **Note 11**).
2. Add appropriate amount of dNTPs, reverse transcription buffer, $MgCl_2$, RNase inhibitor, primer, and AMV reverse transcriptase according to the manufacturer's protocol.
3. Incubate the total volume of 50 μL reaction mixture at 42 °C for 30 min.
4. Using all of RT reaction mixture, prepare 500 μL of PCR reaction with ExTaq polymerase, according to the procedure as described above. Dispense the reaction mixture into ten tubes (50 μL each).
5. To determine the optimum PCR cycles, begin with the 50 μL volume and take 5 μL in each four cycles. PCR amplification should be done in the condition of 98 °C, 10 s; 55 °C, 30 s; and 72 °C, 30 s.
6. From PAGE analysis, determine the appropriate PCR cycle number (Fig. 2a) (*see* **Note 12**). In our case, we chose 11 cycles.
7. Run PCR reaction with the remaining 450 μL of the reaction mixture, and check the PCR product by PAGE (Fig. 2b).

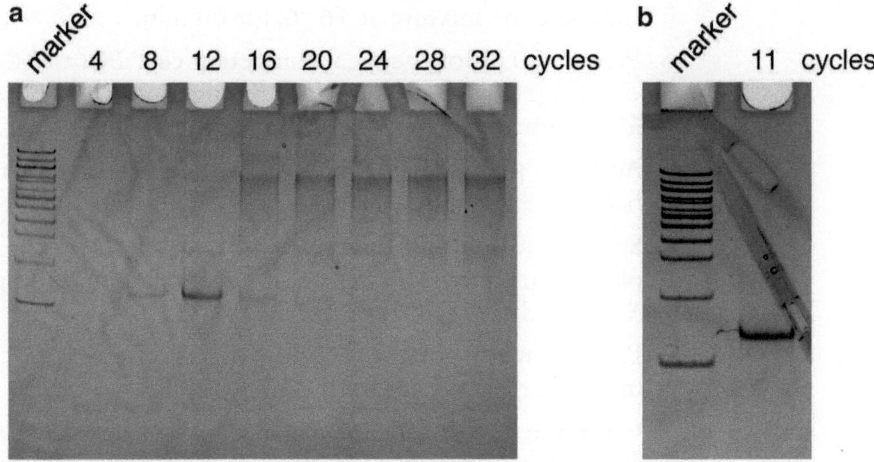

Fig. 2 Native PAGE analysis of RT-PCR products obtained at first round of selection. (**a**) Optimization of PCR cycles using 50 μL of PCR mixture. In each four cycles, 5 μL of the reaction mixture was taken for PAGE analysis. After electrophoresis, the gel was stained with SYBR Green. (**b**) Confirmation of migration of RT-PCR products (11 cycles) obtained in large-scale reaction. After electrophoresis, the gel was stained with SYBR Green

8. Purify the obtained solution by 2-propanol precipitation as described above, and dissolve the obtained pellet in 50 μL of water.

3.1.6 Next Rounds of Selection

1. To proceed to the second round of the selection, prepare 200 μL of transcription mixture and react with the same condition described above using half of the transcribed RNA for loading to selection column. The other half should be stored in −20 °C refrigerator. To cycle selection procedures including negative and positive selection, RT-PCR, and transcription, basically follow the protocol described above (*see* **Notes 13** and **14**).
2. When you confirm the enrichment of RNA bound to the KRAzR-gel after several rounds of selection (*see* **Note 15**), stop the selection cycle after RT-PCR.

3.1.7 Isolation and Identification of Aptamer Sequences

1. Add *BamH*I and *EcoR*I to the obtained PCR product (DNA library), which is dissolved in appropriate buffer according to the manufacturer's protocol and incubate at 37 °C for 3 h.
2. Treat pUC18 in the same condition in a 200 μL tube.
3. After 37 °C incubation, place both the DNA library and pUC18 at 70 °C for 10 min for deactivation of restriction enzymes.
4. Prepare 10 μL of ligation solution including 5 μL of Mighty Mix, about 25 fmol of cleaved pUC18, and about 25–250 fmol of cleaved DNA library.

5. Incubate the mixture at 16 °C for 60 min.
6. Prepare 50 μL of chemical competent cell, *ECOS™ Competent E. coli* DH5a, and add 5 μL of the ligation mix. For the transformation, use general heat shock procedure.
7. Add 450 μL of LB medium to the competent cell, and incubate it at 37 °C for 1 h.
8. Spread the solution directly or after dilution on the LB-agar plates containing 100 μg/mL of ampicillin and 40 μg/mL of X-Gal and incubate at 37 °C overnight.
9. Pick the white colonies (blue-white selection), and check the insertion of DNA library by colony PCR.
10. Grow the *E. coli* having proper length of insert DNA on the vector DNA in 5 mL LB medium containing 100 μg/mL of ampicillin.
11. Extract and purify the plasmid DNA by QIAprep Spin Mniprep Kit.
12. Determine the concentration of obtained plasmid DNAs by measuring the UV_{260}.
13. Read the sequences of obtained RNA aptamers.

3.2 SPR Analysis of Photoresponsive Interaction Between RNA Aptamers and KRAzR Peptide

Basically, follow the instruction manual of Biacore 2000. Use CM5 sensor chip throughout all procedures. Immobilization of KRAzR onto the gold surface was done through amide bond formation.

3.2.1 Ligand Immobilization on Gold Surface

1. Flow HBS-N buffer as running buffer followed by mixture of EDC and NHS to activate carboxylic acid on the surface.
2. Inject immediately 5 mM KRAzR dissolved in coupling buffer, followed by HBS-N buffer for washing.

3.2.2 Sample Preparation

1. To prepare RNA aptamer samples for Biacore analysis, PCR-amplify template DNAs from plasmid DNAs.
2. Transcribe RNA aptamers, and purify them by 8 % denaturing PAGE and 2-propanol precipitation.
3. Dissolve the obtained pellet with binding buffer used in the selection procedure, and prepare 1.0 μM of each sample (*see* **Note 16**).

3.2.3 SPR Analyses

1. Set running buffer (binding buffer) and RNA samples in the Biacore 2000.
2. Flow running buffer for 90 s, and then inject aptamer sample for 120 s for binding to the KRAzR surface (association).
3. Flow the running buffer again for 120 s for dissociation.

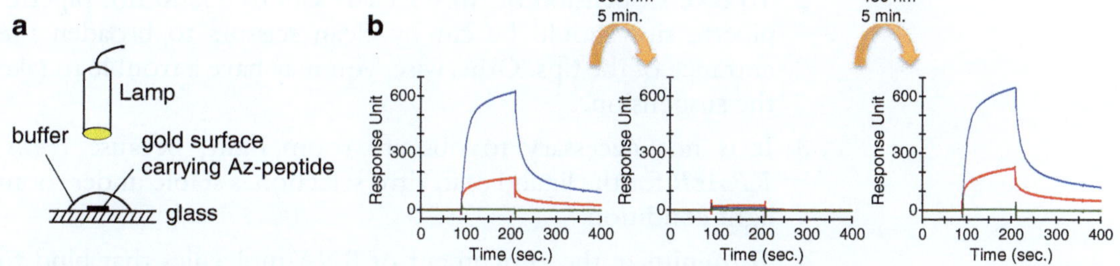

Fig. 3 Photoresponsive interaction between RNA aptamers and KRAzR peptide. (**a**) Illustration of light irradiation to the SPR sensor chip covered with droplet of running buffer. (**b**) SPR sensorgrams before and after light irradiation. Two different RNA aptamers (*red* and *blue line*) were analyzed. As negative control, running buffer was injected instead of RNA aptamers (*green line*)

4. Inject 10 mM NaOH for about 100 s to wash out bound RNAs on the surface.
5. Flow the running buffer again, and proceed to next measurement.

3.2.4 Photoisomerization of KRAzR on Gold Surface

1. To photo-isomerize *trans*-KRAzR to *cis* form, set up a darkroom (*see* **Note 17**).
2. Cover the gold surface of sensor chip with a droplet of running buffer (*see* **Note 18**), and irradiate 360 nm light for 5 min by MAX-302 Xenon Light Source 300 W (Fig. 3a).
3. Blow the droplet away by blowing air with αCLEAN DUSTER ECO, and set the chip back into Biacore 2000 as soon as possible.
4. Analyze the same aptamer with the same procedure as described above.
5. In the case of isomerization from *cis*-KRAzR back to *trans-isomer*, irradiate 430 nm light for 5 min. Photoresponsive binding between RNA aptamers and KRAzR peptide would be observed (Fig. 3b).

4 Notes

1. Through all experimental procedures including RNA handling, contamination of RNase is destructive. So, always take care not to contaminate any dirt or dust. The use of gloves and masks is favorable. Speaking during experiments might cause contamination of RNase. Pipette tips accidentally contacted with skins, clothes, or desks should be discarded immediately. The experiments using RNase such as miniprep and genome extraction should be done in another room or at a space far from your RNA experimental space.

2. To take suspension of Affi-Gel 10® Gel by 1,000 mL pipette, pipette tips should be cut by clean scissors to broaden the entrance of the tips. Otherwise, you may have a trouble to take the suspension.
3. It is not necessary to obscure room light, because *trans-KRAzR* for the ligand of in vitro selection is stable under room light condition.
4. To minimize the enrichment of RNA molecules that bind to the gel itself such as spacer molecules tethered on the gel, negative selection should be done before KRAzR selection.
5. Before large-scale reaction, test with small scale (e.g., 100 μL) and determine the optimal cycle number by PAGE analysis. The excess number of cycles often causes decreased yields of proper product.
6. Determine the concentration of obtained PCR product using UV_{260} measurement. PAGE analysis is also useful to determine the concentration by comparing the band intensity with that of an authentic sample.
7. If you want to confirm the purity of the transcription product, denaturing PAGE analysis is useful.
8. Throughout the procedure, to obtain RNA aptamers against *trans*-KRAzR, you do not need to obscure room light because *trans*-KRAzR is stable under room light with E/Z ratio of 79:21 [12]. However, if you select aptamers against photo-unstable compound such as *cis*-KRAzR, the positive selection procedure should be done in the dark.
9. The glycine-agarose and KRAzR-agarose are reusable after selection. Wash each column with 5-gel volume of 7 M urea for rinsing the remaining molecules, and then equilibrate the gel with 5-gel volume of binding buffer. When you store the gels, put the columns in 4 °C refrigerator.
10. Because most of the RNA sequences are not retained in the column and the recovery rate of bound RNA is very low, use of precipitating agent may increase the recovery rate. But it is not indispensable. The precipitating agent is not harmful for next procedures such as RT-PCR reaction.
11. When you heat your sample containing RNA, magnesium cation must be excluded. Magnesium cation catalyzes RNA-cleaving reaction by coordinating oxygen atom of hydroxyl group and phosphate group on RNA.
12. As seen in Fig. 2a, overcycling of PCR gives decreased proper band and increased by-product, presumably because DNA templates with different sequences at the randomized region would interact with each other and cause undesirable elongation. Therefore, be careful in choosing the proper cycle number.

You may take samples in shorter period of PCR cycles to increase the accuracy of your choice.

13. You do not have to choose the same selection conditions of first round after second round. Increased selection stringency such as reduced gel volume or increased temperature sometimes provides good RNA aptamers with strong affinity.
14. When you stop your experiment, please stop at RT-PCR, transcription, or 2-propanol precipitation. Do not stop your experiment during negative or positive selection because magnesium cation present in binding buffer would cause RNA degradation.
15. When an enrichment of selected RNA aptamers occurs, you will see the reduced PCR cycle number in PAGE analysis of RT-PCR products compared to that in the previous rounds. This is because the amount of RNA bound to KRAzR-agarose gradually increases as the enrichment proceeds.
16. Sample preparation should be done just before the measurements. Do not leave the sample for a long time. Magnesium cation included in binding buffer gradually degrades aptamers.
17. *Cis*-isomer of KRAzR is unstable under room light and isomerizes back to *trans*-isomer for 5 min under the room light. Therefore, when your ligand is photolabile, take care to avoid unwanted light exposure to your ligand.
18. To keep wet condition of the surface, we covered the surface with droplet of running buffer. But that might not be necessary.

Acknowledgements

This work was supported by JSPS grant of Research Fellowship for Young Scientists.

References

1. Mandal M, Breaker RR (2004) Gene regulation by riboswitches. Nat Rev Mol Cell Biol 5:451–463
2. Serganov A, Patel DJ (2007) Ribozymes, riboswitches and beyond: regulation of gene expression without proteins. Nat Rev Genet 8:776–790
3. Winkler WC, Breaker RR (2005) Regulation of bacterial gene expression by riboswitches. Annu Rev Microbiol 59:487–517
4. Serganov A, Patel DJ (2012) Metabolite recognition principles and molecular mechanisms underlying riboswitch function. Annu Rev Biophys 41:343–370
5. Soukup GA, Breaker RR (1999) Engineering precision RNA molecular switches. Proc Natl Acad Sci U S A 96:3584–3589
6. Werstuck G, Green MR (1998) Controlling gene expression in living cells through small molecule-RNA interactions. Science 282:296–298
7. Bauer G, Suess B (2006) Engineered riboswitches as novel tools in molecular biology. J Biotechnol 124:4–11

8. Sinha J, Reyes SJ, Gallivan JP (2010) Reprogramming bacteria to seek and destroy an herbicide. Nat Chem Biol 6:464–470
9. Suess B, Fink B, Berens C, Stentz R, Hillen W (2004) A theophylline responsive riboswitch based on helix slipping controls gene expression in vivo. Nucleic Acids Res 32:1610–1614
10. Suess B, Hanson S, Berens C, Fink B, Schroeder R, Hillen W (2003) Conditional gene expression by controlling translation with tetracycline-binding aptamers. Nucleic Acids Res 31:1853–1858
11. Wilson DS, Szostak JW (1999) In vitro selection of functional nucleic acids. Annu Rev Biochem 68:611–647
12. Hayashi G, Hagihara M, Dohno C, Nakatani K (2007) Photoregulation of a peptide-RNA interaction on a gold surface. J Am Chem Soc 129:8678–8679
13. Hayashi G, Hagihara M, Nakatani K (2009) RNA aptamers that reversibly bind photoresponsive azobenzene-containing peptides. Chemistry 15:424–432

Chapter 4

Identification of RNA Aptamers Against Recombinant Proteins with a Hexa-Histidine Tag

Shoji Ohuchi

Abstract

Artificial riboswitches that respond to the concentrations of intracellular proteins are a promising tool with a variety of applications. They can be designed and engineered using existing RNA aptamers that target proteins. Aptamers are generated via an iterative selection–amplification process, known as systematic evolution of ligands by exponential enrichment (SELEX). This chapter describes a SELEX procedure for the identification of RNA aptamers against hexa-histidine-tagged proteins. For the efficient enrichment of higher affinity aptamers, the selection stringency should be gradually increased. Undesired RNA species that bind to affinity resins can be eliminated from the pool by using a negative selection step and alternating different types of resins.

Key words Aptamer, SELEX, His-tag, In vitro selection, RNA, Artificial riboswitch, Synthetic gene circuit

1 Introduction

Riboswitches are promising gene components for the construction of artificial gene circuits and cell-based biosensors. Using existing aptamers, new artificial riboswitches can be constructed (see later related chapters in this book). Although most artificial riboswitches have been generated to respond to metabolites and small cell-permeable compounds, the generation of artificial riboswitches for intracellular proteins is also possible. Such riboswitches can be used to regulate gene expression according to the intracellular environment; thus, they are valuable for the design of artificial gene circuits with cascade structures [1] (see later related chapters in this book). In addition, artificial riboswitches for disease-related proteins may be applied for novel types of therapy.

Aptamers are generated in vitro from large random sequence libraries based on their high affinity for a target molecule by a process known as systematic evolution of ligands by exponential enrichment

Atsushi Ogawa (ed.), *Artificial Riboswitches: Methods and Protocols*, Methods in Molecular Biology, vol. 1111,
DOI 10.1007/978-1-62703-755-6_4, © Springer Science+Business Media New York 2014

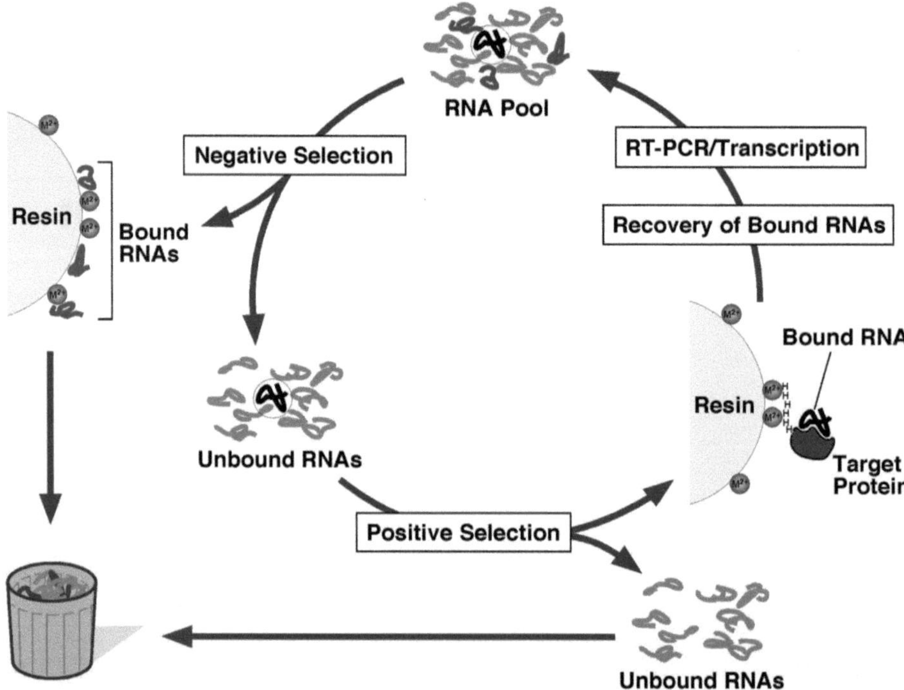

Fig. 1 Schematic diagram of SELEX for the identification of RNA aptamers against a hexa-histidine-tagged protein. Through negative selection, undesired RNA species that bind to the resin (indicated by *dark gray*) are eliminated. The unbound RNA is recovered and used for positive selection. After extensive washing, RNA species that bind tightly to the target (indicated by *black*) are isolated and amplified by RT-PCR. Following in vitro transcription regenerates an enriched RNA pool for the next round of SELEX. These processes are repeated until the desired aptamers are enriched in the pool

(SELEX) [2–7]. SELEX is an iterative selection–amplification process (Fig. 1). Since the establishment of this procedure, a wide variety of aptamers has been generated, and their applications have been extensively studied. In this chapter, I describe a procedure for the identification of RNA aptamers that target the most popular recombinant proteins, hexa-histidine (His)-tagged proteins.

2 Materials

Prepare all solutions under ribonuclease (RNase)-free conditions.

2.1 Random Pool Constructions

1. Random sequence template DNA (R3-N40): 5′-CAAGG AGCGACCAGAGG-N_{40}-TGGCATCCTT CAGCCC-3′ (N_{40} represents a random sequence of 40 nucleotides) (*see* **Notes 1** and **2**).
2. Forward primer (R3f): 5′-<u>TAATACGACT CACTATA</u>GGG CTGAAGGATG CCA-3′ (the T7 promoter region is underlined).
3. Reverse primer (R3r): 5′-CAAGGAGCGA CCAGAGG-3′.

4. TE buffer: 10 mM Tris–HCl, pH 7.5, 1 mM EDTA.
5. Thermostable DNA polymerase for PCR [e.g., *ExTaq* (Takara Bio, Japan)].
6. RiboMAX Large Scale RNA Production System-T7 (Promega).
7. RNase-free water (e.g., MilliQ water).
8. Mixed phenol:chloroform:isoamyl alcohol (PCI) at 25:24:1.
9. Thermocycler (PCR equipment).
10. 0.5× TBE buffer: 44.5 mM Tris–HCl, 44.5 mM borate, and 1 mM EDTA.
11. 8 % polyacrylamide gel in 0.5× TBE buffer.
12. Microcon YM-30 (Millipore).
13. MicroSpin G-25 columns (GE Healthcare).
14. UV spectrometer.
15. Random sequence template DNA for the alternative method (R3-N40+T7p): 5′-CAAGGAGCGA CCAGAGG-N40-TGG CATCCTT CAGCCCTATA GTGAGTCGTA TTA-3′ (the T7 promoter region is underlined; N_{40} represents a random sequence of 40 nucleotides).

2.2 SELEX

1. Buffer A: 20 mM Tris–HCl, pH 7.5, 100 mM potassium acetate, 5 mM magnesium chloride, and 2 mM dithiothreitol (DTT) (*see* **Note 3**).
2. Buffer B: 40 mM Tris–HCl, pH 7.5, 200 mM potassium acetate, 10 mM magnesium chloride, 4 mM DTT, and 0.1 U/μL RNase inhibitor (*see* **Note 4**).
3. 10 mg/mL tRNA (Sigma) in RNase-free water.
4. 1 % (w/v) BSA (Sigma) in RNase-free water.
5. Recombinant His-tagged protein.
6. HisLink Protein Purification Resin (Promega).
7. TALON Metal Affinity Resin (Clontech Laboratories).
8. Buffer C: Buffer A supplemented with 0.1 U/μL RNase inhibitor and 500 mM imidazole (*see* **Note 4**).
9. Co-precipitant for ethanol precipitation [e.g., Pellet Paint Co-Precipitant (Novagen)].
10. Sample tube rotator.
11. Reverse transcriptase [e.g., ReverTra Ace (Toyobo, Japan); SuperScript III (Invitrogen)].

2.3 Pull-Down Assay

1. Denaturing solution: 90 % formamide, 20 mM EDTA, 0.5 % bromophenol blue (BPB), 0.1 % xylene cyanol (XC).
2. 12 % denaturing polyacrylamide gel in 0.5× TBE buffer containing 7 M urea.

3. SYBR Green II (Molecular Probes).
4. Phosphorimager [e.g., FLA 7000 (Fuji Film)].

2.4 Electrophoretic Mobility Shift Assay

1. 6× loading solution: 50 % glycerol, 1 % BPB, and 0.2 % XC.
2. TBM buffer: 89 mM Tris, 89 mM borate, and 1 mM magnesium chloride (*see* **Notes 5** and **6**).
3. 10 % polyacrylamide gel in TBM buffer.

2.5 Filter-Binding Assay

1. Radioisotope (e.g., [^{32}P])-labeled RNA (*see* **Note 7**).
2. Blotting unit [e.g., Minifold Dot-blot System (Whatman)].
3. Nitrocellulose blotting membrane (pore size 0.45 μm) (Whatman).
4. A liquid scintillation counter.

3 Methods

3.1 Random DNA Pool Construction

Two alternative methods are described blow (Fig. 2). The first method (Subheading 3.1.1) is more common. The second method (Subheading 3.1.2) requires longer sequences of synthetic DNA but no enzymatic reaction.

3.1.1 DNA Pool Preparation by Extension Reaction

1. Extension reaction (Fig. 2a): The total reaction volume is 500 μL: 1× PCR buffer (supplemented), 0.2 mM of each dNTP (supplemented), 1 μM R3-N40 (template DNA), 5 μM

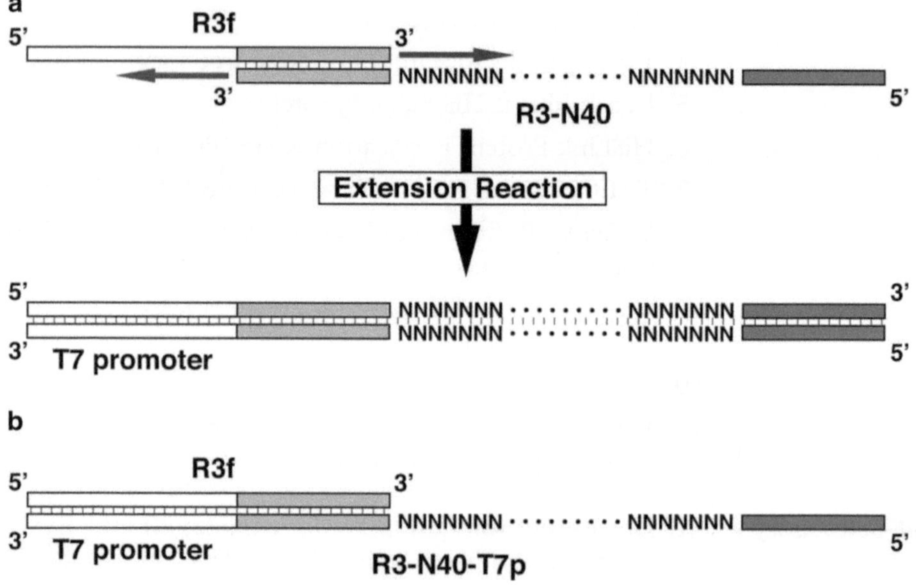

Fig. 2 Procedures for the preparation of the random DNA pool: (**a**) Extension reaction-based procedure. (**b**) Annealing-based procedure. *Boxes* and "N" represent the primer regions and the random sequence, respectively. *Gray arrows* indicate the orientation of the extension reaction

R3f (forward primer), and 50 U *ExTaq*. After denaturing for 1 min at 94 °C, anneal and extend the DNA at 50 °C for 15 s and 72 °C for 3 min, respectively. Analyze the double-stranded DNA (dsDNA) product (3 μL of the reaction) by 8 % PAGE (*see* **Note 8**).

2. Purify the DNA by PCI extraction and ethanol precipitation. Dissolve the DNA in 50 μL of TE buffer.

3.1.2 DNA Pool Preparation by Primer Annealing (Alternative Method)

By using single-stranded DNA template that contains a full promoter region, the DNA pool can be prepared without enzymatic reaction. The annealing of a forward primer to the template makes a transcriptionally competent, double-stranded promoter region (Fig. 2b) [8].

The total volume is 40 μL: 0.1× PCR buffer, 10 μM R3-N30+T7p, and 10 μM R3f. After denaturing for 1 min at 96 °C, anneal the DNA under the following conditions using a thermocycler: 70 °C for 1 min; 60 °C for 1 min; 50 °C for 1 min; and 40 °C for 5 min. All of the annealed DNA will be used for in vitro transcription.

3.2 Preparation of the Random RNA Pool

1. In vitro transcription: Incubate a 100 μL reaction containing 1× transcription buffer (supplemented), 40 μL of the DNA pool prepared by extension reaction (Subheading 3.1.1) or by annealing (Subheading 3.1.2), 5 mM of each rNTP (supplemented), and 10 μL of T7 RNA polymerase enzyme mix at 37 °C for 2–4 h.
2. Digest the template DNA by adding 5 U of RQ1 DNase (supplemented) and incubating at 37 °C for an additional 15 min.
3. Extract the transcript with PCI and purify on Microcon YM-30 and MicroSpin G-25 columns.
4. Measure the RNA concentration based on the absorbance at 260 nm using a UV spectrometer.

3.3 SELEX

3.3.1 Negative Selection

1. Wash 3 μL of empty resin (HisLink or TALON) with 100 μL of buffer A once in a 0.2-mL PCR tube. Resuspend the resin in 100 μL of buffer A, and store the tube at room temperature until use.
2. Dilute 2 nmol (1.2×10^{15} molecules) of the purified RNA pool with water to 48 μL. To facilitate RNA folding, heat the RNA solution at 90 °C for 3 min and chill quickly to 4 °C using a thermocycler. Then, add 50 μL of buffer B and incubate at 37 °C for 10 min.
3. Add 1 μL of 10 mg/mL tRNA and 1 μL of 1 % BSA to the RNA solution (*see* **Note 9**).
4. Remove the supernatant from the resin tube, and add the RNA solution to the resin.

5. Incubate the resin tube at room temperature with gentle rotation for 30 min by using a sample tube rotator (*see* **Note 10**).
6. Spin down the resin, and collect the supernatant. The recovered supernatant containing RNA not bound to the empty resin will be used for positive selection.

3.3.2 Preparation of Target-Immobilized Resin

1. Wash 3 μL of empty resin (HisLink or TALON) with 100 μL of buffer A once in a 0.2-mL PCR tube. Resuspend the resin in 100 μL of buffer A and add 25 pmol of the target protein.
2. Incubate the tube at room temperature with gentle rotation for 30 min using a rotator (*see* **Note 10**).
3. Spin down the resin, and remove the supernatant. To remove un-immobilized proteins, wash the resin with 100 μL of buffer A three times (*see* **Note 11**).

3.3.3 Positive Selection

1. Spin down the target-immobilized resin (Subheading 3.3.2), and remove the supernatant. Add the RNA solution from the negative selection (Subheading 3.3.1) to the resin. Incubate the tube at room temperature with gentle rotation for 30 min using a rotator (*see* **Note 10**).
2. Spin down the resin, and remove the supernatant containing unbound RNA. To remove weakly bound RNA, wash the resin with 100 μL of buffer A four times.
3. Spin down the resin, and remove the supernatant. To release the target protein from the resin, resuspend the resin in 100 μL of buffer C. Incubate the tube at room temperature with gentle rotation for 10 min using a rotator (*see* **Notes 10** and **12**).
4. Collect the supernatant containing the released protein and RNA bound to the protein. Recover the bound RNA by PCI extraction followed by ethanol precipitation with a co-precipitant (*see* **Note 13**).

3.3.4 Reverse Transcription (RT)-PCR

1. Reverse transcription: Dissolve the recovered RNA in 31 μL of RNase-free water. Add 1 μL of 100 μM R3r primer, 10 μL of 5× reaction buffer (supplemented), 5 μL of 10 mM dNTP mixture, 0.5 μL of 10 U/μL RNase inhibitor, and 2.5 μL of 100 U/μL ReverTra Ace. Incubate the mixture at 50 °C for 1 h and then at 99 °C for 5 min to heat inactivate the enzyme (*see* **Note 14**).
2. PCR: The total reaction volume is 300 μL: 50 μL (all) of the reverse transcription product, 1× PCR buffer (supplemented), 0.2 mM of each dNTP (supplemented), 3 μM R3f primer, 3 μM R3r primer, and 0.1 U/μL *ExTaq*. Run eight cycles of PCR as follows: denaturing at 96 °C for 15 s; annealing at 50 °C for 15 s; and extension at 72 °C for 15 s (*see* **Notes 15** and **16**).

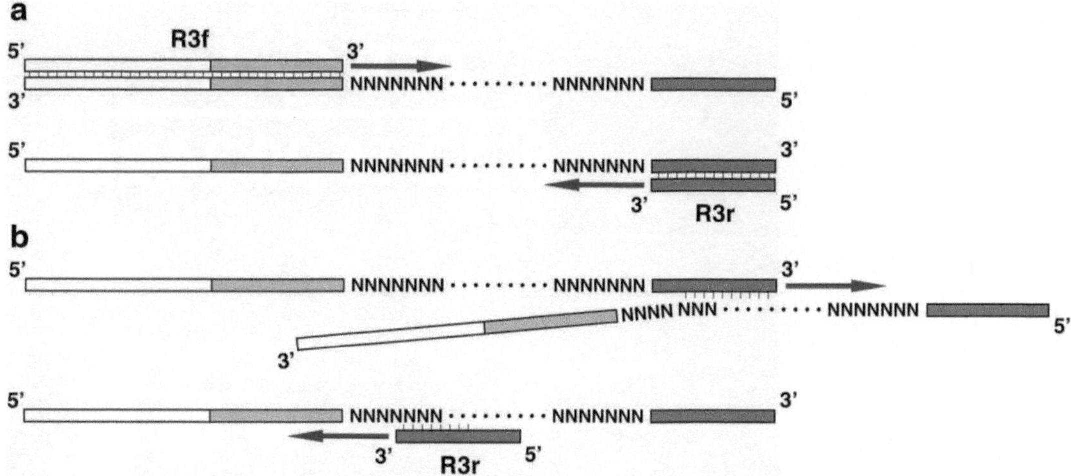

Fig. 3 Desired and undesired reactions during PCR. (**a**) Desired reactions during the extension step. (**b**) Undesired side reactions. Because of the random sequence in the DNA pool, unintended mis-annealing may occur. These side reactions generate longer and shorter by-products. *Boxes* and "N" represent the primer regions and the random sequence, respectively. *Gray arrows* indicate the orientation of the extension reaction

3. Amplification cycle check: Equally divide 50 μL of the PCR mixture into five new PCR tubes (10 μL each). Run 4, 8, 12, 16, or 20 additional cycles of PCR. Analyze the PCR product (3 μL from the tube) by 8 % PAGE, and determine the number of cycles necessary and sufficient for amplification (Figs. 3 and 4) (*see* **Notes 17** and **18**).
4. Run additional cycles of PCR for the remaining PCR mixture, according to the cycle number determined above.
5. Purify the PCR product by PCI extraction and ethanol precipitation. Dissolve the purified DNAs in 20 μL of TE buffer (*see* **Note 19**).

3.3.5 Subsequent Rounds of SELEX

Repeat the process described in Subheadings 3.2–3.3.4, typically for ten rounds. Reaction volumes for in vitro transcription (Subheading 3.2) and RT-PCR (Subheading 3.3.4) can be reduced by half or more. To enrich tightly binding RNA, gradually increase the selection stringency. To achieve this, reducing the target concentration and increasing the number of washes during positive selection are helpful. To avoid enriching RNA that binds tightly to the resin, alternate the type of resins used. Examples of the typical conditions for each round of SELEX are listed in Table 1.

3.4 Analysis of SELEX Progress

The number of SELEX rounds necessary depends on the properties of the target protein and the selection conditions employed. Thus, it is impossible to know how many rounds to perform a priori, and the progress of SELEX should be experimentally confirmed by analyzing the affinity enrichment of the RNA pool (*see* **Note 20**).

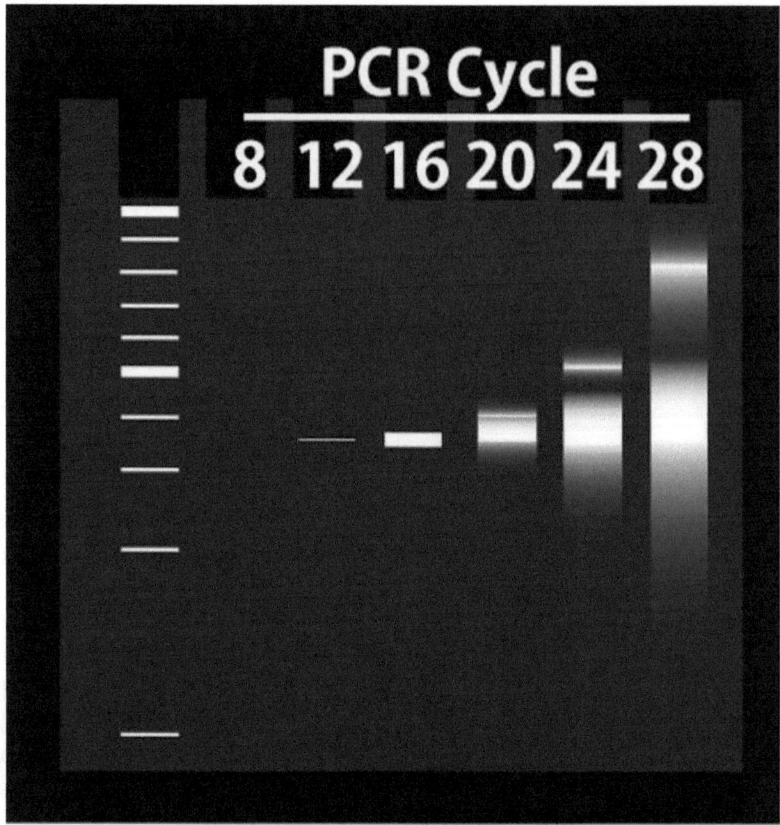

Fig. 4 Optimization of PCR cycle number. Too many PCR cycles generate undesired longer and shorter by-products, which can be recognized by a smear band pattern in PAGE analysis. Therefore, the number of cycles that is necessary and sufficient should be determined. In the case shown, 16–18 cycles are optimal

Table 1
Conditions for positive selection against a His-tagged target protein

Round	Target protein (nM)	Resin	Washing
1	250	HisLink	4
2	250	TALON	5
3	125	HisLink	5
4	125	TALON	6
5	62.5	HisLink	6
6	62.5	TALON	7
7	31.3	HisLink	7
8	31.3	TALON	8
9	15.6	HisLink	8
10	15.6	TALON	10

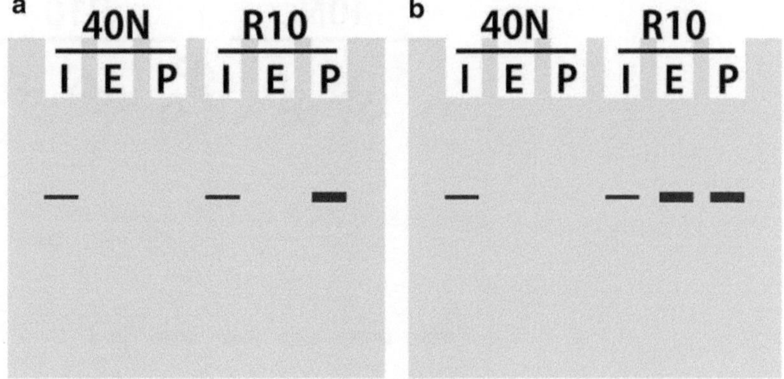

Fig. 5 Examples of pull-down assay results. (**a**) Case demonstrating successful SELEX. If SELEX progresses successfully, the RNA pool (in this case, the Round 10 pool, designated "R10") binds to the protein-immobilized resin (indicated as "P") but not to the empty resin (indicated as "E"). (**b**) Case demonstrating unsuccessful SELEX. If undesired RNA species that bind to resin are enriched, the RNA pool ("R10") binds to the protein-immobilized resin and to the empty resin. "40N" and "I" indicate the control RNA and 1/10 volume of an input sample, respectively

3.4.1 Pull-Down Assay

1. Prepare 6 μL of the resin with 200 pmol of the target protein immobilized as described above (Subheading 3.3.2). Divide it equally into two 0.2-mL PCR tubes. As a resin control, wash 6 μL of empty resin and divide it equally into two 0.2-mL PCR tubes (*see* **Notes 21** and **22**).
2. Dilute 200 pmol of the RNA pool to be tested with water to 96 μL, fold the RNA, and add tRNA and BSA as described above (Subheading 3.3.1). The total volume of the RNA sample is 200 μL. As an RNA control, use 200 pmol of RNA from the initial random pool.
3. Remove the supernatant from the resin tubes, and add 90 μL of each RNA sample onto each resin.
4. Incubate the resin tubes at room temperature with gentle rotation for 30 min using a rotator (*see* **Note 10**).
5. Spin down the resin and wash with 100 μL of buffer A twice. After removing the supernatant, resuspend the resin in 10 μL of the denaturing solution. As an input sample control, transfer 9 μL of the remaining folded RNA to a 0.2-mL PCR tube and add 9 μL of the denaturing solution.
6. Heat the samples at 90 °C for 3 min and chill quickly to 4 °C using a thermocycler. Separate the samples on 12 % denaturing PAGE containing 7 M urea. Soak the gel with SYBR Green II to stain the RNA bands, and analyze the band image using a phosphorimager. If SELEX works successfully, and the desired aptamers are enriched in the pool, the RNA pool binds to the protein-immobilized resin but not to the empty resin (Fig. 5).

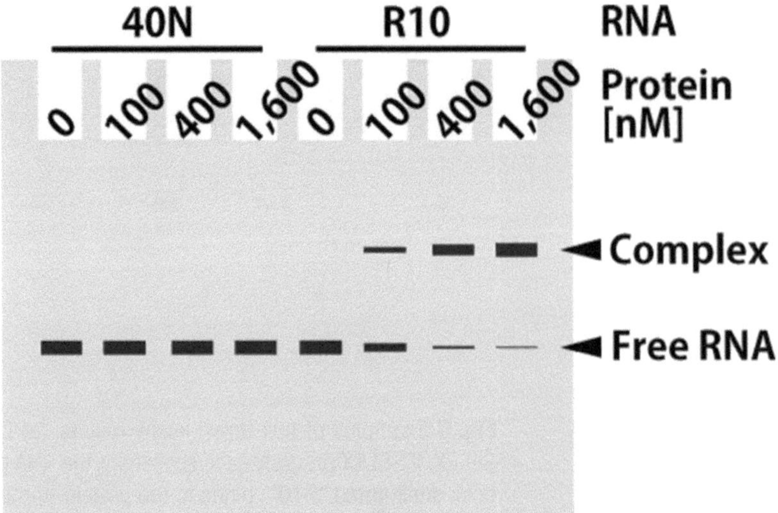

Fig. 6 Example of EMSA results. When SELEX progresses successfully, the RNA pool (in this case, the Round 10 pool, designated "R10") forms complexes with the target in a protein concentration-dependent manner. Not all RNA molecules form complexes even when the protein concentration is high due to mis-folding of the aptamers and the presence of inactive RNA species in the pool. "40N" indicates the control RNA

3.4.2 Electrophoretic Mobility Shift Assay (Alternative Method)

1. Dilute 5 pmol of the RNA pool to be tested with water to 12 μL, fold the RNA, and add tRNA and BSA as described above (Subheading 3.3.1). The total volume of the RNA sample is 25 μL. As an RNA control, use 5 pmol of RNA from the initial random pool (*see* **Note 21**).
2. Dilute 2.5, 10, or 40 pmol of the target protein with buffer A to 12.5 μL.
3. Mix 5 μL of each folded RNA and 5 μL of the diluted protein. As a no-protein control, mix 5 μL of each folded RNA and 5 μL of buffer A. The final concentrations of RNA and protein are 100 nM and 0 nM, 100 nM, 400 nM, or 1,600 nM, respectively.
4. Incubate at 37 °C for 30 min.
5. Chill on ice, and add 2 μL of the 6× loading solution.
6. Run the TBM gel at 75 V for 2 h in a cold room (*see* **Note 23**).
7. Soak the gel with SYBR Green II to stain the RNA bands, and analyze the band image using a phosphorimager. If SELEX works successfully, and the desired aptamers are enriched in the pool, the RNA pool forms a complex with the protein, and mobility-shifted bands can be detected (Fig. 6).

3.4.3 Filter-Binding Assay (Alternative Method)

1. Dilute 0.1 pmol of the radioisotope-labeled RNA pool with water to 120 μL, fold the RNA, and add tRNA and BSA as described above (Subheading 3.3.1). The total volume of the

RNA sample is 250 μL. As an RNA control, use 0.1 pmol of RNA from the initial random pool (*see* **Note 24**).

2. Dilute 2.5, 25, or 250 pmol of the target protein with buffer A to 125 μL.
3. Mix 50 μL of each folded RNA and 50 μL of the diluted protein. As a no-protein control, mix 50 μL of each folded RNA and 50 μL of buffer A. The final concentrations of RNA and protein are 1 nM and 0 nM, 10 nM, 100 nM, or 1,000 nM, respectively.
4. Incubate at 37 °C for 30 min.
5. Chill on ice.
6. Soak the nitrocellulose membrane with buffer A for a few minutes. Place the soaked membrane onto the blotting unit, and assemble the apparatus.
7. Connect to a vacuum pump, and apply the vacuum.
8. Apply each sample to the membrane. Quickly apply 200 μL of buffer A twice to wash the membrane (*see* **Note 25**).
9. Disassemble the blotting unit and turn off the pump. Place the membrane onto a paper towel to dry (*see* **Note 26**).
10. Cut off each blotted position, and evaluate the amount of trapped RNA using a liquid scintillation counter. If SELEX works successfully, and the desired aptamers are enriched in the pool, the amount of RNA trapped increases in a protein concentration-dependent manner (*see* **Note 27**).

3.5 Cloning and Evaluation of Aptamers

3.5.1 Cloning and Sequencing

Once affinity enrichment of the RNA pool is confirmed, clone the variants by TA cloning. Several TA cloning vectors are commercially available [i.e., pGEM-T Easy Vector System (Promega); TA Cloning Kit with pCR (Invitrogen)]. Sequencing more than 30 clones is recommended. Typically, similar or even identical sequences appear multiple times. This sequence convergence indicates successful enrichment of certain clones via SELEX (*see* **Note 28**).

3.5.2 Affinity Estimation of Each Clone

To evaluate the affinity of each clone, perform a binding assay as described for the analysis of SELEX progress (Subheading 3.4). At this stage, determining the precise value of the affinity constant is desirable. Thus, an EMSA (Subheading 3.4.2) or a filter-binding assay (Subheading 3.4.3) with radioisotope-labeled RNA is recommended. Alternatively, other interaction analysis techniques, like surface plasmon resonance analysis or fluorescence correlation spectroscopy, are useful [9–12].

3.5.3 Evaluation of Essential Elements for Binding Activity

Typically, the full-length sequence of the original isolate is not required for the binding activity. In addition, a shorter sequence and simpler structure are beneficial for the design and engineering of artificial riboswitches employing the obtained aptamers.

T06 (71-nt, K_D=8.0 nM)

GAGGAUGGCA GCGGAGAGGA AUGGUGUAGU GUAAGUCAAU UCUACUUUCG UUGCAGCCCA CACCACUCUC C

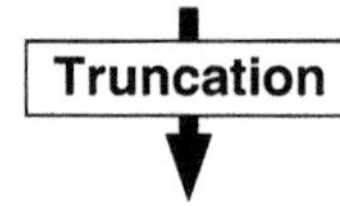

T06-38 (38-nt, K_D=10.5 nM) GUGUAGU GUAAGUCAAU UCUACUUUCG UUGCAGCCCA C

T230-29 (29-nt, K_D=2.9 nM) GC GUAAGUCAAU UCCACUAUCA UUGCUGC

Fig. 7 Generation of a shortened aptamer. Initially, truncations from the 5′- and 3′-ends of the original isolate, an aptamer against T7 RNA polymerase (T06), were examined. A shortened derivative (T06-38) was generated without reducing the affinity. Based on the shortened derivative, secondary SELEX was performed, and further truncation was examined. The resulting 29-nt aptamer (T230-29) is 42 nt shorter than the original aptamer; its affinity for the target protein is higher. In addition, the secondary SELEX revealed the sequence and structural features essential for activity (for details, *see* ref. 13)

To shorten the aptamers, serial truncation of the 5′- and 3′-ends of the original isolate is the best choice for an initial effort (Fig. 7) [13]. Alternatively, two-dimensional (2D) structure predictions facilitate the rational design of shortened aptamers (*see* **Note 29**).

Re-selection from a randomly mutagenized pool, in which ~10–25 % of the nucleotides of the original isolate are randomly replaced with other nucleotides, is also helpful [13, 14]. Based on the sequence comparison of the isolates from this secondary SELEX, essential sequence motifs and/or structural features can be identified. Furthermore, evolved aptamers with higher affinity may be also isolated (Fig. 7).

4 Notes

1. If the several lines of SELEX are performed (either simultaneously or sequentially), use primer sets with different sequences to prevent cross- and carryover contamination.
2. The length of the random region is typically 30–100 nucleotides. Theoretically, libraries with a longer random sequence are more likely to identify aptamers [15]. However, longer random sequences are also more likely to generate by-products (*see* Subheading 3.3.4), and they show lower RT-PCR efficiency.
3. Some proteins show a strong tendency to adsorb RNA nonspecifically. For such proteins, this buffer system may not be suitable for SELEX because the adsorption of RNA severely reduces the enrichment factor in positive selection. To reduce the adsorption, the addition of a nonionic, hydrophilic (high hydrophile–lipophile balance value) surfactant (e.g., 0.05 % Tween 20) may be effective. Such surfactants do not disrupt

protein structure. In addition to a surfactant, higher ionic strength (e.g., <500 mM sodium chloride or potassium chloride) may also be helpful. High ionic strength neutralizes and weakens electrostatic interactions.

4. Buffer B and buffer C should conform to 2× and 1× buffer A, respectively. When buffer A is modified, these buffers should be also modified.
5. TBE buffer (89 mM Tris, 89 mM borate, and 2 mM EDTA) supplemented with 3 mM magnesium chloride can substitute for TBM buffer.
6. TBE buffer can be used for electrophoretic mobility shift assay (EMSA) to analyze the varieties of aptamers. However, in some cases, inclusion of magnesium ion in the electrophoresis buffer is essential [16]. Because it is unknown whether aptamers enriched in the pool require magnesium ion in the buffer, the use of TBM buffer, rather than TBE buffer, is recommended.
7. The radioisotope labeling can be performed by 5′- or 3′-end labeling and internal labeling. At least 1×10^5 cpm/pmol of the relative radioactivity is desired.
8. If the reaction efficiency is not sufficiently high (i.e., less than 50 % of the input DNA is converted to dsDNA), repeat the annealing and extension step several times. The denaturing step, however, should not be repeated. In contrast to a typical PCR, the reaction includes only the forward primer; denaturing the dsDNA products will not reproduce dsDNA. Furthermore, denaturing can enhance the likelihood of side reactions (see below). When repeated extension does not improve the yield of dsDNA products, low-quality template DNA may be at fault. In this case, increase the reaction volume to obtain sufficient amounts of dsDNA.
9. If the target protein or resin strongly adsorbs RNA nonspecifically, the tRNA concentration can be increased up to tenfold.
10. If a rotator is not available, invert the tube every 3 min.
11. When using commercially available His-tagged recombinant proteins, typically, more than 70 % of the input protein is immobilized on the resin through this procedure. However, the immobilization efficiency of protein prepared by an individual researcher may be very low. Thus, a pilot experiment to assess protein immobilization is recommended. If immobilization is inefficient, modification of the buffers, such as the addition of a low concentration of denaturant (e.g., 0.5–2 M urea, guanidine-hydrochloride), may be effective. Alternatively, completely denatured proteins may be efficiently immobilized and refolded on the resin [17].
12. When using commercially available His-tagged recombinant proteins, typically, almost all of the immobilized protein is

released from the resin under these conditions. However, the release efficiency of protein prepared by an individual researcher may be very low. Again a pilot experiment is recommended. If the protein cannot be released even with higher concentrations of imidazole, the bound RNA can be directly recovered by PCI extraction or by heat treatment (i.e., 90 °C for 5 min) of the resin. These recovery procedures are not selective for RNA that binds to protein; undesired RNA species that bind to the resin are also recovered.

13. Typically, the number of RNA molecules at this step is very low. For the efficient recovery of the desired RNA species, the use of a co-precipitant is highly recommended.
14. Reverse transcriptases have a tendency to skip strong stem structures in the template RNA [18]. Such skipping produces shortened cDNA. To avoid skipping, the use of thermostable enzymes, like ThermoScript Reverse Transcriptase (Invitrogen), may be helpful.
15. Typically, eight cycles of PCR is not sufficient amplification during the early rounds of SELEX. However, according to the progress of SELEX, the PCR cycle number required may be lower. If the eight-cycle PCR generates visible amounts of DNA on PAGE analysis, reduce the cycle number.
16. Typically, efficient PCR amplification of very short DNA fragments, like the DNA pool for SELEX, requires a relatively high concentration of primers. To confirm the appropriate PCR conditions for the DNA pool, a pilot experiment using DNA from an initial random pool is recommended.
17. Because of the random sequence in the template DNA, unintended mis-annealing may occur (Fig. 3). These side reactions generate longer and shorter by-products. Typically, the early PCR cycles generate only the desired products. However, the increased concentration of PCR products, that is DNA containing random sequences, increases the likelihood of side reactions. Thus, avoid too many cycles of PCR.
18. A high-concentration agarose gel can be used to confirm the amplification of some PCR products. However, slight size differences between the desired product and undesired by-products cannot be distinguished. Thus, the use of polyacrylamide gels, instead of agarose gels, is recommended.
19. Do not use all the PCR product in the next step (in vitro transcription). A portion of the DNA pool should be stored. Useful aptamers in the pools from earlier rounds may disappear in the pools from later ones.
20. In the author's experience, the progress of SELEX can be speculated from PAGE analysis of RT-PCR; a change in the smear band pattern; and a reduction in the number of PCR cycles

required for sufficient amplification. Although this is not definitive, experienced selectionists may skip the pool analysis step and proceed to the analysis of each clone in the pool.

21. Smaller amounts of protein and RNA are sufficient for the analysis when sensitive detection of RNA is possible. For this purpose, radioisotope labeling (i.e., with [^{32}P]) is least expensive and most sensitive.
22. Alternatively, the detection of RT-PCR products, rather than the RNA itself, is also a sensitive procedure. However, the DNA pool usually has a higher RT-PCR efficiency, and the RT-PCR step may overestimate affinity enrichment.
23. The extent of the mobility shift upon the complex formation depends on the charge and molecular weight of the target protein. Thus, the running time of electrophoresis required for clear discrimination of the complex from free RNA varies from protein to protein.
24. The filter-binding assay tends to show relatively high data dispersions; thus, triplicate experiments are recommended.
25. Keep the membrane wet before washing; drying at this step causes non-ignorable background signal.
26. Do not place the membrane upside down, and do not cover the membrane with the paper towel. Direct contact of the sample surface with the paper towel may remove the trapped RNA.
27. The evaluation can be also performed by exposing the membrane to an imaging plate and analyzing the plate using a phosphorimager.
28. Sequence convergence does not always mean successful enrichment of aptamers. Not only affinity for the target but also efficiency of RT-PCR and transcription can cause enrichment bias during SELEX. In some conditions, parasitic RNA species lacking affinity are efficiently enriched [19]. Thus, the affinity of each clone should be experimentally validated.
29. Several 2D structure prediction programs are freely available [20].

References

1. Ausländer S, Fussenegger M (2013) From gene switches to mammalian designer cells: present and future prospects. Trends Biotechnol 31:155–168
2. Ellington AD, Chen X, Robertson M, Syrett A (2009) Evolutionary origins and directed evolution of RNA. Int J Biochem Cell Biol 41:254–265
3. Joyce GF (2007) Forty years of *in vitro* evolution. Angew Chem Int Ed Engl 46:6420–6436
4. Ellington A, Szostak JW (1990) *In vitro* selection of RNA molecules that bind specific ligands. Nature 346:818–822
5. Tuerk C, Gold L (1990) Systematic evolution of ligands by exponential enrichment: RNA ligands to bacteriophage T4 DNA polymerase. Science 249:505–510
6. Klussmann S (ed) (2006) The aptamer handbook: functional oligonucleotides and their applications. Wiley-VCH, Weinheim

7. Stoltenburg R, Reinemann C, Strehlitz B (2007) SELEX—a (r)evolutionary method to generate high-affinity nucleic acid ligands. Biomol Eng 24:381–403
8. Milligan JF, Groebe DR, Witherell GW, Uhlenbeck OC (1987) Oligoribonucleotide synthesis using T7 RNA polymerase and synthetic DNA templates. Nucleic Acids Res 15:8783–8798
9. Di Primo C, Dausse E, Toulmé JJ (2011) Surface Plasmon resonance investigation of RNA aptamer-RNA ligand interactions. Methods Mol Biol 764:279–300
10. Ishino T, Atarashi K, Uchiyama S, Yamami T, Saihara Y, Yoshida T, Hara H, Yokose K, Kobayashi Y, Nakamura Y (2000) Interaction of ribosome recycling factor and elongation factor EF-G with *E. coli* ribosomes studied by the surface plasmon resonance technique. Genes Cells 5:953–963
11. Werner A, Hahn U (2009) Fluorescence correlation spectroscopy (FCS)-based characterisation of aptamer ligand interaction. Methods Mol Biol 535:107–114
12. Schürer H, Buchynskyy A, Korn K, Famulok M, Welzei P, Hahn U (2001) Fluorescence correlation spectroscopy as a new method for the investigation of aptamer/target interactions. Biol Chem 382:479–481
13. Ohuchi S, Mori Y, Nakamura Y (2012) Evolution of an inhibitory RNA aptamer against T7 RNA polymerase. FEBS Open Bio 2:203–207
14. Ekland EH, Bartel DP (1995) The secondary structure and sequence optimization of an RNA ligase ribozyme. Nucleic Acids Res 23:3231–3238
15. Sabeti PC, Unrau PJ, Bartel DP (1997) Accessing rare activities from random RNA sequences: the importance of the length of molecules in the starting pool. Chem Biol 4:767–774
16. Hellman LM, Fried MG (2007) Electrophoretic mobility shift assay (EMSA) for detecting protein-nucleic acid interactions. Nat Protoc 2:1849–1861
17. Li M, Su ZG, Janson JC (2004) *In vitro* protein refolding by chromatographic procedures. Protein Expr Purif 33:1–10
18. Zhang YJ, Pan HY, Gao SJ (2001) Reverse transcription slippage over the mRNA secondary structure of the LIP1 gene. Biotechniques 31:1286–1294
19. Breaker RR, Joyce GF (1993) Minimonsters: evolutionary byproducts of *in vitro* RNA amplification. In: Fleischaker GR, Colonna S, Luisi PL (eds) Proceedings of the NATO conference on self-reproduction of supramolecular structures. Kluwer Academic Press, Dordrecht, pp 127–135
20. Seetin MG, Mathews DH (2012) RNA structure prediction: an overview of methods. Methods Mol Biol 905:99–122

Chapter 5

Artificial Riboswitch Selection: A FACS-Based Approach

Zohaib Ghazi, Casey C. Fowler, and Yingfu Li

Abstract

Riboswitches have a number of characteristics that make them ideal regulatory elements for a wide range of synthetic biology applications. To maximize their utility, methods are required to create custom riboswitches de novo or to modify existing riboswitches to suit specific experimental needs. This chapter describes such a method, which exploits fluorescence-activated cell sorting (FACS) to quickly and efficiently sort through large libraries of riboswitch-like sequences to identify those with the desired activity. Suggestions for the experimental setup are provided, along with detailed protocols for testing and optimizing FACS conditions FACS selection steps, and follow-up assays to identify and characterize individual riboswitches.

Key words Artificial riboswitch, Aptamer, Biosensor, Fluorescence-activated cell sorting (FACS), High-throughput, Enrichment

1 Introduction

Riboswitches are RNA elements that facilitate genetic control through structural changes that occur in response to directly binding a specific small-molecule ligand [1]. The modular nature of riboswitches, coupled with the simple and direct nature of the regulation that they provide, has sparked the interest of synthetic biologists who have applied riboswitches toward a range of interesting applications [2–6]. One of the principal advantages of riboswitches as regulators in synthetic systems is the potential for novel elements to be created that can exert a desired regulatory effect in response to a hand-selected target molecule. The most critical factor for successfully isolating an effective synthetic riboswitch is that the aptamer domain must possess desirable binding characteristics that suit the intended applications. Creating such an aptamer de novo can be a project in itself, and so riboswitch engineers have generally made use of previously identified aptamers (either created artificially using "in vitro selection" protocols or borrowed from

Atsushi Ogawa (ed.), *Artificial Riboswitches: Methods and Protocols*, Methods in Molecular Biology, vol. 1111,
DOI 10.1007/978-1-62703-755-6_5,

naturally evolved riboswitches) as the foundation for their designs. Many excellent reviews have been published that describe the methodology that can be used to create a novel aptamer with desired properties [7, 8]. Once an appropriate aptamer has been identified, several creative approaches have been successful in applying it toward the production of a functional riboswitch [9–13]. In this chapter, we describe such a method in which fluorescence-activated cell sorting (FACS) is employed to identify active riboswitches from a carefully designed, partially random library of riboswitch candidates [9, 10]. The methodology described here should be generally applicable to bacterial organisms that can take up and maintain plasmid DNA. For some organisms, steps involving the growth and manipulation of the bacteria will need to be tweaked, and for biologically hazardous bacteria, additional safety measures will be required that are not described here.

The distinctive feature of the approach described in this chapter is the use of FACS to identify sequences that produce a regulatory response to the desired target molecule. Flow cytometers are employed for a vast array of purposes in biological research that exploit their analytical capabilities, sorting capabilities, or both. For an extensive review of flow cytometry, consult the following: [14, 15]. With respect to sorting, flow cytometers can be used to sort individual particles in a given size range (typically ~0.2–100 μm, depending upon the instrument being used) based on a number of physical properties. As it is applied here, flow cytometry is used to sort individual bacterial cells on the basis of fluorescence intensity. A library of sequences, carefully designed to have the potential to act as riboswitches, is cloned upstream of a fluorescent protein on a plasmid that is transformed into the host bacteria. When its expression is activated, this fluorescent protein represents the dominant source of fluorescence, and thus the bacteria can be sorted based on their expression of this gene. To identify bacteria in which the fluorescent protein is regulated by the target molecule, a procedure is used in which bacteria are sorted under two different conditions: one where the riboswitch target is present at high levels and the other where it is absent or present at low levels (*see* Fig. 1).

The focus of this chapter is to provide detailed protocols for conducting FACS-based riboswitch selections. This method is only as good as the starting library, and the success of this protocol hinges completely on the presumption that it contains active riboswitches. While a limited description of the basics of creating a library is provided in Subheading 3.1, there are endless options for the design of the library, and a detailed description of these possibilities falls outside the scope of this chapter.

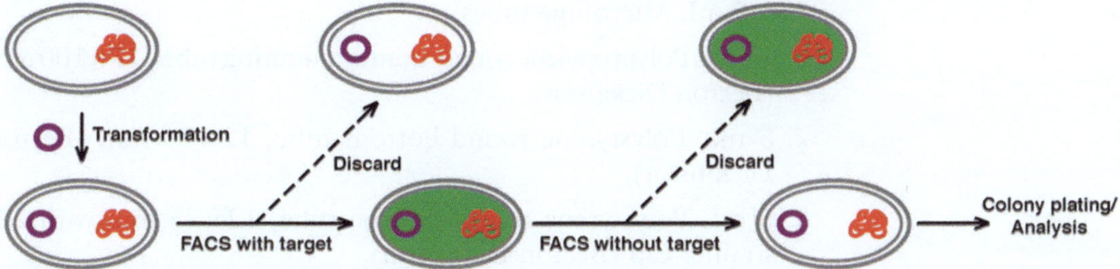

Fig. 1 Isolation of functional riboswitches in *E. coli* using FACS. Following the transformation of the plasmid library (*small circle*), which contains riboswitch candidates placed within the 5′ UTR of an FP gene. *E. coli* cells are sorted based on fluorescence intensity to isolate cells that are highly fluorescent in the presence of the target molecule that regulates the riboswitch, but less so in its absence

2 Materials

2.1 Growth Media

1. Luria Broth (LB).
2. LB + 15 % glycerol.
3. LB + 30 % glycerol.
4. Growth medium to create low target metabolite levels (see below).
5. Growth medium to create high target metabolite levels (see below).

The selection of the growth media that will be used for the culture steps preceding FACS and fluorescence assays is not trivial. The protocol requires that two culture conditions be created: one where the riboswitch ligand is absent or present at low levels and another where this molecule is highly abundant. Some strategies for creating these conditions are described in Subheading 3.1.1. In some instances, such as when the target molecule is synthetic, convenient, chemically undefined media such as LB might be appropriate. Typically, we suggest using chemically defined media to more carefully control the levels of the target molecule. Nutrients such as amino acids and vitamins can be added to simple media formulations to permit more robust growth and gene expression [2].

2.2 Fluorescence-Activated Cell Sorting

1. Agar.
2. Antibiotics.
3. Sterile polystyrene petri dishes, 100 × 15 mm (Thermo Fisher Scientific). Prepare LB agar plates supplemented with the appropriate antibiotic.
4. 10× Dulbecco's phosphate-buffered saline (DPBS), supplemented with calcium and magnesium (Invitrogen). Make up a 1× solution, filter, and autoclave.

5. 1.5 mL Microfuge tubes.
6. 14 mL Polypropylene round-bottom culturing tubes, 17×100 mm (Becton Dickinson).
7. 5 mL Polystyrene round-bottom tube, 12×75 mm (Becton Dickinson).
8. 5 mL Polystyrene round-bottom tube, 12×75 mm with cell strainer cap (Becton Dickinson).

2.3 96-Well Plate Fluorescence Assay

1. 96-well, flat bottom polystyrene non-treated sterile assay plates (Corning).
2. 96-well, half-area flat bottom polystyrene NBS plates (Corning).
3. ABgene 96-well, square-well storage plates (Thermo Fisher Scientific).

3 Methods

3.1 Considerations Before Beginning the Project

3.1.1 The Target Molecule

Pioneering projects to create synthetic riboswitches have often selected a target molecule based primarily on the availability of a high-quality, well-characterized RNA aptamer. While this is a reasonable approach for method development, future endeavors should generally be conducted with specific downstream applications in mind. There are, however, certain factors that can make the development of a synthetic riboswitch for a given ligand particularly challenging or even impossible using the methodology described here (and, in general, other methods as well). One consideration is that RNA has limited chemical diversity, and, although it has shown the potential to tightly and specifically bind a very wide range of molecules, certain types of ligands will intrinsically present a greater challenge for aptamer development. For example, molecules with significant negative charge and highly hydrophobic molecules are likely to be difficult ligands for RNA.

A second consideration is that this protocol depends on the ability to predictably manipulate the cellular concentration of the target molecule. To achieve this for molecules that are not native to the cell, the primary concern is that the target ligand must be able to permeate host cells and maintain suitable levels. In certain instances, strains featuring mutations to broad-specificity efflux pumps might assist in obtaining elevated concentrations of molecules that are excreted. For cellular metabolites, a generic approach that is effective for many molecules is to employ mutant strains in which the synthesis of the target molecule has been disrupted. The molecule (or a precursor) is then added to the growth medium at controlled levels to produce the "high" and "low" concentrations required for the selection steps. For certain molecules it might prove difficult or even impossible to manipulate their cellular levels in a controlled fashion over the desired range, which would preclude the use of the strategy described in this chapter.

A final consideration is that all riboswitches are only responsive over a limited concentration range. The binding characteristics of the aptamer will be the primary factor in dictating this range, and it is important for the target molecule concentration range that is relevant for your purposes to be compatible with the aptamer that is selected. Importantly, in vitro binding constants will not always hold true within the context of a cell, and the kinetics of RNA folding, transcription, and translation have been shown to impact the responsive range of naturally occurring riboswitches [16–18]; these factors should be considered but can be difficult to predict. The unpredictability of these issues highlights the advantage of exploiting the power of selection to identify the best solution to this complicated problem.

3.1.2 Library Design

Once an aptamer has been chosen, the next step before beginning the riboswitch selection is to design a plasmid-borne library of candidate riboswitches. The only region of the plasmid that will vary between individual members of the library is the "riboswitch region," which resides in the 5′ untranslated region (UTR) of a fluorescent protein. To drive the expression of the transcript that contains the riboswitch region and the fluorescent protein, it is advisable to use a well-characterized "constitutive" promoter, which is as stable as possible under the selection conditions that will be used. Two important considerations for the library plasmid are its origin of replication (copy number) and the strength of the promoter. For each of these two factors, the required signal strength for the FACS steps must be balanced with the potential complications and intrusiveness associated with protein overexpression. In addition, for riboswitches with high-affinity aptamers that detect low levels of the target molecule, it is possible that, within the aptamer's dynamic range, the target molecule may not be in excess over the transcript when using high copy plasmids and strong promoters.

The methodology described in this chapter is intended to be generally applicable for a range of riboswitch designs and mechanisms. The focus is therefore on setting up and performing the selection that allows for the identification of active riboswitches from a starting library and on follow-up experiments to identify and characterize individual sequences. There are limitless options for the design of the starting library, which could target riboswitches that act at the level of transcription [19, 20], translation [10–13, 21], or RNA processing [22–24]. Specific library designs are not discussed here. Generally, the library will consist of a combination of invariant and random residues in the 5′ UTR shortly upstream of a fluorescent protein (FP). The aptamer domain, which is typically held largely constant to maintain the binding

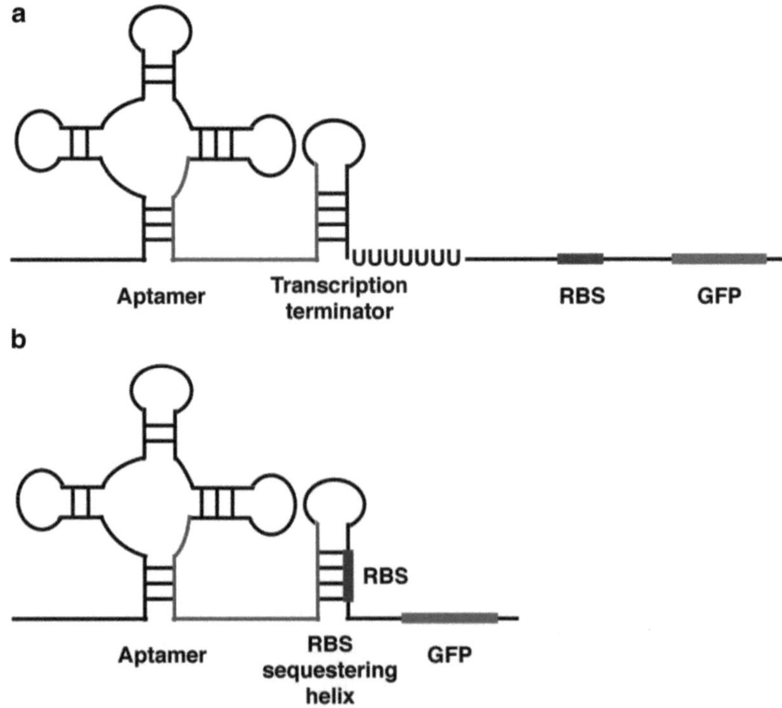

Fig. 2 Two representative library designs, one with a transcription terminator (**a**) and the other with an RBS sequestering helix (**b**). Random or partially random residues (*grey segment*) are introduced at key positions that will allow for the binding event to be transduced into a change in the expression of downstream genes. More specific examples of library designs can be found in the literature (see main text) (color figure online)

capabilities of the library, will be arranged in close proximity to a regulatory element that is capable of hindering the expression of the downstream fluorescent protein via its ribosomal binding site, an intrinsic transcriptional terminator, or a self-cleaving ribozyme. Random residues are introduced at key positions that are likely to transduce the structural changes that occur in response to aptamer–ligand binding to the disruption or the activation of the regulatory element. While many of the sequences in this library might be inert, a low number of active sequences can be identified using the selection and screening steps outlined in this chapter. A schematic demonstrating two generic library designs is shown in Fig. 2. For specific examples of designs that have been used previously, *see* refs. 10–13, 19–25.

3.1.3 The Fluorescent Protein

Decades of intense research has yielded an abundance of FPs with diverse properties. Different FPs can vary greatly in their excitation/emission spectra, extinction coefficient and quantum yield, maturation speed and efficiency, turnover rate, environmental sensitivity, photostability, oligomerization state, etc. A number of excellent reviews have been published summarizing these factors and how they apply to individual FPs that are available; this material should be carefully reviewed before making a final choice [27–30]. Many FP reviews focus on microscopy applications; this information is useful and relevant, but be mindful of the particular requirements of your experiments. For example, factors such as photostability and oligomeric state are critical for many imaging experiments but are not significant factors for FACS.

The optics of the flow cytometer that will be employed for this project is an essential consideration when selecting an FP; the optimal excitation/emission spectrum of the FP must match up with the lasers and filter sets of this instrument (see below). The signal-to-background ratio is another important consideration. The signal generated by the FP will depend on a number of factors including the efficiency of its expression and maturation within your strain and the extinction coefficient and quantum yield of the FP. The background will be dictated by the level of autofluorescence of your strain, which will generally be greater in the blue region of the spectrum and less pronounced in the red range. The amount of time required for the FP to fold and mature and the length it will persist in the population (turnover time) are issues that are particularly relevant for this project. Slow-maturing FPs create a significant delay between the time that regulatory decisions are made by potential riboswitches and the appearance of the corresponding signal. FPs are generally very stable and can persist for hours or even days within cells. Signal contamination by FPs synthesized at previous steps of the protocol has the potential to significantly interfere with the selection process described in this chapter. Dilution can be effectively used to combat this problem; however, FPs with a shorter half-life are generally preferable. The caveats of this are that FPs must persist for at least a few hours to accommodate the lag between growth and the time samples are processed by FACS and that FPs with a shorter half-life will yield a lower signal due to their decreased abundance.

3.1.4 Instrument Considerations

High-speed cell sorters are available from several manufacturers, such as the FACSAria™ (BD Biosciences), EPICS® ALTRA™ (Beckman Coulter), or the MoFlo® (Dako) cell sorters. Flow cytometers, particularly those with sorting capabilities, are tremendously valuable for a wide range of applications and are very complex and expensive instruments. Accordingly, many research institutes have a limited number of flow cytometers that are shared in common facilities, often operated exclusively by specially trained personnel. For those that do not have ready access to a capable instrument that allows the sorting of bacteria, many centralized facilities permit access to users from other institutes at hourly rates. While most modern instruments are compatible with the experiments described here, there are several issues to consider before beginning your experiment. We strongly suggest speaking with an experienced director or operator from the facility you intend to use about the specific details of your experiment before getting started.

With regard to the instrument that will be used, the main factors that must be considered when planning your experiment are the difficulties associated with sorting particles as small as bacterial cells, the optics of the instrument, the sorting rates, and any pertinent safety/containment issues (particularly for those working with BSL-2 or BSL-3 organisms). Most commercial instruments can be optimized for small-particle analysis, such as bacteria; however, they approach the limit of detection, as historically they have been used to detect much larger particles (i.e., mammalian cells). As much smaller particles, the probability of bacteria not being in the proper position or orientation within the stream to be properly assessed by the instrument is much higher. From our experience, using more dilute samples (and consequently sorting at slower rates) is very important for dealing with this issue. The choice of the FP that will be used for your project (see previous section) needs to be coordinated with the lasers and filter sets that are available with the instrument that will be used. While most instruments will be compatible with many green and red FPs, suitable lasers for more exotic FP selections are optional additions and not always available. The sorting rate achievable for modern instruments analyzing bacteria depends on a variety of factors; however, most flow cytometers can efficiently sort at rates in the 5,000–10,000 events per second range or higher for eukaryotic cells. The more pressing issue is for the present method is not how quickly the bacteria can be sorted, but how quickly they can be sorted without a significant drop off in purity. Be sure to describe the strain of bacteria that will be used and its safety requirements to facility personnel prior to the project; many facilities will not be equipped to deal with pathogenic species.

3.2 Establishing Controls and FACS Parameters

Prior to beginning selection steps, it is important to test the feasibility of your particular project, to experience running your samples on the flow cytometer that will be employed, to gain insight into the properties of your samples, and to determine the gating and other settings that will best suit your selection. This section presents a set of experiments that will help accomplish these goals. Depending upon the success of these experiments, additional tests along these same lines may be required in order to prepare for the selection.

At a minimum, these tests require two plasmids to be stably introduced in the strain that will be used for the selection:

1. The library plasmid containing the FP you have chosen but lacking any potential regulatory elements in its 5′ UTR (FP-positive control plasmid).
2. A similar plasmid that contains a different antibiotic resistance cassette and that lacks a fluorescent protein (FP-negative control plasmid).

In addition, we strongly suggest also applying a third plasmid. This plasmid will simulate the FP expression levels of a riboswitch from your library that is locked in the "expression off" state (FP-repressed control plasmid). For example, for a riboswitch library targeting a transcriptional termination mechanism, this plasmid will contain the intact transcriptional terminator that will be a part of the library upstream of the FP. Since some leak in expression will occur with even very efficient riboswitches, this control plasmid will provide a more realistic representation of the fluorescence intensity of riboswitches in the "off" state. Like the FP-negative control plasmid, the FP-repressed control plasmid should parallel the FP-positive control plasmid (same basic vector with the same promoter, same origin of replication, etc.) but should have a different antibiotic resistance cassette.

1. Transform your selected host strain with the FP-positive and FP-negative control plasmids described above. Plate onto LB agar plates supplemented with the appropriate antibiotics, and grow overnight at 37 °C.
2. Inoculate a 2 mL culture of the selected growth medium (*see* Subheading 2.1) supplemented with antibiotics with a single colony picked from each construct transformed in **step 1** and grow overnight at 37 °C for ~16–18 h with shaking (250 rpm).
3. The following day, subculture both samples 1:1,000 into 2 mL of fresh medium and incubate at 37 °C for ~6–8 h (*see* **Note 1**) with shaking to late log phase (OD_{600} ~ 0.8).
4. Measure the OD_{600} of both cultures, and infer the approximate concentration of bacteria (# bacteria/mL culture) (*see* **Note 2**).
5. Transfer 1 mL of each culture from **step 4** into a 1.5 mL microfuge tube and centrifuge for 10 min at 5,000 × *g* to pellet the bacteria. Wash several times with an equal volume of DPBS (*see* Subheading 2.2). Resuspend in 1 mL of DPBS.
6. Using the approximate cell concentrations determined in **step 4**, prepare ~5 mL of the following cell suspensions in DPBS at a concentration of ~10^6 to 10^7 cells/mL (*see* **Note 3**):
   (a) Cells containing the FP-positive control plasmid.
   (b) Cells containing the FP-negative control plasmid.
   (c) A 50:50 mixture of the FP-positive and FP-negative cells.
   (d) A 95:5 mixture of the FP-positive and -negative cells (FP-positive:FP-negative).
   (e) A 95:5 mixture of the FP-positive and -negative cells (FP-negative:FP-positive).
7. Put all samples on ice, and transport them to the flow cytometer. Keep samples on ice both before and after FACS analysis.

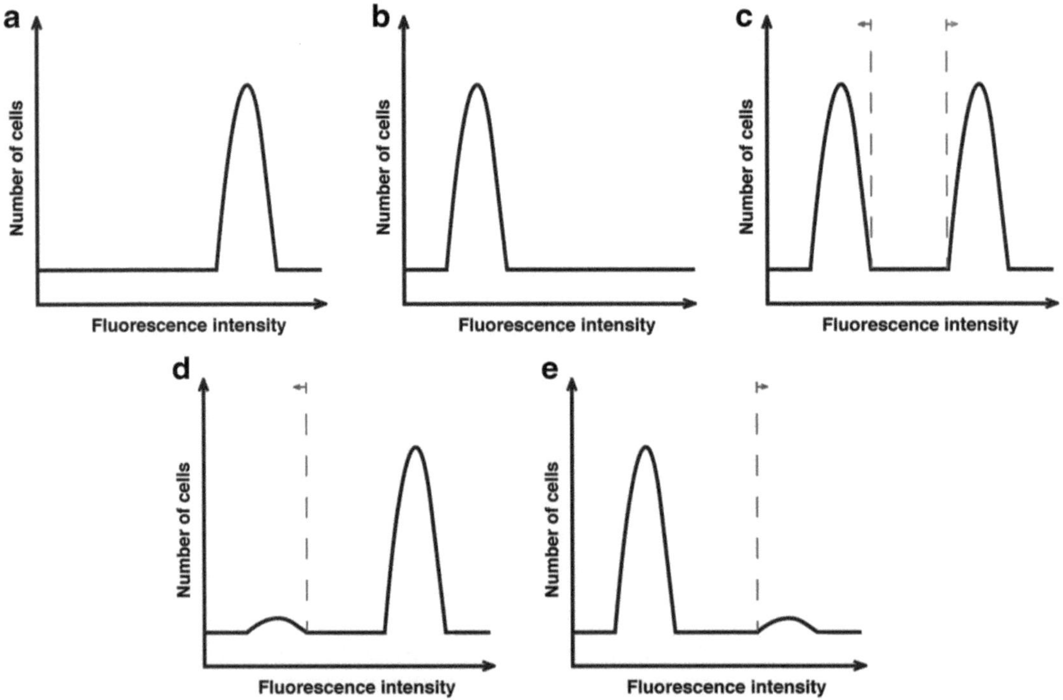

Fig. 3 Schematic illustrations of the expected fluorescence histograms for (**a**) cells containing the FP-positive control plasmid, (**b**) cells containing the FP-negative control plasmid, (**c**) a 50:50 mixture of the FP-positive and FP-negative cells, (**d**) a 95:5 mixture of the FP-positive and FP-negative cells, and (**e**) a 95:5 mixture of the FP-negative and FP-positive cells. Each *dash line* with an *arrow* indicates an example of the gating threshold settings that can be used for sorting, with the *arrow* indicating the cells to be kept from the sort

Keep a small volume of samples **c**, **d**, and **e**. They will be required for **step 13**.

8. Beginning with sample **a**, analyze the bacteria using your flow cytometry system. As a first step you will need to determine the forward scatter and side scatter profile of your bacteria (*see* **Note 4**). To assist in determining these parameters, the fluorescence intensity profile can be used, since intact bacteria should represent the dominant fluorescent signal for sample **a**.
9. Examine the fluorescence histograms for each of the five samples that have been prepared. Figure 3 provides an illustration that depicts the expected profiles for each of these samples. *See* **Note 5** if distinct peaks are not observed for your two control populations (in sample **c**).
10. The goal of each of the following steps in this section is to gauge your ability to sort mixtures of the control populations. Begin with the 50:50 mixture. Establish gating (threshold fluorescence values) for a sort such that only the highly fluorescent cells are kept (*see* Fig. 3c). You might wish to try a few different gating settings to vary the stringency of your selection.

Collect a minimum of several thousand positive cells for each sort, and store samples on ice. Be sure to carefully label all collection tubes.

11. Repeat the above step, this time sorting to keep only the low-fluorescence cells (*see* Fig. 3c).
12. Move on to the mixtures containing 95 % of one control and only 5 % of the other. Sort as described in **step 10**, keeping only the rare population of cells in each instance (Fig. 3d, e). Adjusting the stringency of the gating settings for these populations can be particularly informative of the gating that will be required to get the desired purity for future sorts.
13. For mixtures **c**, **d**, and **e**, as well as for each of the sorted samples described above, plate equal dilutions on two sets of LB agar plates: one set containing the antibiotic to which the FP-positive control plasmid is resistant and the other set containing the antibiotic for FP-negative selection. Aim to plate several hundred total bacteria per plate; plating multiple dilutions is advised.
14. Count the total number of colonies on all plates from **step 13**, and record the CFU numbers.
15. Determine the percentage of each control that has been recovered from each sort. Comparing these figures to the initial populations, the efficiency of each sort can be ascertained. For populations with clearly distinct, non-overlapping fluorescence intensity peaks (such as those shown in Fig. 3) it should be possible to separate the cells with >95 % purity using conservative gating.
16. *Optional*: It is strongly recommended that the process described above be repeated using the FP-repressed control plasmid (described in the opening of this section) in place of the FP-negative control. In this instance, the two populations are likely to have more similar levels of fluorescence, and mixtures may not yield two completely distinct histogram peaks. This process should provide excellent insight into appropriate gating settings for the selection steps. If these populations cannot be separated from one another, the design of the library, the choice of the FP, or the inhibitory element used to control expression might need to be altered prior to beginning the selection.

3.3 Artificial Riboswitch Selection

A thorough description of the possibilities and considerations for the design and construction of a library of "riboswitch candidates" would provide subject matter for a lengthy manuscript and is beyond the scope of this chapter. To create a large library, a high-efficiency cloning method in conjunction with highly competent host cells will be required. Once the design and molecular cloning steps have been sorted out, single colonies containing individual library sequences should be pooled and stored for future use. As a general guide, we suggest a protocol along the following lines: Transform your plasmid DNA library into your chosen host strain, and plate the transformants to yield ~1,000 colonies per 10 cm LB agar plate. Add 2–3 mL of medium to each plate, and scrape the colonies off the agar and into the liquid. Mix well, and transfer a portion to a common pool with the bacteria from other plates. This pool should be mixed well, diluted to an appropriate concentration, and multiple aliquots should be stored in LB + 15 % glycerol (or another storage solution) at −80 °C for future use. For large library sizes, several of these pools can be created, stored, and grown separately (to maintain maximum diversity) and combined immediately prior to selection (**step 6** below).

The following protocol assumes a selection for a riboswitch that activates the expression of the downstream gene(s) in response to high ligand levels. With obvious adjustments, the same methodology can be used to identify riboswitches that repress the expression of the regulated gene(s) at high target molecule concentrations. The method requires two iterations of FACS for each selection cycle: one to identify clones that yield relatively low levels of fluorescence when the target molecule is present at low levels (or absent) and the other to identify clones with relatively high FP expression when the target molecule is present at high concentrations. The order of these selection steps can be reversed from that described below (*see* **Note 6**).

1. Inoculate 2 mL starter cultures for host cells carrying the FP-positive control plasmid and those carrying the FP-negative control plasmid (*see* Subheading 3.2). Allow these cultures to grow to stationary phase overnight. These populations will be required as controls to configure the FACS settings.
2. Thaw an aliquot of your host cells carrying the plasmid library (see above) on ice for ~10–15 min.
3. Dilute the library significantly into the growth medium selected for your experiment wherein the target molecule is present at low levels (or absent). Under these conditions, the riboswitches are expected to be in the unbound state with the expression of the FP repressed. Grow the culture at 37 °C with shaking until late log phase. [The dilution factor of the starting library should be such that the population has undergone many generations and the culture can grow at least ~5–8 h before reaching late log phase (*see* **Note** 7)].
4. Concurrent with **step 3**, subculture the two control populations that are described in **step 1** and grow to late log phase in the same growth medium used to culture the library.

5. Collect each of the cultures described in **steps 3** and **4** and place on ice for 10 min.
6. Pellet a fraction of each of the cell populations by centrifugation at 5,000×*g* for 10 min at 4 °C. The number of cells that need to be collected depends on the size of the library (*see* **Note 8**). Wash 3× with an equal volume of ice-cold DPBS and resuspend in DPBS at a final concentration of ~10^6–10^7 cells/mL (*see* **Note 3**). Bring samples to the flow cytometer.
7. Using the FP-positive and the FP-negative populations, establish instrument settings for the flow cytometer to create an appropriate forward and side scatter window for your bacteria and to create an optimal spread in fluorescence between the two controls.
8. Using the settings established in **step 7**, examine the fluorescence spread of your library population. Note the spread of the population as well as the fluorescence intensity of the library relative to those of the positive and negative controls.
9. Establish the gating settings for sorting your library to keep the cells with a relatively low fluorescence value and to discard cells with a relatively high fluorescence value. This is a crucial and difficult decision that should be guided by the desired characteristics of the riboswitches you seek, the observations noted in **step 8** and your experiences from the experiments described in Subheading 3.2. If the gating is not stringent enough the enrichment will be poor, and if the gating is too stringent desirable sequences will be lost. Unless the fluorescence distribution of the cells dictates otherwise, the fraction of the population of cells that is kept at any given FACS step should be small. This fraction should not drop below ~0.5 %, however, since FACS is disposed to yielding a small portion of erroneously processed particles that will likely dominate the extremities of the spectrum.
10. Sort the desired number of cells (*see* **Note 8**), keeping your samples on ice both before and after sorting.
11. When sorting is complete, dilute the surviving population into a rich growth medium and grow for several hours at 37 °C with shaking to revitalize cells after the stress of sorting and to recover cell numbers.
12. As with the starting library, this pool should be mixed well, diluted to an appropriate concentration, and multiple aliquots should be stored in LB+15 % glycerol (or another storage solution) at −80 °C for future use. This population can be described as Round 1A.
13. Repeat the entire above procedure using the Round 1A population as the starting library. For this iteration, grow the

cells under conditions that result in a relatively high target molecule concentration. Riboswitches are now expected to be in the bound state, which activates gene expression. Therefore the population with a relatively high level of fluorescence should be isolated and the low-fluorescence cells discarded. Collect, grow, and store the isolated population as described above. This population is dubbed Round 1B.

3.4 Identifying and Characterizing Individual Riboswitches

The two-step FACS selection described in Subheading 3.3 can be repeated any number of times prior to checking the progress of the selection and seeking individual riboswitch candidates. For most projects, two to three rounds are likely to be sufficient; however, this will depend on factors such as the frequency of desirable sequences in the library, the size of the library, and the extent of purification that is achieved at each FACS step. Importantly, the pool of clones that remains after each FACS step must be saved (as described in the previous section), and so it is always possible to return to any stage in the selection process. One addition that can help improve the final outcome of your project is to identify the most active riboswitch sequences from a given selection (populations 2A, 2B, 3A, #B, etc) and then to build a second library based on these sequences with random residues introduced at key locations [9]. The process of identifying and improving upon promising sequences can allow you to more shrewdly survey the almost limitless sequence space that exists in a riboswitch-sized oligonucleotide library.

Even after your library has been significantly enriched, it will still contain a large number of unique sequences with varying levels of activity. The procedure described below allows for the identification of the most active individual sequences from your final, most enriched population.

1. Thaw an aliquot of the population of cells that you wish to survey on ice for 10–15 min.
2. Create several serial tenfold dilutions of the thawed cells in LB, plate onto LB agar plates containing the appropriate antibiotic, and incubate overnight at 37 °C. The goal of this step is to generate plates containing large numbers of well-separated, single colonies.
3. Fill a 96-well plate with 200 μL of LB plus the appropriate antibiotic in each well. Inoculate each well with a single colony from **step 2**. Set aside a few wells and inoculate them with cells carrying the FP-positive and the FP-negative control vectors to serve as controls for subsequent fluorescence assays. Grow overnight (~16 h) with shaking at 37 °C.
4. The following day, fill a 96-well storage plate (ABgene) with 100 μL per well of LB + 30 % glycerol. Using a multichannel pipette, add 100 μL per well of the plate from **step 3** to the

corresponding well of the storage plate and mix well. Store indefinitely at –80 °C. Repeat **steps 3** and **4** to create multiple plates in order to screen larger numbers of clones. Alternate media can be used to create the frozen stocks if required (*see* **Note 9**).

5. Prepare duplicate 96-well plates, one containing 250 μL of the medium that creates low target metabolite levels and the other containing 250 μL of the high target metabolite concentration medium.
6. Pin the frozen stock plate prepared in **step 4** to inoculate each of the duplicate plates from **step 5** and grow for several hours (~5–8 h is suggested for most situations) at 37 °C with shaking.
7. Measure the OD_{600} for the plate using a plate-based spectrophotometer.
8. Using a multichannel pipette, transfer 100 μL of sample from each plate to a 96-well half-area flat-bottom NBS microplate.
9. Analyze the fluorescence of the duplicate plates using a TECAN Safire fluorimeter or a similar instrument. The excitation/emission settings should mirror those that are optimal for the FP you have chosen.
10. Normalize the fluorescence values for cell growth by dividing each fluorescence value by the corresponding OD_{600} value for the same sample.
11. Examine the normalized fluorescence values from your control wells. The values should be similar between replicates and between duplicate plates.
12. Calculate activation ratios for each well by dividing the corrected fluorescence values from the plate grown in high levels of the target molecule to those from the plate grown in low levels of the target molecule. An activation ratio of 1 indicates that the target molecule does not influence FP expression. Higher activation ratios indicate greater levels of (possible) riboswitch activity.
13. Identify a handful of clones that represent the most active sequences observed. Streak these sequences out on fresh LB agar plates from the frozen stock. Assay these clones using a similar methodology to that described above but on a lower throughput scale using multiple replicates of each sample in order to confirm the observed activity.
14. Purify plasmid DNA from the most active clones, and sequence the relevant region of these vectors using standard techniques.
15. These sequences can be used as the basis to create a new library (as described in the introduction to this section) or can be further analyzed as described below.

16. A number of experiments will be required to confirm that the regulation observed for these samples is due to a riboswitch mechanism, to characterize the regulatory activity, and to sort out the mechanism of regulation. Some of these experiments may include the following:
    - Creating point mutants that are known to disrupt the binding capabilities of the aptamer: This mutant should lose regulatory activity, which would imply that the regulation is indeed due to the riboswitch binding its ligand.
    - Assaying for regulatory activity over a broad range of target molecule levels to generate a dose–response curve: This will indicate the dynamic range of the riboswitch. To ensure the specificity of your riboswitch, this can also be repeated using a similar concentration range of a close chemical derivative of the target molecule.
    - A myriad of experiments can be performed to help identify the level at which the riboswitch is regulating the downstream gene (transcription, translation, RNA processing). These can include mutagenesis, in vitro transcription or translation, analyzing transcript levels from cells, and checking for regulation of co-transcribed genes with distinct ribosome-binding sites.
    - In-line probing [26] has been a very successful technique for monitoring the structural changes undergone by RNA in response to binding a ligand. Comparing the RNA degradation patterns at different target molecule concentrations using this technique can be instrumental in identifying the residues within the riboswitch that undergo significant structural changes in response to target binding. This information is key for piecing together a detailed molecular mechanism.

4 Notes

1. Throughout this document, late log phase is suggested as the end point for key culture steps. For many of these same steps, the amount of time that the culture is allowed to grow prior to collection is also an important factor (*see* **Note** 7), which can create difficulties, particularly for conditions/strains that support fast growth. To accommodate both of these stipulations, the volume of the culture and the dilution factor of the starter culture are variables that can be manipulated. Alternately, another subculture step can be introduced part way through the growth stage to dilute the bacteria and allow for additional time and growth to occur before stationary phase begins.
2. OD_{600} measurements are a useful means of gauging the relative concentration of bacteria in a culture medium. The OD_{600}

can be used to approximate the number of bacteria/mL. To accomplish this, plate several dilutions of the bacterial culture at a given OD_{600} and count the colony-forming units (CFU) at each dilution. As an approximate guide, for normal *E. coli* cultures growing in LB, an OD_{600} of 1.0 is often estimated to indicate ~10^9 bacteria/mL.

3. Speak with the operator of the FACS instrument you will be using regarding sample preparation. Samples should generally be diluted to ~106 to 107 cells/mL using 1× DPBS (or a similar solution), but we strongly suggest optimizing the sort rates to best suit the sorting speed and purity required for your individual setup and experimental needs. Immediately prior to sorting, samples can be strained through a polystyrene cell strainer cap (*see* Subheading 2.2) to reduce issues with aggregation, however this is not generally required for bacterial cells.
4. In flow cytometry, the measures of forward and side scatter are useful parameters to identify particles with the approximate size and makeup of those that you intend to sort. Forward scatter profiles are typically proportional to the diameter of the cell, whereas side scatter profiles highlight the internal complexity or granularity of a cell. Prior to sorting, it is important that the instrument has demonstrated that it is capable of identifying the particles (bacterial cells) that you wish to sort. The flow cytometer's settings should be adjusted such that only particles with the correct forward/side scatter profile are examined and sorted.
5. The fluorescence profiles of the positive and negative controls will, in most cases, represent the highest and lowest levels of fluorescence achievable using your current setup. If there is not a generous spread in the fluorescence values of these populations (~2 orders of magnitude or more), the FP is not a dominant enough source of fluorescence under these conditions and the selection methodology described will not be possible. To overcome the weak signal various changes can be applied, for example increasing expression of the FP (by increasing the plasmid copy number and the strength of the promoter or by altering the RBS region to a more efficient sequence) or selecting an alternate FP that produces a stronger signal or that incurs less interference from autofluorescence.
6. The order of the two FACS steps (to select for low FP expression in the absence of the ligand and for high expression in the presence of the ligand) can be switched. One consideration is that most FPs are very stable and will persist in the population for many hours of growth. When selecting for highly fluorescent cells first, one must be sure to dilute this population significantly and allow many generations to pass before selecting for cells with a low level of fluorescence to avoid desirable sequences from being lost. If multiple cycles of the two-step

FACS protocol are used, this is a consideration regardless of the order of the steps in a given cycle.

7. It is important for sufficient time to pass and for many growth generations to occur during the culture phase prior to FACS. One reason for this is to allow for the dilution of the FP and of the target molecule from the frozen stock. It is important that the target molecule equilibrates at the desired levels created by the growth medium conditions and for these levels to dictate the regulatory decisions of any riboswitches in the population. Following these regulatory decisions, the FP must have time to mature and accumulate. This is essential in order for the fluorescence level of cells in the population to reflect any riboswitch activity of the individual plasmids they harbor.
8. The number of cells that should be sorted depends on the size of the library and the limitation imposed by sorting speeds (*see* **Note 3**). To comfortably survey your library, ~100 cells should be sorted per sequence in the library. For example, for a library size of 10^5 sequences, it might be possible to sort, 10^7 cells in a single session, which would provide good library coverage. If multiple cycles of selection are conducted, somewhat lower numbers of cells can be sorted for later rounds when the library diversity has presumably declined. For very large library sizes that push the limits of this method, the ratio of cells sorted to library size can be lowered; however, it should be noted that as this ratio drops, the probability of stochastically losing desirable clones increases. Such large libraries will also require a different strategy from that described in the introduction of Subheading 3.3 to select and pool the library. Regardless of the number of cells that you intend to sort, always bring a generous excess of sample with you beyond this amount.
9. Just as for the FACS experiments, it is also important when screening individual clones to ensure that the levels of the target molecule and the FP from the previous step do not affect the present experiment (*see* **Note 7**). This can be a challenge in the screening step due to the low volumes involved in the 96-well plate format. Accordingly, if growth in LB creates high levels of the target metabolite or the FP that affect these assays, individual clones should be grown and stored in alternate media that mitigate these effects.

Acknowledgments

The riboswitch work in the Li Lab has been supported by the Natural Science and Engineering Research Council of Canada (NSERC).

References

1. Roth A, Breaker RR (2009) The structural and functional diversity of metabolite-binding riboswitches. Annu Rev Biochem 78:305–334
2. Fowler CC, Brown ED, Li Y (2010) Using a riboswitch sensor to examine coenzyme B(12) metabolism and transport in *E. coli*. Chem Biol 17:756–765
3. Topp S, Gallivan JP (2007) Guiding bacteria with small molecules and RNA. J Am Chem Soc 129:6807–6811
4. Blount KF, Breaker RR (2006) Riboswitches as antibacterial drug targets. Nat Biotechnol 24: 1558–1564
5. Sudarsan N, Hammond MC, Block KF et al (2006) Tandem riboswitch architectures exhibit complex gene control functions. Science 314:300–304
6. Jin Y, Watt RM, Danchin A, Huang J (2009) Use of a riboswitch-controlled conditional hypomorphic mutation to uncover a role for the essential csrA gene in bacterial autoaggregation. J Biol Chem 284:28738–28745
7. Djordjevic M (2007) SELEX experiments: new prospects, applications and data analysis in inferring regulatory pathways. Biomol Eng 24: 179–189
8. Stoltenburg R, Reinemann C, Strehlitz B (2007) SELEX—a (r)evolutionary method to generate high-affinity nucleic acid ligands. Biomol Eng 24:381–403
9. Fowler CC, Brown ED, Li Y (2008) A FACS-based approach to engineering artificial riboswitches. ChemBioChem 9:1906–1911
10. Lynch SA, Gallivan JP (2009) A flow cytometry-based screen for synthetic riboswitches. Nucleic Acids Res 37:184–192
11. Lynch SA, Desai SK, Sajja HK et al (2007) A high-throughput screen for synthetic riboswitches reveals mechanistic insights into their function. Chem Biol 14:173–184
12. Topp S, Gallivan JP (2008) Random walks to synthetic riboswitches—a high-throughput selection based on cell motility. ChemBioChem 9:210–213
13. Muranaka N, Abe K, Yokobayashi Y (2009) Mechanism-guided library design and dual genetic selection of synthetic OFF riboswitches. ChemBioChem 10:2375–2381
14. Shapiro HM (2003) Practical flow cytometry, 4th edition. Wiley-Liss, New York, NY
15. Macey MG (2007) Flow cytometry: principles and applications. Humana Press, Totowa, NJ
16. Wickiser JK, Winkler WC, Breaker RR et al (2005) The speed of RNA transcription and metabolite binding kinetics operate an FMN riboswitch. Mol Cell 18:49–60
17. Wickiser JK, Cheah MT, Breaker RR et al (2005) The kinetics of ligand binding by an adenine-sensing riboswitch. Biochemistry 44: 13404–13414
18. Gilbert SD, Stoddard CD, Wise SJ et al (2006) Thermodynamic and kinetic characterization of ligand binding to the purine riboswitch aptamer domain. J Mol Biol 359: 754–768
19. Trausch JJ, Ceres P, Reyes FE, Batey RT (2011) The structure of a tetrahydrofolate-sensing riboswitch reveals two ligand binding sites in a single aptamer. Structure 19: 1413–1423
20. Wachsmuth M, Findeiß S, Weissheimer N et al (2013) De novo design of a synthetic riboswitch that regulates transcription termination. Nucleic Acids Res 41:2541–2551
21. Suess B, Fink B, Berens C et al (2004) A theophylline responsive riboswitch based on helix slipping controls gene expression in vivo. Nucleic Acids Res 32:1610–1614
22. Win MN, Smolke CD (2007) A modular and extensible RNA-based gene-regulatory platform for engineering cellular function. Proc Natl Acad Sci U S A 104:14283–14288
23. Wieland M, Hartig JS (2008) Improved aptazyme design and in vivo screening enable riboswitching in bacteria. Angew Chem Int Ed 47:2604–2607
24. Ogawa A, Maeda M (2007) Aptazyme-based riboswitches as label-free and detector-free sensors for cofactors. Bioorg Med Chem Lett 17:3156–3160
25. Beisel CL, Smolke CD (2009) Design principles for riboswitch function. PLoS Comput Biol 5:e1000363
26. Regulski EE, Breaker RR (2008) In-line probing analysis of riboswitches. Methods Mol Biol 419:53–67
27. Tsien RY (1998) The green fluorescent protein. Annu Rev Biochem 67:509–544
28. Shaner NC, Steinbach PA, Tsien RY (2005) A guide to choosing fluorescent proteins. Nat Methods 2:905–909
29. Telford WG, Hawley T, Subach F et al (2012) Flow cytometry of fluorescent proteins. Methods 57:318–330
30. Chudakov D, Matz M, Lukyanov S et al (2010) Fluorescent proteins and their applications in imaging living cells and tissues. Physiol Rev 90:1103–1163

Chapter 6

FRET-Based Optical Assay for Selection of Artificial Riboswitches

Svetlana V. Harbaugh, Molly E. Chapleau, Yaroslav G. Chushak, Morley O. Stone, and Nancy Kelley-Loughnane

Abstract

Artificial riboswitches are engineered to regulate gene expression in response to a variety of non-endogenous small molecules and, therefore, can be useful tools to reprogram cellular behavior for different applications. A new synthetic riboswitch can be created by linking an in vitro-selected aptamer with a randomized expression platform followed by in vivo selection and screening. Here, we describe an in vivo selection and screening technique to discover artificial riboswitches in *E. coli* cells that is based on TEV protease–FRET substrate reporter system.

Key words Aptamer, Riboswitch, In vivo selection, TEV protease, FRET-based fusion protein

1 Introduction

The ability to create artificial riboswitches that control gene expression in response to desired small molecules has important implications for the development of chemical and synthetic biology. Riboswitches can be powerful tools and have been used to study microbial genetics [1], to identify new antimicrobial targets in a variety of bacterial species [2, 3], and to develop logic gates for reprogramming cell behavior and engineering metabolic pathways [4–6]. New riboswitches can be created by linking an in vitro-selected aptamer that binds a target molecule with a randomized expression platform followed by in vivo selection and screening. While methods to discover aptamers that selectively recognize and bind a desired compound are well established [7, 8], the next challenge has been to develop and optimize a method for converting these aptamers into artificial riboswitches that are active in cells. Since the starting point for in vivo selection is large libraries of riboswitches, there are some key requirements for genetic (in vivo) selection in order to

Atsushi Ogawa (ed.), *Artificial Riboswitches: Methods and Protocols*, Methods in Molecular Biology, vol. 1111,
DOI 10.1007/978-1-62703-755-6_6, © Springer Science+Business Media New York 2014

result in functional switches. Ideally, a method for discovering riboswitches should be rapid, sensitive, inexpensive, and of high throughput. The method should have the ability to quantitatively distinguish the switches with the best performance characteristics from those switches that only weakly activate or repress gene expression. Since engineered riboswitches ultimately control gene expression, the efficiency of the selection technique generally depends on the choice of reporter gene system.

We have developed a FRET-based optical assay for monitoring riboswitch activation in *E. coli* cells [9]. This assay is based on expression of two genes, tobacco etch virus (TEV) protease [10] and its optical engineered protein substrate, within the same cell. The protein substrate is a fluorescence resonance energy transfer (FRET) construct between two fluorescent protein conjugates coupled with a peptide linker containing a TEV protease cleavage site (17 amino acid peptide sequence). The first protein in the pair is enhanced green fluorescent protein (eGFP), which acts as the electron energy donor; the second is a non-fluorescent mutant of yellow fluorescent protein called resonance energy-accepting chromoprotein (REACh), which acts as an electron energy acceptor [11]. Use of this type of FRET pair eliminates acceptor fluorescence, and therefore little to no fluorescence is observed prior to cleavage [9].

In order to evaluate the performance of the reporter system we placed a previously developed theophylline synthetic riboswitch [12] upstream of the TEV protease gene. Activation of the riboswitch initiated translation of TEV protease, which cleaved the FRET linker, allowing detection of eGFP fluorescence (Fig. 1). Previously, Gallivan and co-workers showed that riboswitch incorporation can significantly reduce the expression of downstream genes; for instance, in the theophylline synthetic riboswitch, analyte induces translation of only ~16 % of the transcripts while ~84 % remain inactive [12]. In this case, a highly sensitive enzymatic reporter system is advantageous for riboswitch discovery, since only a small amount of enzyme is required for cleavage of the substrate and achievement of a measurable optical signal. Using our reporter system, we were able to observe a detectable increase in fluorescence intensity as early as 30 min post analyte addition [9], when presumably only a small amount of TEV protease was present in the system. In comparison with other enzymatic reporters such as the widely used β-galactosidase and luciferase, our system does not require addition of substrate; fusion protein substrate can be directly expressed in cells. Moreover, we have demonstrated superiority of TEV protease–FRET substrate system over direct coupling of the riboswitch with fluorescent protein in terms of sensitivity. When the eGFP gene was placed downstream of theophylline riboswitch, we observed a very modest increase (~1.4-fold) in fluorescence intensity of cells in response to analyte [9]. In contrast, riboswitch activation of TEV protease gene expression followed by

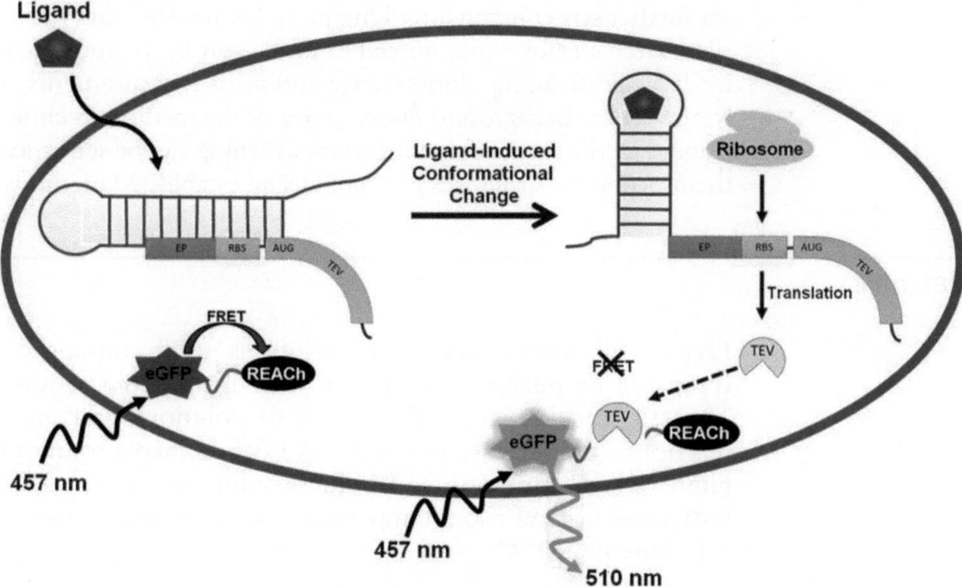

Fig. 1 TEV protease–FRET substrate reporter system for monitoring riboswitch performance in *E. coli* cells. In the absence of ligand, extensive pairing of riboswitch with ribosome-binding site (RBS) prevents translation of TEV protease. Ligand binding induces conformational change of the riboswitch, unpairing RBS and initiating translation of the TEV protease mRNA. The produced protease cleaves the FRET-based protein substrate, yielding an increase in fluorescent emission at 510 nm upon excitation at 457 nm

cleavage of FRET protein resulted in an 11.3-fold increase in fluorescence intensity in response to theophylline.

Here we describe a screening protocol that is based on TEV protease–FRET substrate reporter system and was used to identify artificial riboswitches that activate protein translation in *E. coli* cells in the presence of 2,4-dinitrotoluene [13]. The in vivo selection presented here focuses on isolation of synthetic riboswitches that upregulate gene expression at the posttranscriptional level based on the availability of the ribosome-binding site (RBS). However, it is also possible to adapt this technique to discover riboswitches that downregulate gene expression or that act at the transcriptional level. The following procedure assumes that the user already has an aptamer (from literature or from performing previously described aptamer selection methods [7, 8]) that binds to the desired cell-permeable and nontoxic compound. The first step in riboswitch selection is construction of a riboswitch library by incorporating a randomized expression platform (8–15 degenerate bases) between the aptamer sequence and RBS coupled with the TEV protease-encoding sequence. The next step is co-transformation of created plasmid library and a plasmid encoding FRET-based TEV protease substrate into *E. coli* cells. Then, all obtained clones must be visually screened for fluorescence increase in the presence of the compound of interest. Colonies that appear to show increased fluorescence in the presence of the desired compound are subjects for further screening in liquid medium. Clones showing the highest riboswitch performance in cell cultures can be further characterized by performing fluorescence intensity measurements in cell lysates where background fluorescence of the medium is eliminated. Candidate riboswitches identified by screening can be sequenced and their secondary structures predicted using established methods [14].

2 Materials

Prepare all media and buffer solutions using ultrapure water (prepared by purifying deionized water to obtain a sensitivity of 18 MΩ cm at 25 °C). For antibiotic solutions and enzymatic reactions, use nuclease-free water. Autoclave media solutions, and filter-sterilize antibiotic and inducer solutions. Store media and buffer solutions at room temperature. Store antibiotic and inducer solutions at −20 °C.

2.1 Construction of Plasmid Containing FRET-Based Protein

1. MAX Efficiency® DH5α™ Competent Cells (Invitrogen, Carlsbad, CA).
2. Plasmid containing YFP coding sequence (we have used pET21a(+):YFP plasmid vector, created in the lab).
3. QuikChange II Site-Directed Mutagenesis Kit (Agilent Technologies/Genomics).
4. Plasmid containing eGFP coding sequence (we have used pIVEX2.3d:eGFP plasmid vector—a generous gift from Dr. Vincent Noireaux from the University of Minnesota).
5. PfuUltra High-fidelity DNA Polymerase (Agilent Technologies/ Genomics).
6. *NdeI*, *HindIII*, and *BamHI* restriction enzymes and supplemented reaction buffers (Invitrogen, Carlsbad, CA).
7. T4 DNA Ligase and supplemented 5× T4 DNA Ligase Buffer (Invitrogen, Carlsbad, CA).
8. *E. coli* expression vector (we have used pHWG640 plasmid vector—a generous gift from Dr. Josef Altenbuchner from the Institute of Industrial Genetics at the University of Stuttgart, Germany).
9. UltraPure™ DNase/RNase-Free Distilled Water (Invitrogen, Carlsbad, CA).
10. Ampicillin: 50 mg/mL stock solution in water, final concentration 100 μg/mL of media for all procedures.
11. Chloramphenicol: 34 mg/mL stock solution in 100 % ethanol, final concentration 25 μg/mL of media for all procedures.
12. LB agar: 35 g/L Difco LB Agar (Becton, Dickinson and Company, Franklin Lakes, NJ).

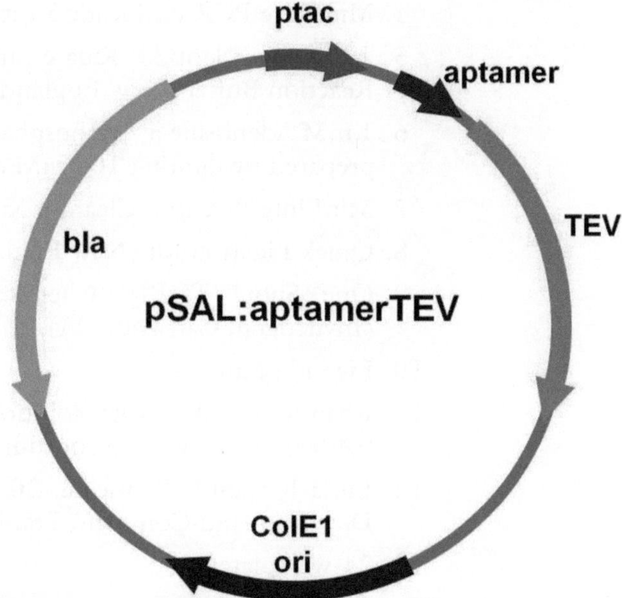

Fig. 2 Plasmid map of pSAL:aptamer TEV used as a template to construct riboswitch library. The plasmid contains aptamer sequence placed upstream of TEV protease encoding gene under the control of ptac promoter, β-lactamase encoding gene responsible for resistance to ampicillin

13. Sterile Petri dishes, 100 mm.
14. Agarose.
15. 1× Tris–acetate–EDTA (TAE) buffer prepared from TAE 50× solution.
16. 2-Log DNA Ladder (0.1–10.0 kb) and Gel Loading Dye, Blue (6×) (New England Biolabs, Ipswich, MA).
17. SYBR® Safe DNA Gel Stain (Invitrogen, Carlsbad, CA).
18. MinElute PCR Purification Kit (Qiagen, Valencia, CA).
19. QIAEX II Gel Extraction Kit (Qiagen, Valencia, CA).
20. QIAprep Spin Miniprep Kit (Qiagen, Valencia, CA).

2.2 Construction of a Riboswitch Library and FRET-Based Selection

1. *E. coli* expression plasmid containing the aptamer of interest coupled to TEV protease reporter gene. (Our construct of the aptamer upstream of TEV protease gene was made by GenScript (Piscataway, NJ). The construct was cloned into pSAL plasmid vector (Fig. 2)—generous gift from Dr. Justin Gallivan from Emory University, Atlanta, GA.)
2. Phusion DNA polymerase (New England Biolabs, Ipswich, MA).
3. *Dpn*I restriction endonuclease (Promega, Fitchburg, WI).

4. MinElute PCR Purification Kit (Qiagen, Valencia, CA).
5. T4 Polynucleotide Kinase and 1× Polynucleotide Kinase Reaction Buffer (New England Biolabs, Ipswich, MA).
6. 1 mM Adenosine 5′-triphosphate disodium salt solution (ATP) prepared by diluting 100 mM ATP solution.
7. MinElute Reaction Cleanup Kit (Qiagen, Valencia, CA).
8. Quick Ligation Kit (New England Biolabs, Ipswich, MA).
9. One Shot® TOP10 Chemically Competent *E. coli* cells (Invitrogen, Carlsbad, CA).
10. Ligand of interest.
11. Rhamnose: 30 % stock solution in water, final concentration 0.4 % in media for all procedures.
12. Luria-Bertani (LB) media: 20 g/L Difco LB Broth (Becton, Dickinson and Company, Franklin Lakes, NJ).
13. 24-well plates.
14. Absorbance and fluorescence microplate reader.
15. Dark Reader blue light transilluminator, amber screen, viewing glasses (Clare Chemical Research, Dolores, CO).
16. Incubator with shaker.

2.3 Analysis of Individual Riboswitch Clones

1. Baffled culture flasks, 500 mL.
2. Lysis buffer: 100 mM Tris–HCl, pH 8.0, 150 mM NaCl, 1 mM $MgCl_2$, 0.1 mg/mL of lysozyme (Sigma, St. Louis, MO), 1 μL/mL of benzonase nuclease (Sigma, St. Louis, MO), 8 mM iodoacetamide (Sigma, St. Louis, MO).
3. Centrifuge.

3 Methods

3.1 Creation of Plasmid Containing Gene Encoding FRET-Based Protein–TEV Protease Substrate

1. Perform sequential site-directed mutagenesis of YFP gene with QuikChange II Site-Directed Mutagenesis Kit. To create a YFP mutant (Y145W), use *E. coli* cloning vector containing YFP coding sequence as a template, forward (5′-CACAAGC TGGAGTAC AACTGGAACAGCCACAACGT CTATATC-3′) and reverse (5′-GATATAGACGTTGTGGCTG TTCCAGTT GTACTCCAGCTTGTG-3′) primers. Perform PCR reaction, treatment with *Dpn*I endonuclease, and transformation into competent *E. coli* cells according to the manufacturer's protocol. Pick six to eight obtained clones for plasmid isolation using QIAprep Spin Miniprep Kit. Verify sequences of purified plasmids by DNA sequencing, and identify plasmids containing Y145W-YFP. To create a YFP mutant (Y145W/H148V),

use *E. coli* cloning vector containing Y145W-YFP coding sequence as a template, forward (5′-CTGGAGTACAAC TGGAACAGCGT GAACGTCTATATCATGGCCG-3′) and reverse (5′-CGGCC ATGATATAGACGTTCACGCTGTTCC AGTTGTAC TCCAG-3′) primers. Perform all manipulations which are described above. Resulting plasmid vector should contain sequence encoding a double mutant of YFP, Y145W/ H148V-YFP, called REACh.

2. Amplify the coding sequence of REACh (Y145W/H148V-YFP) using forward (5′-GTGGATCC**GAAAACCTGTAC TTCCAGAGCGGCACC**GTGAGCAAGGGCGAA-3′) and reverse (5′-CGTAAGCTTTTA*ATGGTGATGGTGATGGTG* CTTGTACAGCTCGTCC-3′) primers. Amplify the coding sequence of eGFP using forward (5′-CGTCATATGGTGA GCAAGGGCGAGGAG-3′) and reverse (5′-CGTGGATCC **GGCCTT CTTGTACAGGCT**CTTGTACAGCTCGTC-3′) primers. (PCR will allow the addition of specific restriction sites (underlined) to the ends of amplified products, a nucleotide sequence (bold) encoding linker peptide and a nucleotide sequence (italic) encoding 6× His-tag at the C-terminus of REACh.) Perform both PCR reactions in 50 μL volume using PfuUltra High-fidelity DNA polymerase according to the manufacturer's protocol. Analyze PCR-amplified products by standard agarose gel electrophoresis (*see* **Note 1**). Purify PCR products using MinElute PCR Purification Kit.
3. Digest the PCR products amplified from eGFP and REACh and plasmid vector (pHWG640) with *Nde*I–*BamH*I, *BamH*I–*Hind*III, and *Nde*I–*Hind*III restriction enzymes, respectively. Set up restriction enzyme digestion reactions according to the manufacturer's protocol. Separate the digested fragments on an agarose gel, and purify the DNA fragments using QIAEX II Gel Extraction Kit.
4. Ligate digested PCR products amplified from eGFP and REACh into linearized plasmid vector (pHWG640).

Ligation reaction (*see* **Note 2**):

Insert:vector molar ratio	5:1
Digested vector	3–20 fmol
Digested PCR product 1	15–100 fmol
Digested PCR product 2	15–100 fmol
Total DNA	0.01–0.1 μg
5× Ligase reaction buffer	2 μL
T4 DNA ligase	1 μL (1 unit)

(continued)

(continued)

Water	To 10 μL final
Temperature	25 °C
Time	3 h

For optimal transformation, dilute the ligation reaction fivefold with 1× TE buffer. (This method of construction introduces 17 amino acid linker peptide (SLYKKAGSENLYFQSGT), which joins the C-terminus of eGFP to the N-terminus of REACh and also places 6× His-tag at the C-terminus of REACh. The linker contains a TEV protease cleavage site (underlined) and flanking spacer arms to increase TEV protease accessibility.)

5. Transform *E. coli* DH5α competent cells with 10 ng/μL of the ligation reaction mixture and grow overnight on LB-agar plates supplemented with antibiotic (25 μg/mL of chloramphenicol).
6. Pick the colonies from the plate. Transfer each colony into 5 mL of LB media supplemented with antibiotic and grow overnight cultures. Collect cells; purify plasmids using QIAprep Spin Miniprep Kit.
7. Verify sequences of purified plasmids by DNA sequencing. The plasmid map of the resulting construct (pHWG640:eGFP-TL-REAChHis) is shown in Fig. 3.

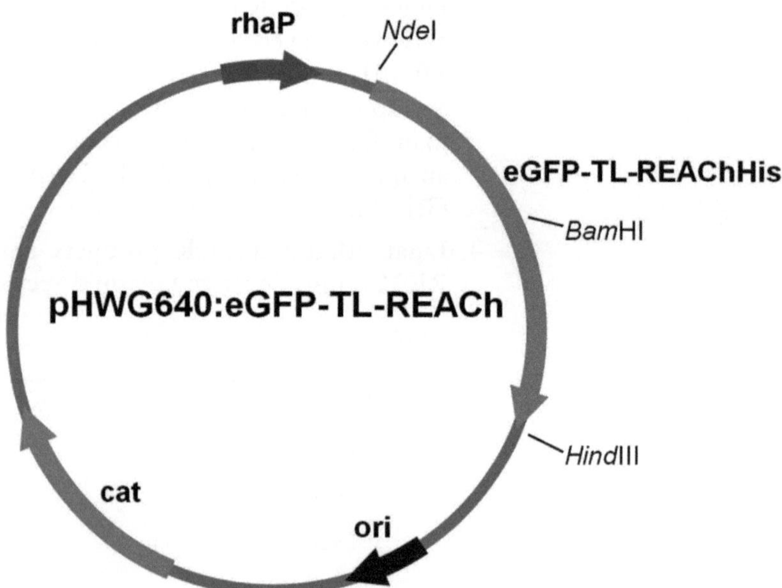

Fig. 3 Plasmid map of pHWG640:eGFP-TL-REACh. The plasmid contains gene encoding fusion protein composed of eGFP- and REACh-encoding sequences coupled with a linker (TL) containing TEV protease cleavage site. The fusion protein gene placed downstream of Rhamnose-inducible promoter (rhaP). The plasmid carries gene (cat) responsible for resistance to chloramphenicol

3.2 Construction of a Riboswitch Plasmid Library and an In Vivo Selection

The schematic illustration of the construction of a riboswitch plasmid library is shown in Fig. 4.

1. Design PCR primers for incorporation of 10–15 degenerate bases between the aptamer and a region of 5–7 constant bases located immediately before the start codon of a reporter gene, TEV protease-encoding sequence (*see* **Note 3**).
2. Incorporate randomized sequence (expression platform) between the aptamer and the start codon of a TEV protease gene by performing PCR reaction. Use *E. coli* expression plasmid containing the aptamer of interest coupled to TEV protease reporter gene as a template and primers designed in **step 1**. Perform PCR reaction in 50 μL volume using Phusion DNA polymerase:

Component	Volume/reaction (μL)	Final concentration
Nuclease-free water	29	
Template DNA (10 ng/μL)	1	10 ng/reaction
Forward primer (5 pmol/μL)	2	0.2 μM
Reverse primer (5 pmol/μL)	2	0.2 μM
dNTP mixture (2.5 mM each)	4	0.2 μ
5× Phusion HF or GC buffer	10	1×
DMSO	1.5	3 %
Phusion DNA polymerase	0.5	1.0 units/50 μL PCR

Set up thermocycling conditions according to the manufacturer's protocol. Use 5 μL of the reaction mixture to analyze PCR-amplified product by standard agarose gel electrophoresis (*see* **Note 1**).

3. To digest the template plasmid, add 5 units of *Dpn*I restriction endonuclease to the PCR mixture and incubate at 37 °C for 60 min. Purify PCR product using MinElute PCR Purification Kit.
4. To add 5′-phosphates to the PCR product, use T4 Polynucleotide Kinase. Use up to 300 pmol of 5′ termini in 50 μL reaction containing 1× T4 Polynucleotide Kinase Buffer, 1 mM ATP, and 10 units of T4 Polynucleotide Kinase. Incubate at 37 °C for 30 min (*see* **Note 4**). Purify the phosphorylated DNA with MinElute Reaction Cleanup Kit.

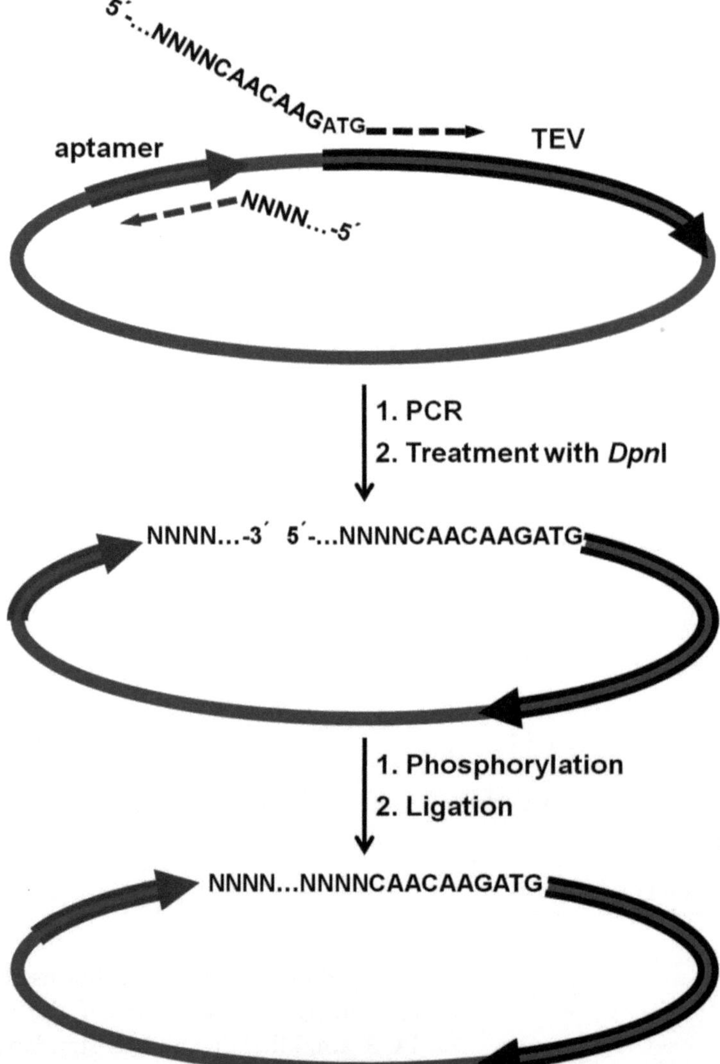

Fig. 4 Construction of a riboswitch plasmid library. The library is constructed by performing PCR using template plasmid with an aptamer sequence upstream of a reporter gene and forward and reverse primers containing degenerate bases at 5′-ends. The resulting PCR product is treated with *DpnI* to digest the template plasmid followed by phosphorylation of 5′-ends and ligation of linear DNA into circular plasmid

5. Self-ligate linear DNA using Quick Ligation Kit:

Linear DNA	50 ng (adjust volume to 10 μL with nuclease-free water)
2× quick ligation buffer	10 μL
Quick T4 DNA ligase	1 μL
Temperature	25 °C
Time	5 min

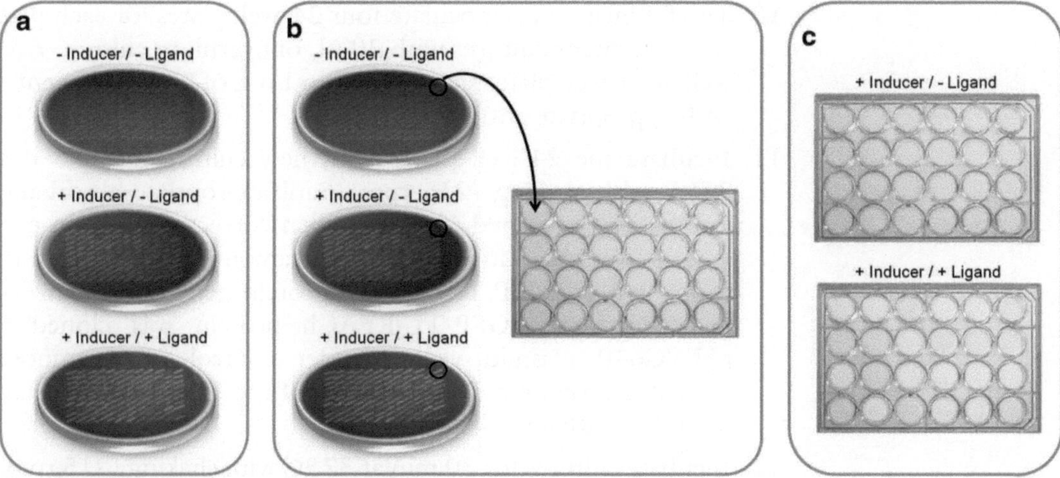

Fig. 5 In vivo screening for functional riboswitches. (**a**) Obtained after transformation *E. coli* colonies are picked and streaked on three LB-agar plates: without inducer and ligand; with inducer (expression of the fusion protein-encoding gene) and without ligand (no riboswitch activation); with inducer (expression of the fusion protein-encoding gene) and with ligand (riboswitch activation and TEV protease production). (**b**) Colonies that show increased fluorescence in the presence of ligand are selected for screening in liquid medium (an example is *circled*). (**c**) The gene expression of the selected riboswitch clones is quantified in the presence and absence of ligand

6. Transform *E. coli* TOP10 chemically competent cells with 5 ng of the ligation mixture and 5 ng of *E. coli* expression vector containing sequence encoding eGFP-TL-REACh protein (TL-17 amino acid peptide sequence containing TEV protease cleavage site). Plate the transformed cells on LB-agar plates supplemented with appropriate antibiotics (ampicillin and chloramphenicol) and grow overnight at 37 °C (*see* **Note 5**).
7. Pick all identified colonies from the plate, and streak each colony onto the same section of each of the three plates: LB-agar plate supplemented with antibiotics, LB-agar plate supplemented with antibiotics and inducer (0.4 % Rhamnose) for eGFP-TL-REACh gene expression, and LB-agar plate supplemented with antibiotics, inducer, and ligand (Fig. 5a). (The number of plates will depend on the number of obtained colonies. In our work, we have analyzed 1,000 clones.) Grow cells overnight at 37 °C.
8. Visualize fluorescent colonies using Dark Reader blue light transilluminator and amber screen or viewing glasses. Pick colonies that show increased fluorescence in the presence of ligand, and inoculate them in 24-well plates with each well containing 1 mL of LB supplemented with appropriate antibiotics (Fig. 5b). (The number of plates will depend on the number of colonies assayed in **step 7**.)
9. Incubate the 24-well plates overnight with shaking (37 °C, 215 rpm).

10. The following day, inoculate four 24-well plates for each plate with overnight cultures with 20 μL of overnight culture. Each well of 24-well plates should contain 1 mL of LB supplemented with appropriate antibiotics.
11. Incubate the 24-well plates with new cultures for 2.5 h at 37 °C with shaking (215 rpm) until appropriate absorbance at 600 nm is reached (0.3–0.5 when corrected to 1-cm path length cuvette). Add an appropriate amount of an inducer for expression of eGFP-TL-REACh protein gene. (In our work, gene encoding eGFP-TL-REACh protein was cloned in pHWG640 plasmid vector under control of Rhamnose-inducible promoter. For induction of the expression, we have used 0.4 % Rhamnose.)
12. Incubate cultures for 30 min at 37 °C with shaking (215 rpm). Then, divide each group of four plates with the same cultures on two sets of two. Add ligand to the first set of plates (this set will contain cells with activated riboswitches). Add equivalent volume of the solvent where ligand is dissolved to the second set of plates (this set will contain cells with non-activated riboswitch) (Fig. 5c). Measure the absorbance at 600 nm and fluorescence at 510 nm (excitation 457 nm) of the cell cultures in each plate.
13. Incubate the 24-well plates for 6 h at 37 °C with shaking (215 rpm). Measure the absorbance at 600 nm and fluorescence at 510 nm (excitation 457 nm) of the cell cultures in each plate. To normalize the data, divide fluorescence intensity by absorbance for each well.
14. Compare the ratios of the normalized data for cultures grown in the presence of ligand to those grown in the absence of ligand (the "activation ratio") to identify functional riboswitches.

3.3 Analysis of Individual Clones for Riboswitch Activation

1. Using 10 μL of selected culture, inoculate 10 mL of LB supplemented with appropriate antibiotics. Incubate overnight at 37 °C with shaking.
2. The following day, using 1.5 mL of fresh overnight culture, inoculate six new flasks containing 150 mL of LB supplemented with appropriate antibiotics.
3. Incubate flasks with new cultures at 37 °C with shaking (220 rpm) until absorbance at 600 nm is reached 0.4–0.5. Add an appropriate amount of an inducer for expression of eGFP-TL-REACh protein gene.
4. Incubate cultures for 30 min at 37 °C with shaking (220 rpm). Then, divide flasks into two sets of three. Add ligand to the first set of flasks (this set will contain cells with activated riboswitches). Add equivalent volume of the solvent (vehicle)

to the second set of flasks (this set will contain cells with non-activated riboswitch). Measure the absorbance at 600 nm and fluorescence at 510 nm (excitation 457 nm) of the cell cultures in each flask (it will be time 0 h). Record the absorbance at 600 nm and fluorescence at 510 nm (excitation 457 nm) for each cell culture at time 1, 2, 3, 4, 5, and 6 h after riboswitch activation (ligand addition). At each time point, collect 10 mL of each culture for preparation of cellular lysates. Harvest the cells by centrifugation at 3,200 rpm for 10 min at 4 °C. Discard the supernatant. Store cell pellets at −80 °C until ready for processing.

5. For preparation of cellular lysates, thaw cell pellets on ice for approximately 15–20 min, vortex, and resuspend in 0.4 mL of lysis buffer. Incubate cell suspensions on ice for 30 min and mix two to three times by gently swirling. Remove cell debris by centrifugation at 13,200 rpm for 30 min at 4 °C. Transfer supernatant from each sample into new tubes, and measure fluorescence at 510 nm (excitation 457 nm). Determine protein concentrations of the clarified cellular lysates using BCA assay; follow the protocol provided by the manufacturer. To normalize the data, divide fluorescence intensity by total protein concentration for each sample.
6. Compare the ratios of the normalized data for lysates from cultures grown in the presence of ligand to those grown in the absence of ligand (the "activation ratio") to determine the riboswitch performance at different time points (time course).

4 Notes

1. 1 % Agarose gel electrophoresis: Add 0.5 g agarose to 50 mL 1× TAE. Heat the solution to boiling in the microwave or on the hotplate stirrer to dissolve the agarose. Cool the solution to approximately 50 °C, and add 3 μL of SYBR® Safe DNA Gel Stain (Invitrogen, Carlsbad, CA). Pour the melted agarose into a gel tray with the well comb in place. Cool gel at room temperature for 20–30 min, until it is completely solidified. Remove the comb, and place the tray with the agarose gel into the electrophoresis box. Fill gel box with 1× TAE buffer until the gel is covered. Load a 2-Log DNA Ladder (New England Biolabs, Ipswich, MA) and samples mixed with gel loading dye (New England Biolabs, Ipswich, MA). Run the gel at 80–130 V until the dye line is approximately 75–80 % of the way down the gel. Using any device that has blue light or UV excitation, visualize the DNA fragments.
2. If the DNA fragments are too dilute, the reaction can be scaled up to 20 μL ligation volume.

3. In our previous work, a riboswitch plasmid library was created by incorporation of randomized expression platform between aptamer and RBS upstream of TEV protease gene. However, replacement of fixed RBS sequence with randomized allows to optimize the strength of the RBS to achieve the high level of reporter protein expression [15]. The constant region between the RBS and the start codon (usually five to seven bases) is important to provide suitable spacing for optimal ribosome binding. The constant sequence in our library was "CAACAAG" that was previously used by Gallivan and co-workers [15].
4. Since T4 Polynucleotide Kinase Buffer contains 1 mM dithiothreitol for optimal activity of T4 Polynucleotide Kinase, fresh buffer is required to perform a phosphorylation (the reduced dithiothreitol in older buffer lowers activity). The phosphorylation efficiency can be improved by heating to 70 °C for 5 min, then chilling on ice prior to kinase addition, and adding PEG-8000 to 5 % (w/v).
5. In our work, we used 50 μL of transformation reaction per each LB-agar plate.

Acknowledgement

This research was funded by the Air Force Office of Scientific Research (AFOSR).

References

1. Jin Y, Watt RM, Danchin A, Huang JD (2009) Use of a riboswitch-controlled conditional hypomorphic mutation to uncover a role for the essential csrA gene in bacterial autoaggregation. J Biol Chem 284:28738–28745
2. Lee ER, Blount KF, Breaker RR (2009) Roseoflavin is a natural antibacterial compound that binds to FMN riboswitches and regulates gene expression. RNA Biol 6:187–194
3. Kim JN, Blount KF, Lim J, Link KH, Breaker R (2009) Design and antimicrobial action of purine analogues that bind guanine riboswitches. ACS Chem Biol 4:915–927
4. Jise AM, Soukup GA, Breaker RR (2001) Cooperative binding of effectors by an allosteric ribozyme. Nucleic Acids Res 29: 1631–1637
5. Win MN, Smolke CD (2008) Higher-order cellular information processing with synthetic RNA devices. Science 322:456–460
6. Sharma V, Nomura Y, Yokobayashi Y (2008) Engineering complex riboswitch regulation by dual genetic selection. J Am Chem Soc 130: 16310–16315
7. Tuerk C, Gold L (1990) Systematic evolution of ligands by exponential enrichment: RNA ligands to bacteriophage T4 DNA polymerase. Science 249:505–510
8. Roth A, Breaker RR (2004) Selection in vitro of allosteric ribozymes. In: Sioud M (ed) Ribozymes and siRNA protocols, methods in molecular biology, 2nd edn. Humana Press, Totowa, NJ, pp 145–164
9. Harbaugh S, Kelley-Loughnane N, Davidson M, Narayanan L, Trott S, Chushak YG, Stone MO (2009) FRET-based optical assay for monitoring riboswitch activation. Biomacromolecules 10:1055–1060
10. Van der Berg S, Löfdahl P-A, Härd T, Berglund H (2006) Improved solubility of TEV protease by directed evolution. J Biotechnol 121:291–298
11. Ganesan S, Ameer-beg SM, Ng TT, Vojnovic B, Wouters FS (2006) A dark yellow fluorescent protein (YFP)-based resonance energy-accepting

chromoprotein (REACh) for Förster resonance energy transfer with GFP. Proc Natl Acad Sci U S A 103:4089–4094

12. Lynch SA, Desai SK, Sajja HK, Gallivan JP (2007) A high-throughput screen for synthetic riboswitches reveals mechanistic insights into their function. Chem Biol 14:173–184
13. Davidson ME, Harbaugh SV, Chushak YG, Stone MO, Kelley-Loughnane N (2013) Development of a 2,4-dinitrotoluene-responsive synthetic riboswitch in *E. coli* cells. ACS Chem Biol 8:234–241
14. Zuker M (2003) Mfold web server for nucleic acid folding and hybridization prediction. Nucleic Acids Res 31:3406–3415
15. Lynch SA, Gallivan JP (2009) A flow cytometry-based screen for synthetic riboswitches. Nucleic Acids Res 37:184–192

Chapter 7

Portable Two-Way Riboswitches: Design and Engineering

Ye Jin and Jian-Dong Huang

Abstract

Riboswitches are RNA-based regulatory devices that mediate ligand-dependent control of gene expression. However, there has been limited success in rationally designing riboswitches. Moreover, most previous riboswitches are confined to a particular gene and only perform one-way regulation. Here, we describe a library screening strategy for efficient creation of ON riboswitches of *lacI* of *Escherichia coli*. An ON riboswitch of *lacI* is then integrated with the *lac* promoter, generating a hybrid device to achieve portable sequential OFF-and-ON gene regulation.

Key words Riboswitch, LacI, LacP, Portable two-way regulation

1 Introduction

Inducible promoters are the major tools for specifically controlling target gene expression. However, there are only limited numbers of inducible promoters which, on some occasions, are insufficient for independent control of multiple genes by addition of distinct inducers. A promising alternative is riboswitch that is an RNA-based, ligand-dependent genetic control element located in noncoding regions of messenger RNAs. Unlike inducible promoters, these RNA-based control elements are potentially numerous as a result of high-throughput screen for ligand-specific aptamers (the sensor domain of a riboswitch) [1–5].

Advancements have recently been made in the ability to convert ligand-specific aptamers to functionally active riboswitches by various random screening approaches [6–8] or rational design [9–11]. However, library screening is labor intensive and time consuming, and successful rational design is not guaranteed even though a long list of design principles are followed due to incomplete understanding of RNA-based gene regulation mechanisms. In addition, a common disadvantage of riboswitches as well as inducible promoters is that they only allow for one-way gene regulation.

Atsushi Ogawa (ed.), *Artificial Riboswitches: Methods and Protocols*, Methods in Molecular Biology, vol. 1111,
DOI 10.1007/978-1-62703-755-6_7, © Springer Science+Business Media New York 2014

Once activated, a target gene cannot be turned off unless promoter inducers or riboswitch ligands are removed. The one-way regulation mode makes the existing regulation systems unsuitable for quick gene control in live animals or in vitro studies where intact cell cultures are required. Thus, there is a great need for a two-way regulation system, in which target gene can be switched off and on in response to distinct effector molecules.

Here we solve the above questions by engineering an ON theophylline-responsive riboswitch of the *lacI* gene and then integrating it with the Lac-inducible promoter system (abbreviated as LacP), in which the LacI repressor binds as a homotetramer to two LacI-binding sites (LacIbs) positioned immediately downstream of the *lac* promoter. First, we use a library screening strategy to create ON theophylline-responsive riboswitches of the *lacI* gene that encodes the LacI transcriptional regulator. Then, by combining an ON riboswitch of LacI with two LacIbs, we establish a portable hybrid device that acts as a versatile key for controlling gene expression without the need for further rational design or library screening. Portability of this device has been verified previously by applying it to *rpoS* (encoding RpoS, a master regulator of acid resistance) and *csrB* (encoding a small noncoding RNA CsrB) [12]. This portable device regulates target genes in a two-way manner, switching off the targets in response to theophylline and restoring the target expression in response to isopropyl β-D-1-thiogalactopyranoside (IPTG).

2 Materials

All media and buffers are prepared using sterilized distilled deionized water (ddH_2O). All containers and pipette tips are sterilized by heating (121 °C, 15 min) and stored at room temperature.

2.1 Incubation of Bacteria

1. Bacterial strain: *E. coli* strain MG1655 (F- lambda- *ilvG- rfb*-50 *rph*-1).
2. Growth medium: Luria-Bertani (LB) medium supplemented with 2 mM magnesium sulfate. The antibiotics 50 μg/ml ampicillin, 50 μg/ml kanamycin, and 12.5 μg/ml chloramphenicol are used for selection when appropriate.
3. Glycerol.
4. Temperature-controlled microplate shaker.
5. Temperature-controlled flask shaker.
6. Sterilized 96-well microtiter plate.
7. Spectrophotometer and cuvettes.
8. Flasks (500 ml).
9. Refrigerators at 4 °C and freezers at −20 and −80 °C.

2.2 Recombineering

1. Ultrapure LB agar.
2. Ultrapure agarose (MB grade).
3. Micropulser electroporator.
4. 50-ml centrifuge tubes.
5. 1.5-ml microfuge tubes.
6. Controlled low-speed centrifuge (3,000 × *g*) at 4 °C.
7. Bench top centrifuge (max speed 16,000 × *g*).
8. Petri dishes (100 mm × 15 mm).
9. Pipettors of various volumes (10, 200, 1,000 μl).
10. Sterile micropulser electroporation cuvettes (0.1 cm gap) (Bio-Rad) (*see* **Note 1**).
11. 42 °C water bath.
12. Ice buckets.
13. Plasmid isolation kit (Qiagen).
14. Gel extraction kit for DNA purification from agarose (Qiagen) (*see* **Note 2**).
15. pCm: A plasmid carrying a self-driven chloramphenicol resistance gene (*cat*) flanked by two loxP sites.
16. Materials for polymerase chain reaction (PCR): Standard Taq polymerase, dNTP mixture, chimeric primers for PCR amplification of recombineering substrates, PCR thermal cycler.
17. Agarose gel electrophoresis apparatus.
18. Materials required to express the Red functions: pSIM6 plasmid, *E. coli* MG1655, ampicillin, chloramphenicol, distilled deionized water at 4 °C.

2.3 X-Gal Screening and Verification

1. 20 mg/ml X-gal (5-bromo-4-chloro-3-indolyl-beta-D-galactopyranoside) in DMSO, stored at −20 °C.
2. OFF plate: LB agar supplemented with 2 mM caffeine, 1 mM magnesium sulfate, 60 μg/ml X-gal, and 12.5 μg/ml chloramphenicol.
3. ON plate: LB agar supplemented with 2 mM theophylline, 1 mM magnesium sulfate, 60 μg/ml X-gal, and 12.5 μg/ml chloramphenicol.
4. X-gal plate: LB agar supplemented with 60 μg/ml X-gal and 12.5 μg/ml chloramphenicol.
5. Beta-galactosidase activity kit (Thermo Scientific), stored at −20 °C.
6. 100 mM Theophylline in ddH_2O, stored at room temperature.
7. 100 mM Caffeine in ddH_2O, stored at room temperature.
8. 1 M IPTG in ddH_2O, stored at −20 °C.

2.4 Assessment of Riboswitches

1. A computerized program for structural analysis of RNAs (RNAstructure) [13].
2. Reagents for acid resistance assay: Acidified LB medium (pH 2.0), neutral LB.

3 Methods

To construct a riboswitch-LacP hybrid device, the first task is engineering ligand-responsive riboswitches for the *lacI* gene which encodes the LacI transcriptional regulator in the Lac system. To do this, we establish a chromosomal riboswitch library, with the aid of the recombineering technique, to screen for active riboswitches for *lacI*. We use the theophylline-responsive aptamer as the sensor domain of riboswitches. The aptamer is linked to a chloramphenicol resistance gene (*cat*) by ligation PCR, generating a riboswitch cassette library containing a *cat*-aptamer-linker-random fragment on the chromosome. The *cat* gene facilitates recombineering of the cassette into the genome, and the 5-nt random sequence allows construction of a riboswitch library. The cassette is integrated, using the recombineering technique, into the genome of the *E. coli* MG1655 strain right upstream of the ribosome-binding site (RBS) of target genes, resulting in numerous mutant strains, each of which carries a riboswitch candidate (functional or nonfunctional) on the chromosome. In theory, the mutant library contains 4^5 riboswitch candidates. An effective ON riboswitch of *lacI* is selected from the library by screening the color of colonies grown on agar supplemented with X-gal and theophylline or caffeine (an analogue of theophylline, serving as a negative control). Next, the ON riboswitch of *lacI* is integrated with the LacP system, generating a hybrid device that performs portable, two-way gene control.

3.1 Construction of a Randomized Riboswitch Cassette (Fig. 1)

1. A *cat* fragment is PCR amplified using pCm as a template. The resulting DNA product carries the *cat* gene to confer chloramphenicol resistance. Upstream of *cat* is a constitutive promoter to drive expression of the chloramphenicol resistance gene. The *cat* gene is followed by a 3′ untranslated region, which serves as a linker between the *cat* gene and the downstream theophylline-responsive aptamer (*see* **Note 3**).
2. The theophylline-responsive aptamer is synthesized by annealing complementary primers. Sequence of the theophylline aptamer is GAAATACCAGCATCGTCTTGATGCCCTTGGCAGTTC (5′–3′). Design a forward primer GAAATACCAGCATCGTCTTGATGCCCTTGGCAGTTC (5′–3′) and a reverse primer GAACTGCCAAGGGCATCAAGACGATGCTGGTATTTC (5′–3′) which is the reverse complement of the forward primers. The two primers are denatured at 94 °C for 30 s and then annealed

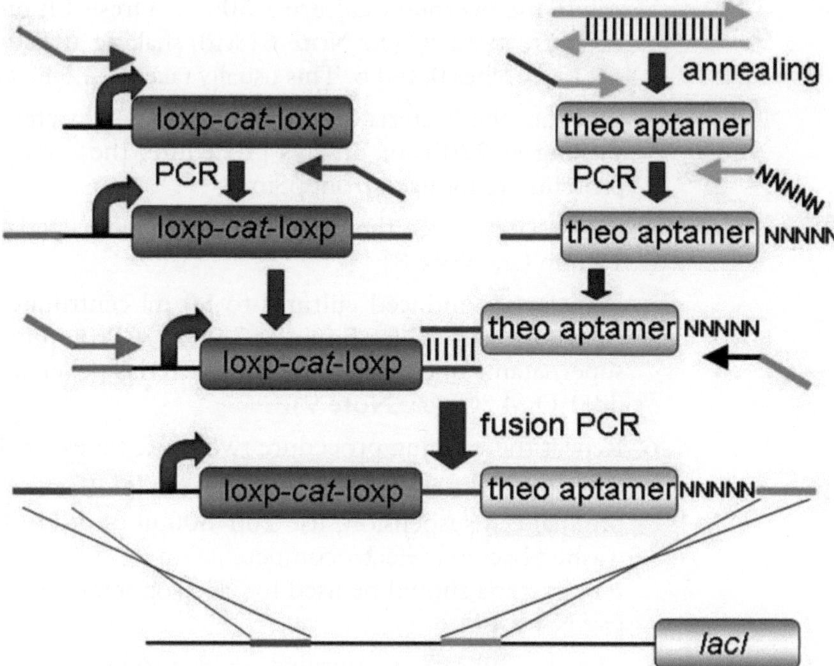

Fig. 1 Construction of a theophylline-aptamer cassette. NNNNN denotes a 5-nt random sequence which is added to the theophylline aptamer by PCR. The loxp-*cat*-loxp cassette is linked to the theophylline aptamer followed by the 5-nt random sequence by fusion PCR. The resulting cassette is flanked with arms (45 nt) homologous to the regions surrounding the ribosome-binding site of the *lacI* gene

at 55 °C for 30 s in a PCR thermal cycler, generating the double-stranded aptamer DNA (*see* **Note 4**).

3. A 5-nt random sequence is then linked to 3′ of the theophylline aptamer by PCR. To do this, a second reverse primer is designed for the aptamer by adding five random nucleotides. In companies like Techdragon, *N* denotes random nucleotides. The aptamer DNA is PCR amplified using the above-described forward primer and the second reverse primer, resulting in a theophylline aptamer fragment followed by a 5-nt random sequence (*see* **Note 5**).
4. The PCR-amplified *cat* cassette is linked to the aptamer with the 5-nt random sequence by ligation PCR. Primers are such designed that the resulting PCR products are flanked with arms (45 nt) homologous to the regions surrounding the RBS of the *lacI* gene. Separate the PCR products with agarose electrophoresis, and purify the products with gel extraction kit (*see* **Note 2**).

3.2 Construction of Randomized Riboswitch Libraries for lacI

1. Chemically transform *E. coli* MG1655 with plasmid pSim6 using conventional molecular techniques. pSim6 expresses the λ recombination proteins at 42 °C.
2. Incubate an isolated colony of MG1655 carrying pSim6 in 5 ml of LB plus 50 μg/ml ampicillin at 32 °C. Next morning,

dilute the overnight culture 1:500 into fresh LB medium and incubate at 32 °C (*see* **Note 6**) with shaking at 220 rpm, until OD_{600} reaches 0.4–0.6. This usually takes 2–2.5 h (*see* **Note 7**).

3. Incubate the bacteria for 15 min in a 42 °C water bath, with shaking at 220 rpm. In this procedure, the λ recombination proteins are induced from pSim6.
4. Immediately place the flask containing the bacteria on ice for 15 min (*see* **Note 8**).
5. Transfer the induced cultures to 50 ml centrifuge tubes and centrifuge at 4 °C for 5 min at 3,000 × *g*. Promptly decant the supernatant, and resuspend the bacterial pellet in 50 ml of ddH_2O (4 °C) (*see* **Note 9**).
6. Repeat the washing procedure two more times (*see* **Note 10**).
7. Resuspend the bacteria in ddH_2O at 4 °C. For each 50 ml of original cell suspension, use 200–300 μl of ddH_2O. Keep the washed bacteria (electrocompetent) on ice. The electrocompetent bacteria should be used for electroporation within 30 min (*see* **Note 11**).
8. Mix 1–5 μl of the purified PCR products (approximately 100 ng, salt-free) from Subheading 3.1, **step 4**, and the electrocompetent cells in microcentrifuge tubes on ice (*see* **Note 12**).
9. Meanwhile, dry the electroporation cuvettes (*see* **Note 13**).
10. Transfer the mixtures to electroporation cuvettes (0.1-cm gap) on ice, and perform electroporation at 1.8 kV (*see* **Note 14**).
11. Immediately add 1 ml of LB (room temperature) to the cuvette, and transfer the bacterial suspension to a microcentrifuge tube. Incubate the bacteria at 32 °C with shaking at 220 rpm for 60 min (*see* **Note 15**).

3.3 Identification of an ON Riboswitch of lacI (Fig. 2)

1. Spin down the bacteria using a bench-top microcentrifuge at 16,100 × *g* for 30 s. Spread the pellet on OFF plates. Incubate the plates overnight at 32 °C.
2. Next morning, search for colonies that are bluer than others. LacI is a repressor of *lacZ*, and therefore bluer color represents the off status of *lacI*. As the random library contains 4^5 (=1,024) riboswitch candidates, it is recommended that at least 1,000–2,000 colonies be checked for *lacI* expression (*see* **Note 16**).
3. Individually streak the blue colonies on ON plates. Incubate the plates overnight at 32 °C. As controls, the streaking is also conducted on OFF plates for the same set of colonies (*see* **Note 17**).
4. Next morning, search for colonies that are whiter on theophylline agar than on caffeine agar. The selected colonies carry ON riboswitch that turns on *lacI* in response to theophylline.

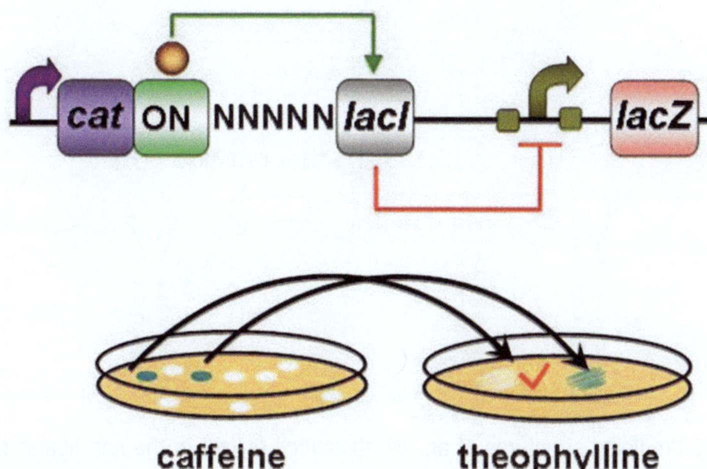

Fig. 2 Creation of ON riboswitches for the *lacI* gene using a library screening-based method. A random cassette mixture containing a theophylline-responsive aptamer and five random bases is "recombineered" into the chromosome of MG1655, generating a riboswitch library for *lacI*. Cells are plated on X-gal plates without theophylline but with 2 mM caffeine. Relatively blue colonies are searched for and transferred to X-gal plates with 2 mM theophylline. Those that turn into white are considered to carry ON riboswitches of *lacI*. The *blue arrow* denotes activating effects; the *blunt red arrow* denotes repressing effects. Reproduced with permission from Oxford University Press [12] (color figure online)

5. Compare the effectiveness of the identified ON riboswitches of *lacI* by beta-galactosidase assay in the presence of either caffeine or theophylline, and select the riboswitch that displays the widest regulation range.
6. Determine the sequence of the 5-nt random region of the selected riboswitch, and predict the secondary structure of the riboswitch using RNAstructure [13]. Choose the optimal structure with the minimum free energy for the structure analysis. To predict riboswitch structures in the ligand-bound form, utilize the "force pair" function of RNAstructure to force the theophylline-binding pocket to form (Fig. 3).

3.4 Construction of a Riboswitch-LacP Hybrid Device

Although the above-described riboswitches fine-tune *lacI* in response to theophylline, they have the following limitations. First, they are not applicable to genes other than *lacI* and therefore lack portability. Second, they perform either ON or OFF function and thus are "one-way" switches. We next demonstrate how to utilize the ON riboswitch of *lacI* in combination with the LacI protein and LacIbs to solve these problems. The transcriptional repressor LacI inhibits *lacZ* transcription by binding to two LacIbs upstream of *lacZ*. Inducible gene control built on binding of LacI to LacIbs has been widely used for various non-*lacZ* genes in prokaryotic and

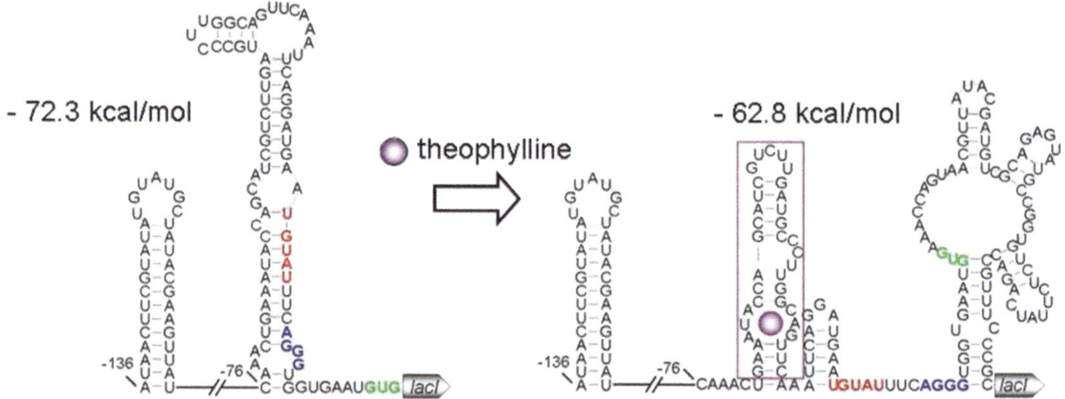

Fig. 3 Predicted structures of an ON riboswitch of *lacI*. In the non-ligand-bound state, there is a 9 bp stem structure immediately upstream of the ribosome-binding site and is likely to block the gene translation. In the theophylline-bound state, the long stem does not form and consequently permits the gene translation. The random sequence is shown in *red*; the ribosome-binding site in *blue*; and the start codon in *green*. Reproduced with permission from Oxford University Press [12] (color figure online)

eukaryotic cells [14–18], showing that the inducibility of the LacI-responsive promoters are less likely to be affected by downstream targets than that of riboswitches. Thus, the ON-*lacI*-LacIbs hybrid device should be capable of regulating any target gene when integrated upstream of the RBS. Next we showcase the portability and two-way control of this hybrid device by directly applying it to the *rpoS* gene without adjusting the sequence of the device.

1. PCR amplify the ON-*lacI*-LacIbs cassette (approximately 2,500 bp) from the genome of the strain carrying the ON riboswitch of *lacI*. The cassette is composed of the *cat* gene, the ON-*lacI* riboswitch, the *lacI* gene, and the *lacI*–*lacZ* intergenic region containing the LacP with two LacIbs. The *cat* gene facilitates integration of this cassette immediately upstream of any target gene via recombineering (*see* **Note 18**).
2. Delete the *lacI* gene from MG1655 by recombineering, generating a *lacI* null mutant. Design primers specific to the loxP-cat-loxP cassette. Primers are preceded by 45-nt overhangs at the 5′ ends. The resulting PCR products encompass the loxP-*cat*-loxP fragment flanked with homology (45 nt) to the regions immediately flanking *lacI*. This fragment is inserted into the genome of MG1655 harboring pSim6, replacing the *lacI* gene, by recombineering. Recombinants are selected for chloramphenicol resistance (encoded by the cat gene) and verified by the blue color of colonies on X-gal plates (*see* **Note 19**).
3. Insert the riboswitch cassette upstream of the RBS of the *rpoS* gene (or any gene of interest) of the *lacI* null mutant by recombineering, generating a new strain named ON-*lacI*-*rpoS* (Fig. 4a). Verify the mutation by sequencing the 5′ region upstream of *rpoS*.

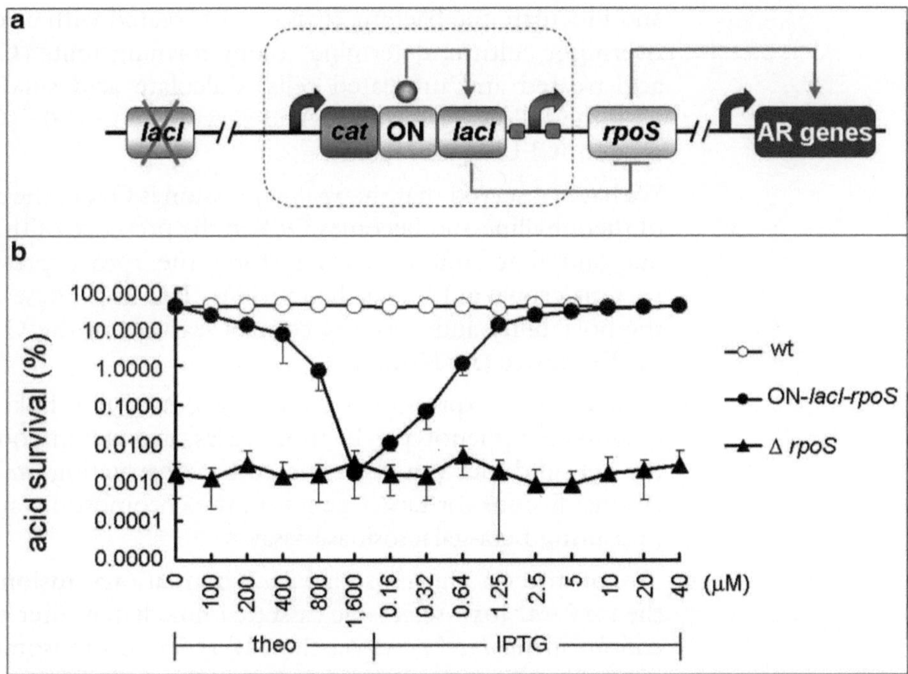

Fig. 4 Gene control mediated by the hybrid riboswitch device. (**a**) Constitution of the riboswitch hybrid device. The *boxed* is a portable riboswitch cassette ON-*lacI*-LacIbs (LacI-binding sites) integrated immediately upstream of ribosome-binding site of the *rpoS* gene that is a master regulator of acid resistance (AR). (**b**) Tunable and two-way control of *rpoS* via the ON-*lacI*-LacIbs hybrid device, as revealed by acid survival. The strain ON-l*acI*-rpoS that carries the ON-*lacI*-LacIbs device upstream of the ribosome-binding site becomes increasingly sensitive to acid with theophylline (0–1,600 μM). When increasing concentrations of IPTG are added to cells grown in media with 1,600 μM theophylline, acid survival is increased and restored to wild-type levels in the presence of 40 μM IPTG. Error bars, SD. Reproduced with permission from Oxford University Press [12]

3.5 Evaluation of the Hybrid Device for Portable, Two-Way Gene Control

1. RpoS is an alternative sigma factor of RNA polymerase, playing a critical role in survival of a diverse number of stresses such as acid shock [19, 20]. Since *rpoS* is required for acid resistance, we assay survival of the strain ON-*lacI-rpoS* under acidic conditions to test if the hybrid riboswitch device controls the *rpoS* expression in response to the ligand theophylline.
2. Before the acid resistance assay, add 10 μl of the overnight culture of the strain ON-*lacI-rpoS* grown into 2.5 ml of each of the following fresh LB medium: (a) LB plus 2 mM caffeine, (b) LB plus 2 mM theophylline, and (c) LB plus 2 mM theophylline and 1 mM IPTG. Incubate the bacteria aerobically at 37 °C with shaking at 220 rpm.
3. To assay the acid resistance of bacteria, treat the overnight cultures (in stationary phase) with acidic LB medium (pH 2.0) for 2 h and then serially (1:10) dilute the bacteria in neutral LB medium. Inoculate 20 μl of the dilutions on LB agar which has been dried at 37 °C for 2 h to ensure that the drops are inoculated to soak quickly into the agar. Meanwhile, serially dilute

and inoculate the bacteria that are not treated with acid. After overnight culture, determine colony-forming units (CFU) of acid-treated and untreated cells. Calculate acid survival (%) with the following formula: Acid survival (%) = 100 × (CFU of treated/CFU of untreated).

4. We have observed that the *rpoS* expression is ON in the absence of theophylline and becomes OFF in the presence of theophylline and that adding IPTG restores the *rpoS* expression of bacteria grown in LB plus theophylline (Fig. 4b). These confirm the portability and two-way control features of the ON-*lacI*-LacIbs device (*see* **Note 20**).
5. Unlike *rpoS*, expression of many genes is not linked to a quantifiable phenotype. In those cases, expression control by the hybrid device can be evaluated by constructing *lacZ* translational fusions for target genes on the chromosome and then measuring beta-galactosidase assay.
6. To construct a chromosomal *lacZ* translational fusion, insert the loxP-*cat*-loxP selectable cassette immediately after the stop codon of the *lacZ* gene on the MG1655 chromosome using recombineering (as described above). Next, PCR amplify the *lacZ*-loxP-*cat*-loxP cassette, and insert it (in frame) immediately prior to the stop codon of the target gene on the chromosome. The inserted *lacZ* fragment starts from the eighth codon of *lacZ* gene and is co-transcribed and translated with the fused genes. Quantify the expression of gene-*lacZ* fusions using a beta-galactosidase assay kit (*see* **Note 21**).

4 Notes

1. Micropulser electroporation cuvettes (0.1 cm gap) (Bio-Rad) can be reused although they are claimed to be disposable. After use, rinse the cuvette surface and their gaps with ddH_2O and store in 75 % ethanol at room temperature.
2. DNA used for electroporation must be free of salt; otherwise, the electric shock will be discharged very fast, resulting in a small explosion. The explosion will kill most competent cells. Therefore, the PCR products must be purified using the gel extraction kit to remove the salts.
3. In our design, the riboswitch cassette is preceded by a self-driven *cat* cassette. This cassette carries a strong constitutive promoter to drive the expression of the downstream riboswitch. The *cat* gene also makes it possible to select for clones with the riboswitch cassette on LB agar plus chloramphenicol. The *cat* gene can be replaced by any other antibiotic resistance gene as long as the drug resistance cassette carries a constitutive promoter.

4. In our design, we utilize the theophylline-responsive aptamer as the sensing domain of riboswitches. The theophylline aptamer can be replaced by any other ligand-specific aptamers.
5. The size of a riboswitch library is 4^N where N is the number of the randomized nucleotides fused to the ligand-specific aptamer.
6. All bacteria harboring pSim6 must be incubated at 32 °C or lower temperature. Higher temperature will cause the loss of pSim6.
7. If OD_{600} is higher than 0.6, the recombineering efficiency will be reduced dramatically.
8. Keep bacteria at 0–4 °C during the entire process of preparation of electrocompetent bacteria for recombineering, as low temperature is critical for recombineering efficiency.
9. In preparation of electrocompetent bacteria for recombineering, mix bacterial cells by vigorously shaking the centrifuge tubes after centrifugation. Add 30 ml of chilled ddH_2O to the bacterial pellets and shake until the pellets on the tube bottom disappear. Then add chilled ddH_2O to 50 ml. Do the shaking quickly as the bacteria are exposed to room temperature during this step. Immediately insert the entire tubes into ice after shaking to keep the tubes cold.
10. In preparation of electrocompetent bacteria for recombineering, bacteria must be washed three times with chilled ddH_2O. Two-time washing reduces recombineering efficiency significantly.
11. For recombineering, electrocompetent bacteria must be freshly prepared. No successful cases of recombineering have ever been reported with electrocompetent bacteria from frozen stock.
12. If more than 5 μl of DNA is used for recombineering, recombineering efficiency will be reduced.
13. Water on cuvettes will cause small explosion in electroporation. Before electroporation, dry by draining the cuvettes upside down on a paper towel.
14. Bubbles in cuvette gaps will cause small explosion in electroporation. Avoid bubbles by gently transferring electrocompetent bacteria to electroporation cuvettes and gently striking the cuvettes on the bench if bubbles occur.
15. Add recovery LB medium (room temperature) to bacterial cells immediately after electroporation. One-minute delay can cause a threefold reduction in recombineering efficiency.
16. To ensure that each member of a riboswitch library is examined for its regulatory capacity, no less than 4^N bacterial clones should be checked. N represents the number of the randomized nucleotides fused to the ligand-specific aptamer. We introduce a 5-nt random sequence to construct a riboswitch library

for the *lacI* gene. However, the number of the randomized nucleotides does not have to be five. If shorter random sequence is employed, the library size is smaller; if longer random sequence is used, more clones have to be screened due to the larger size of the random library.

17. For library screening, incubate bacteria carrying the riboswitch library on X-gal agar plate for 12 h. Longer incubation makes it difficult to discriminate whiter colonies from bluer ones.
18. The LacP harbors a cAMP receptor protein (CRP)-binding site, and therefore the downstream gene can be activated by the cAMP–CRP complex. As cAMP–CRP is inactivated by glucose, the ON-*lacI*-LacIbs cassette confers glucose (in addition to theophylline) repression of gene expression.
19. The native *lacI* gene will mask the riboswitch regulation of the second copy of *lacI* and, therefore, must be deleted.
20. The riboswitch hybrid device not only regulates gene expression in a two-way manner but also fine-tunes gene expression. For example, with the strain ON-*lacI-rpoS* that carries the hybrid device immediately upstream of *rpoS*, acid survival (representing the expression levels of *rpoS*) decreases with increasing concentrations of theophylline and increases with increasing concentrations of IPTG.
21. The ON riboswitch of *lacI* regulates the *lacI* expression at the translational level. However, the hybrid riboswitch device regulates its target expression at the transcriptional level as the LacI is a transcriptional repressor. Thus, in addition to the translational *lacZ* fusion described in Subheading 3, techniques quantifying RNA levels such as real-time PCR and northern blot can be used for evaluating the gene control by the hybrid device.

References

1. Ellington AD, Szostak JW (1990) In vitro selection of RNA molecules that bind specific ligands. Nature 346:818–822
2. Tuerk C, Gold L (1990) Systematic evolution of ligands by exponential enrichment: RNA ligands to bacteriophage T4 DNA polymerase. Science 249:505–510
3. Chushak Y, Stone MO (2009) In silico selection of RNA aptamers. Nucleic Acids Res 37:e87
4. Collett JR, Cho EJ, Lee JF et al (2005) Functional RNA microarrays for high-throughput screening of antiprotein aptamers. Anal Biochem 338:113–123
5. Li Y, Lee HJ, Corn RM (2006) Fabrication and characterization of RNA aptamer microarrays for the study of protein-aptamer interactions with SPR imaging. Nucleic Acids Res 34: 6416–6424
6. Lynch SA, Desai SK, Sajja HK et al (2007) A high-throughput screen for synthetic riboswitches reveals mechanistic insights into their function. Chem Biol 14:173–184
7. Lynch SA, Gallivan JP (2009) A flow cytometry-based screen for synthetic riboswitches. Nucleic Acids Res 37:184–192
8. Muranaka N, Abe K, Yokobayashi Y (2009) Mechanism-guided library design and dual genetic selection of synthetic OFF riboswitches. Chembiochem 10:2375–2381
9. Suess B, Fink B, Berens C et al (2004) A theophylline responsive riboswitch based on helix slipping controls gene expression in vivo. Nucleic Acids Res 32:1610–1614

10. Jin Y, Watt RM, Danchin A et al (2009) Use of a riboswitch-controlled conditional hypomorphic mutation to uncover a role for the essential csrA gene in bacterial autoaggregation. J Biol Chem 284:28738–28745
11. Win MN, Smolke CD (2007) A modular and extensible RNA-based gene-regulatory platform for engineering cellular function. Proc Natl Acad Sci U S A 104:14283–14288
12. Jin Y, Huang JD (2011) Engineering a portable riboswitch-LacP hybrid device for two-way gene regulation. Nucleic Acids Res 39:e131
13. Reuter JS, Mathews DH (2010) RNAstructure: software for RNA secondary structure prediction and analysis. BMC Bioinformatics 11:129
14. Studier FW, Moffatt BA (1986) Use of bacteriophage T7 RNA polymerase to direct selective high-level expression of cloned genes. J Mol Biol 189:113–130
15. Luukkonen BG, Seraphin B (1998) Construction of an in vivo-regulated U6 snRNA transcription unit as a tool to study U6 function. RNA 4: 231–238
16. Grilly C, Stricker J, Pang WL et al (2007) A synthetic gene network for tuning protein degradation in Saccharomyces cerevisiae. Mol Syst Biol 3:127
17. Cronin CA, Gluba W, Scrable H (2001) The lac operator-repressor system is functional in the mouse. Genes Dev 15:1506–1517
18. Ellis T, Wang X, Collins JJ (2009) Diversity-based, model-guided construction of synthetic gene networks with predicted functions. Nat Biotechnol 27:465–471
19. Small P, Blankenhorn D, Welty D et al (1994) Acid and base resistance in Escherichia coli and Shigella flexneri: role of rpoS and growth pH. J Bacteriol 176:1729–1737
20. Fang FC, Libby SJ, Buchmeier NA et al (1992) The alternative sigma factor katF (rpoS) regulates Salmonella virulence. Proc Natl Acad Sci U S A 89:11978–11982

Chapter 8

Generation of Orthogonally Selective Bacterial Riboswitches by Targeted Mutagenesis and In Vivo Screening

Helen A. Vincent, Christopher J. Robinson, Ming-Cheng Wu, Neil Dixon, and Jason Micklefield

Abstract

Riboswitches are naturally occurring RNA-based genetic switches that control gene expression in response to the binding of small-molecule ligands, typically through modulation of transcription or translation. Their simple mechanism of action and the expanding diversity of riboswitch classes make them attractive targets for the development of novel gene expression tools. The essential first step in realizing this potential is to generate artificial riboswitches that respond to nonnatural, synthetic ligands, thereby avoiding disruption of normal cellular function. Here we describe a strategy for engineering orthogonally selective riboswitches based on natural switches. The approach begins with saturation mutagenesis of the ligand-binding pocket of a naturally occurring riboswitch to generate a library of riboswitch mutants. These mutants are then screened in vivo against a synthetic compound library to identify functional riboswitch–ligand combinations. Promising riboswitch–ligand pairs are then further characterized both in vivo and in vitro. Using this method, a series of artificial riboswitches can be generated that are versatile synthetic biology tools for use in protein production, gene functional analysis, metabolic engineering, and other biotechnological applications.

Key words Orthogonal riboswitches, Artificial riboswitches, Aptamer domain, Synthetic ligands, Gene expression tools, In vivo screening, In vitro transcription, Isothermal titration calorimetry, Synthetic biology

1 Introduction

Riboswitches are modular noncoding regions of mRNAs that function as genetic switches to turn genes either on or off (*see* refs. 1, 2 for recent reviews). They consist of an aptamer domain, which binds a small-molecule ligand with both high affinity and selectivity, and a regulatory expression platform. Ligand binding to the aptamer domain induces a conformational change in the downstream expression platform, thereby modulating gene expression.

Atsushi Ogawa (ed.), *Artificial Riboswitches: Methods and Protocols*, Methods in Molecular Biology, vol. 1111,
DOI 10.1007/978-1-62703-755-6_8, © Springer Science+Business Media New York 2014

The majority of riboswitches are found in the 5′ untranslated region (UTR) of bacterial mRNAs, although thiamine pyrophosphate (TPP)-dependent riboswitches have also been found in archaea, fungi, and plants [3]. Over 20 classes of bacterial riboswitch have been identified to-date, which collectively respond to a diverse array of ligands including complex coenzymes (e.g., TPP, adenosylcobalamin, flavin mononucleotide, *S*-adenosylmethionine); purines and their derivatives (e.g., adenine, guanine, 2′-deoxyguanosine, adenosine triphosphate, 7-aminomethyl-7-deazaguanine (preQ$_1$)); amino acids (glycine, glutamine, lysine); sugars (glucosamine-6-phosphate); and simple inorganic ions (Mg^{2+}, F^-) [1, 2]. Typically, these ligands regulate genes required for their own biosynthesis, catabolism, and/or transport with the mechanism of gene regulation utilized being largely dependent upon the organism from which the riboswitch originates. Riboswitches in Gram-negative bacteria (e.g., *Escherichia coli*) typically operate at the translational level, whereas transcriptional mechanisms dominate in Gram-positive bacteria (e.g., *Bacillus subtilis*), and in eukaryotes alternative splicing is affected [2].

The relative simplicity of their protein-free mechanism of action, with sensor and regulatory domains contained within a single RNA molecule, makes riboswitches an attractive platform for the development of new gene expression tools. In fact, with their diversity in both ligand recognition and regulatory mechanisms, natural riboswitches appear to represent a ready-made gene expression toolkit that could be applied to the fields of protein production, gene functional analysis, metabolic engineering, and synthetic biology. However, since the majority of riboswitches respond to essential metabolites (e.g., amino acids, coenzymes, purines, and sugars) their use may be limited by regulation of intracellular metabolite concentrations, and the exogenous addition of these compounds is likely to negatively impact normal cellular function. Therefore, there is a clear need to develop artificial riboswitches that can regulate gene expression in response to synthetic nonnatural ligands, which do not affect normal cellular function.

Like natural riboswitches, artificial riboswitches must bind their ligands with high affinity and selectivity; ligand binding should lead to a conformational change that results in modulation of gene expression, and this regulation should be fast and dose-dependent with a dynamic range at least equivalent to that exhibited by natural switches. A number of strategies have now been validated for the production of artificial riboswitches. The most common approach is to take RNA aptamers, generated in vitro (e.g., through SELEX protocols [4]), and connect them to an expression platform, either by rational design [5–7] or through selection/screening for in vivo functionality [5–9]. However, recently our laboratory has pioneered a novel strategy for the production of artificial orthogonal riboswitches, which involves reengineering the aptamer domain of natural riboswitches so that they respond to synthetic ligands rather than the native metabolites [10]. Applications such as the production of designer auxotrophs [11], bioremediation [12], development of gene expression tools for pathogenic organisms [13, 14], and simultaneous control of multiple genes within the same bacterial cell [15] are beginning to be explored.

The major advantage of in vitro generation of aptamer domains is that SELEX can, in theory, be used to produce aptamers with high affinity and selectivity against any ligand of choice. However, in practice, only a limited number of aptamers have been applied to the development of riboswitch technologies in bacteria: the theophylline aptamer [5–9, 11, 13, 14, 16, 17] and, more recently, an atrazine aptamer [12]. This is largely because the de novo development of aptamers in vitro can be both costly and time consuming and, more critically, does not guarantee in vivo functionality [18]. Functionality is fundamentally dependent on the bioavailability of the ligand. Even the validated synthetic ligands theophylline and atrazine exhibit poor cell permeability, such that far higher concentrations of ligand are required to elicit a response in vivo than would be predicted from binding affinities determined in vitro [11, 12, 18]. Also, when challenged with the intracellular metabolome, the selectivity of the aptamer domain for the synthetic ligand is often weaker than observed in vitro [19]. Finally, in vivo functionality is strongly dependent upon how the aptamer and expression platform are linked, and this connection is not easy to rationally design. Consequently, the majority of functional in vitro-engineered artificial riboswitches have only been isolated following further extensive in vivo genetic selections or screens [5–9, 12].

To address the limited diversity of functional synthetic aptamers, our laboratory has developed a novel strategy for reengineering the aptamer domains of natural riboswitches to produce artificial riboswitches that respond to diverse synthetic ligands [10]. We have demonstrated the application of this approach with the adenine-responsive *add* translational-ON switch from *Vibrio vulnificus*, producing an orthogonally selective artificial riboswitch for inducible gene expression in *E. coli* [10]. In this chapter, we present the protocol that we use to change the selectivity of translational riboswitches for use as gene expression tools in bacteria [10]. An overview of this protocol is presented in Fig. 1. In principle, this strategy is applicable to any riboswitch, for use in any bacterial organism. However, for simplicity, the method presented here focuses on *E. coli*, a common model organism in the laboratory.

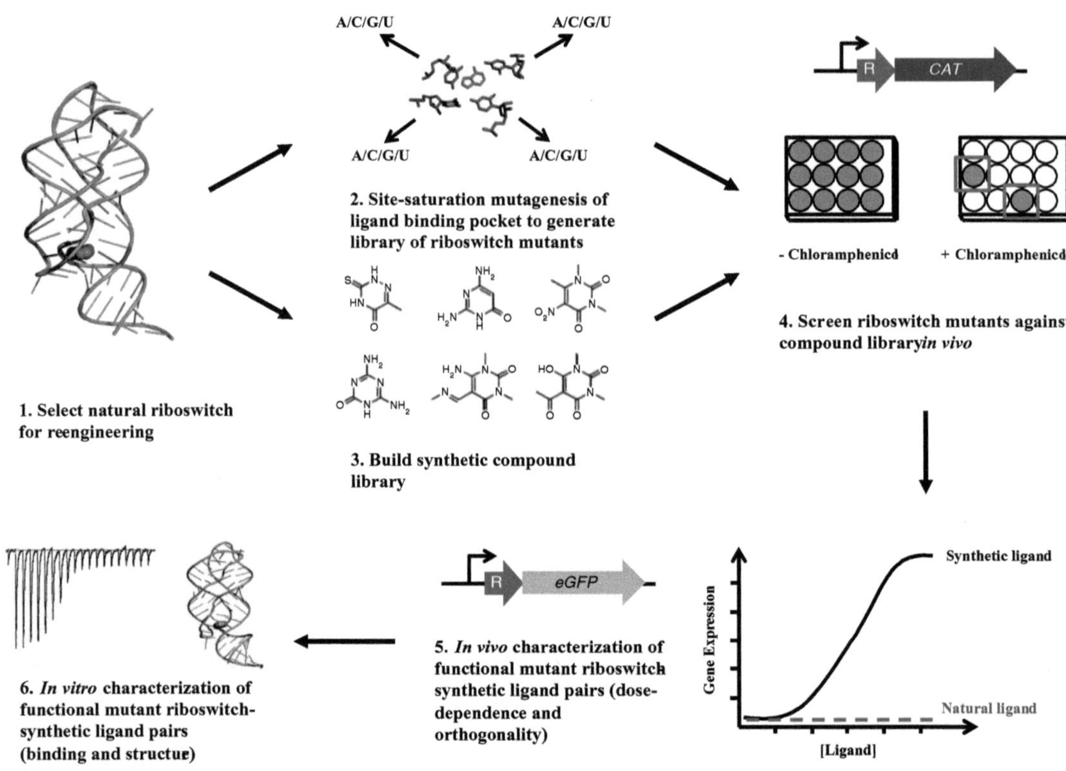

Fig. 1 Schematic overview of the strategy to engineer orthogonally selective artificial riboswitches. Individual steps are explained in detail in the subsequent protocol

In order to overcome the challenge of linking the aptamer domain and expression platform in a functional conformation, our approach starts with an entire natural riboswitch (the aptamer domain and expression platform) for which in vivo functionality has been validated and for which an X-ray crystal structure of the ligand-bound aptamer domain is available. Based on the crystal structure, the ligand-binding pocket is mutated so that it no longer accepts the natural ligand. These riboswitch mutants are fused upstream of a chloramphenicol resistance gene (*CAT*) and then screened in the presence of chloramphenicol against a library of suitable nonnatural compounds for ligand-dependent growth (ON switches) or growth inhibition (OFF switches). By starting with an in vivo functionality screen, only those mutant riboswitch–ligand pairs with the potential to be suitable for biological applications are carried forward. Selected candidates are then cloned upstream of an enhanced green fluorescent protein (*eGFP*) gene to allow for characterization of ligand selectivity/orthogonality and dose-dependent response in vivo. Finally, in vitro characterization of ligand binding by biophysical methods (e.g., isothermal titration calorimetry (ITC) and X-ray crystallography) is carried out to confirm orthogonality and to inform subsequent rounds of riboswitch–ligand design.

2 Materials

2.1 Construction of Mutant Riboswitch Chloramphenicol-Based Reporter Plasmids

1. Standard reagents and equipment for molecular cloning.
2. A high-copy-number vector carrying an ampicillin resistance gene, e.g., pBluescript II KS(+) (Stratagene) (*see* **Note 1**).
3. DNA fragments containing (1) the *lac* promoter/operator (*see* **Note 2**), (2) the natural riboswitch, and (3) the *CAT* gene (*see* **Note 3**).
4. QuikChange Site-directed Mutagenesis Kit (Stratagene).
5. Mutagenic primers.

2.2 Building a Screening Library of Nonnatural Compounds

1. Access to CAS SciFinder (American Chemical Society) (*see* **Note 4**).
2. Commercial databases, e.g., Aurora Fine Chemicals, TimTec, Enamine, Bionet.
3. Chemical database management software, e.g., ChemFinder (CambridgeSoft) (*see* **Note 4**).
4. Dimethyl sulfoxide (DMSO).

2.3 Screening of Mutant Riboswitches Against a Library of Nonnatural Compounds

1. 37 °C incubator for plates.
2. 37 °C shaker incubator for tubes, flasks, and 96-well plates.
3. Anthos Zenyth 3100 plate reader (or a suitable alternative capable of measuring absorbance at 620 nm).
4. Sterile 90 mm Petri dishes.
5. Sterile 20 ml culture tubes.
6. Sterile 250 ml flasks.
7. Sterile non-treated polystyrene 96-well microplates with 400 μl wells with flat, clear bottoms (Nunc).
8. LB-agar: 1 % (w/v) bactotryptone, 0.5 % (w/v) yeast extract, 0.5 % (w/v) NaCl, and 2.5 % (w/v) agar in dH_2O, sterilized by autoclaving.
9. LB media: 1 % (w/v) tryptone, 0.5 % (w/v) yeast extract, 0.5 % (w/v) NaCl in dH_2O, sterilized by autoclaving.
10. 5× M9 salts (1 l): 64 g $Na_2HPO_4{\cdot}7H_2O$, 15 g KH_2PO_4, 2.5 g NaCl, and 5 g NH_4Cl in dH_2O, sterilized by autoclaving.
11. M9 minimal media (1 l): 1× M9 salts, 2 mM $MgSO_4$, 0.1 mM $CaCl_2$, 0.4 % (w/v) glucose, 2 mg/ml casamino acids, and 0.1 mg/ml thiamine prepared in sterile dH_2O.
12. 100 mg/ml ampicillin stock prepared in dH_2O, filter sterilized, and stored at −20 °C.
13. 34 mg/ml chloramphenicol stock prepared in 70 % ethanol, filter sterilized, and stored at −20 °C.

14. 1 M isopropyl β-D-THIOGALACTOPYRANOSIDE (IPTG) stock prepared in dH_2O, filter sterilized, and stored at −20 °C.
15. *E. coli* TOP10F′ chemically competent cells (Life Technologies) (*see* **Note 5**).
16. The parental and mutant riboswitch chloramphenicol reporter plasmids (*see* Subheading 3.1).
17. DMSO.
18. 5 mM natural riboswitch ligand stock prepared in DMSO and stored at −20 °C.
19. Screening library of nonnatural/synthetic compounds (*see* Subheading 3.2).

2.4 In Vivo Characterization of Functional Mutant Riboswitch–Synthetic Ligand Pairs

1. Standard reagents and equipment for molecular cloning.
2. 37 °C incubator for plates.
3. 37 °C shaker incubator for tubes, flasks, and 96-well microplates.
4. Anthos Zenyth 3100 plate reader (or a suitable alternative capable of measuring absorbance at 620 nm and fluorescence with excitation at 485 nm and emission at 535 nm).
5. Sterile 90 mm Petri dishes.
6. Sterile 20 ml culture tubes.
7. Sterile non-treated polystyrene 96-well microplates with 400 μl wells with black sides and flat, clear bottoms (Greiner Bio-One).
8. LB-agar (*see* Subheading 2.3, **item 8**).
9. LB media (*see* Subheading 2.3, **item 9**).
10. 5× M9 salts (*see* Subheading 2.3, **item 10**).
11. M9 minimal media (*see* Subheading 2.3, **item 11**).
12. 100 mg/ml ampicillin stock (*see* Subheading 2.3, **item 12**).
13. 34 mg/ml chloramphenicol stock (*see* Subheading 2.3, **item 13**).
14. 1 M IPTG stock (*see* Subheading 2.3, **item 14**).
15. A high-copy-number vector carrying an ampicillin resistance gene, e.g., pBluescript II KS(+) (Stratagene).
16. DNA fragments containing (1) the *lac* promoter/operator (*see* **Note 2**), (2) the natural riboswitch, and (3) the *eGFP* gene.
17. QuikChange Site-directed Mutagenesis Kit (Stratagene).
18. Mutagenic primers.
19. *E. coli* TOP10F′ chemically competent cells (Life Technologies) (*see* **Note 5**).
20. DMSO.

21. 5 mM natural riboswitch ligand stock prepared in DMSO.
22. 5 mM synthetic ligand stocks prepared in DMSO from the screening compound library (*see* Subheading 3.2).

2.5 Preparation of DNA Template to Produce Aptamer RNA for In Vitro Biophysical Characterization

1. PCR thermocycler.
2. MaXtract high-density tubes, 1.5 ml (Qiagen).
3. 3 M sodium acetate (pH 5.2).
4. 70 and 100 % ethanol solutions, chilled to –20 °C.
5. Phenol:chloroform:isoamyl alcohol mixture (25:24:1).
6. Chloroform.
7. Nuclease-free H_2O (Ambion), not diethylpyrocarbonate (DEPC) treated (*see* **Note 6**).
8. Phusion High-Fidelity DNA Polymerase, including supplied 5× Phusion HF buffer (New England BioLabs).
9. Deoxyribonucleoside 5′ triphosphate (dNTP) mix: 2 mM each of dATP, dCTP, dGTP, and dTTP.
10. Suitable 5′ and 3′ DNA oligonucleotide primers, 100 μM each in dH_2O.
11. Single-stranded sense DNA oligonucleotide template, 100 μM in dH_2O.

2.6 Preparation of Aptamer RNA by In Vitro Transcription for In Vitro Biophysical Characterization

1. 37 °C incubator.
2. 12 ml Slide-A-Lyzer dialysis cassettes, 3,000 MWCO (Thermo Scientific).
3. Disposable syringes and 18-guage needles.
4. Vivaspin 6 centrifugal concentrators, 3,000 MWCO (Sartorius).
5. Nuclease-free H_2O (Ambion), not DEPC treated (*see* **Note 6**).
6. DEPC-treated dH_2O.
7. 10× transcription buffer: 300 mM Tris-HCl (pH 8.0), 100 mM dithiothreitol (DTT), 34 mM spermidine, 0.1 % Triton X-100 prepared in nuclease-free H_2O and stored at –20 °C.
8. 1 M $MgCl_2$ prepared in nuclease-free H_2O.
9. 1 M DTT prepared in nuclease-free H_2O and stored at –20 °C.
10. Nucleoside 5′ triphosphate (NTP) mix: 25 mM each of ATP, CTP, GTP, and UTP prepared in nuclease-free H_2O and stored at –20 °C.
11. 0.5 M EDTA (pH 8.0) prepared in nuclease-free H_2O.
12. Riboswitch buffer: 100 mM KCl, 50 mM HEPES (pH 7.5), 20 mM $MgCl_2$ in DEPC-treated H_2O and stored at 4 °C.
13. DNase I, RNase-free (New England BioLabs).

14. Bacteriophage T7 RNA polymerase purified in-house or purchased from Ambion (*see* **Note 7**).
15. Purified DNA template prepared in nuclease-free H_2O (*see* Subheading 3.5).

2.7 Purification of Aptamer RNA for In Vitro Biophysical Characterization

1. ÄKTA FPLC system (GE Healthcare) at 4 °C.
2. HiLoad 26/600 Superdex 200 PG size-exclusion column (GE Healthcare) (*see* **Note 8**).
3. RNase-free disposable tubes for FPLC fraction collector.
4. Disposable syringes and 18-guage needles.
5. 3 ml Slide-A-Lyzer dialysis cassettes, 3,000 MWCO (Thermo Scientific).
6. Vivaspin 6 centrifugal concentrators, 3,000 MWCO (Sartorius).
7. 0.5 ml Amicon Ultra centrifugal concentrators, 3,000 MWCO (Millipore).
8. 0.2 μm bottle-top filter units (Nalgene).
9. 0.2 μm syringe filter units (Millipore).
10. DEPC-treated dH_2O.
11. Riboswitch buffer: 100 mM KCl, 50 mM HEPES (pH 7.5), 20 mM $MgCl_2$ in DEPC-treated dH_2O.
12. Crystallography buffer: 100 mM KCl, 50 mM HEPES (pH 7.5), 5 mM $MgCl_2$ in nuclease-free dH_2O (0.2 μm filtered and vacuum degassed).
13. 2.5 ml in vitro transcription reaction (*see* Subheading 3.6).

2.8 In Vitro Biophysical Characterization of Aptamer–Ligand Interactions: Isothermal Titration Calorimetry

1. A VP-ITC (GE Healthcare) or other appropriate microcalorimeter.
2. Origin software (OriginLab) or similar software capable of nonlinear curve fitting, including "Origin for ITC" custom software (GE Healthcare) for use with a VP-ITC microcalorimeter.
3. Thermovac degassing unit (MicroCal).
4. 2.5 ml long needle Hamilton filling-syringe.
5. Disposable syringe fitted with tubing.
6. KimWipes.
7. DEPC-treated H_2O.
8. 12 % NaOH solution.
9. 5 % RNaseZap (Invitrogen) solution.

10. Dialysis buffer: 100 mM KCl, 50 mM HEPES (pH 7.5), 20 mM $MgCl_2$ in DEPC-treated dH_2O (*see* Subheading 3.7, **step 8**).
11. 10 μM aptamer RNA (*see* Subheadings 3.5–3.7) in dialysis buffer.
12. 100 μM ligand in dialysis buffer.

3 Methods

The following protocols assume that the user has a good working knowledge of standard molecular cloning techniques. All in vitro procedures (*see* Subheading 3.5 onwards) should be carried out in an RNase-free environment (*see* **Note 9**), including the use of sterile reagents and consumables. Gloves should be changed frequently to avoid contamination from ribonucleases found on the hands and in the environment. RNA samples should be kept on ice wherever possible.

3.1 Construction of Mutant Riboswitch Chloramphenicol-Based Reporter Plasmids

The X-ray crystal structures of aptamer domains in complex with their cognate ligand have been determined for most classes of riboswitch. In general, aptamer domains form tight binding pockets that completely surround the ligand. With the notable exception of adenosylcobalamin riboswitches [20, 21], the majority of riboswitches appear to selectively bind the ligand through specific hydrogen bonding interactions mediated by conserved nucleotides within the ligand-binding pocket, with the binding of planar ligands further stabilized through base-stacking interactions [2, 22]. When changing the specificity of natural riboswitch aptamers, a good initial strategy is to specifically target the conserved nucleotides responsible for forming the aptamer–ligand hydrogen bonds.

1. Select a suitable riboswitch for reengineering (*see* **Note 10**).
2. Identify key ligand-binding nucleotides from the X-ray crystal structure of the aptamer–ligand complex, and select nucleotides for mutagenesis (*see* **Note 11**).
3. Using established methods, construct a parental chloramphenicol reporter plasmid (Fig. 2a) by cloning the following elements, in order, into a suitable high-copy-number vector (*see* **Note 1**).
   (a) The IPTG-inducible *lac* promoter and operator derived from the 5′ UTR of the *lacZ* gene from *E. coli* (*see* **Note 2**).
   (b) The natural riboswitch aptamer domain and expression platform up to the native translational start site.
   (c) The open reading frame of the *CAT* gene, which confers chloramphenicol resistance (*see* **Note 3**).

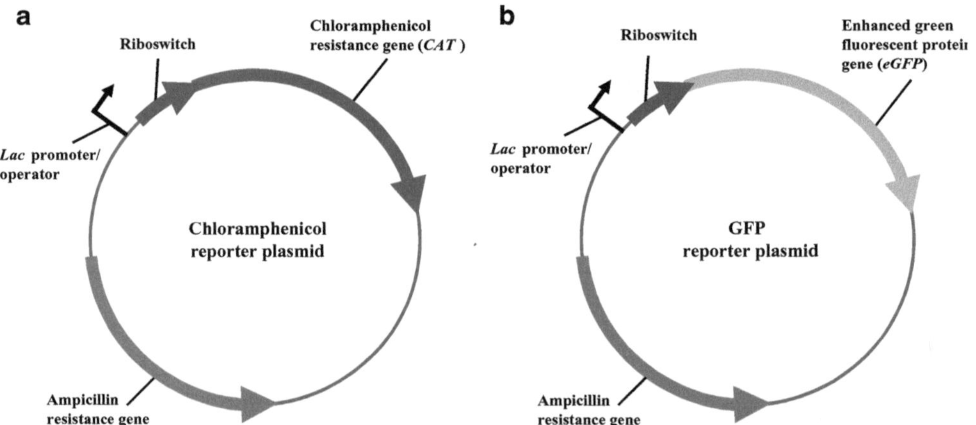

Fig. 2 Schematic representation of the reporter plasmids. (**a**) Chloramphenicol reporter plasmid for in vivo screening. (**b**) GFP reporter plasmid for quantitative in vivo characterization

The final cassette should be sequenced to ensure that the correct construct has been successfully assembled.

4. Using the parental chloramphenicol reporter plasmid as a template and the QuikChange Site-directed Mutagenesis Kit (Stratagene) with appropriate mutagenic primers perform site-directed mutagenesis to mutate each selected ligand-binding nucleotide to all other nucleotide possibilities (*see* **Note 12**). Final clones should be sequenced to ensure that the desired mutations have been introduced.

3.2 Building a Screening Library of Nonnatural Compounds

The choice of the screening library size and composition directly affects the probability of successfully identifying functional mutant riboswitch–synthetic ligand combinations. In practice, we have found that a pool of 15 riboswitch mutants (based upon saturation mutagenesis of two nucleotides in the ligand-binding pocket), screened against a carefully selected library of no more than 100 compounds (1,500 different riboswitch–ligand scenarios in total), should be sufficient to find functional mutant–ligand pairs. When designing the library it is important to consider that in order to be fit for purpose, the compounds should have suitable physicochemical properties (e.g., stable, at least modestly soluble in water, cell permeable) and they should be accessible in terms of availability and cost.

1. Using the chemical structure of the natural ligand as the query, use structural chemical research tools (e.g., CAS SciFinder) and search commercial databases (e.g., Aurora Fine Chemicals, TimTec, Enamine, Bionet) using chemical database software (e.g., ChemFinder) to generate a list of analogues.
2. Filter out any compounds that are natural metabolites, since these compounds would be expected to encounter similar

problems with regard to disrupting normal cellular function as the natural ligand.

3. Filter out any compounds with a molecular weight of more than twice that of the natural ligand since it is unlikely that the ligand-binding pocket will be able to accommodate compounds that vary dramatically in size from the natural ligand.
4. Filter out any compounds that do not contain hydrogen bond acceptor or donor groups since hydrogen bonds are crucial for the specific recognition of ligands by riboswitches (*see* **Note 13**).
5. Check for commercial availability of the refined library using online chemical databases (e.g., ChemBuyersGuide, ChemExper, eMolecules), and filter out any that are not generally kept in stock (*see* **Note** 14).
6. Decide upon a price threshold, and filter out compounds above this threshold (*see* **Note** 15).
7. Purchase the remaining compounds.
8. Prepare 5 mM stocks of each compound in DMSO and store at –20 °C.

3.3 Screening of Mutant Riboswitches Against a Library of Nonnatural Compounds

To identify functional mutant riboswitch–ligand pairs, the parental and mutant riboswitches are screened against each of the compounds in the library, for compound-dependent changes in *CAT* expression, by simply assaying growth in the presence and absence of chloramphenicol. Functional mutant riboswitch–synthetic ligand pairs should grow in the presence and absence of chloramphenicol (ON switches) or only in the absence of chloramphenicol (OFF switches). Furthermore, to satisfy the requirement for orthogonality, these responses should only be observed for the synthetic ligand and not for the natural ligand (e.g., *see* Fig. 3). The following screen should be carried out in duplicate to reduce the possibility of selecting false positives.

1. Using established methods, transform the parental and mutant riboswitch reporter plasmids into *E. coli* TOP10F′ (*see* **Note 5**) cells.
2. Streak each strain to single colonies on LB-agar, supplemented with 100 μg/ml ampicillin, and incubate overnight at 37 °C.
3. Inoculate 5 ml of LB, supplemented with 100 μg/ml ampicillin, with individual colonies from each strain and incubate at 37 °C overnight with shaking.
4. Dilute 2.5 ml of the overnight cultures 20-fold into 50 ml M9 minimal media, supplemented with 100 μg/ml ampicillin and 1 mM IPTG (to induce transcription of the chloramphenicol reporter cassette), and incubate for 1 h at 37 °C with shaking.

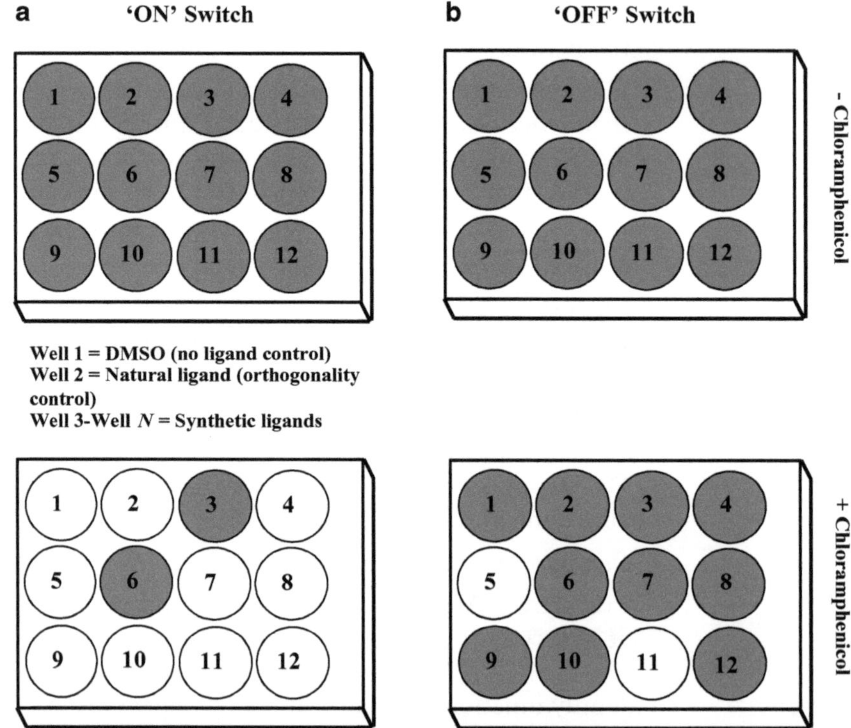

Fig. 3 Schematic representation of the expected results of chloramphenicol screening. *Shaded circles* indicate bacterial growth, and *white circles* indicate growth inhibition. (**a**) "ON" riboswitches. Wells *1* and *2* represent the no ligand and natural ligand controls and should grow only in the absence of chloramphenicol. Wells *3–N* represent different synthetic ligands. Functional riboswitch–ligand pairs are identified by growth in both the absence and presence of chloramphenicol, e.g., wells *3* and *6*. (**b**) "OFF" riboswitches. Wells *1* and *2* represent the no ligand and natural ligand controls and should grow in both the absence and presence of chloramphenicol. Wells *3–N* represent different synthetic ligands. Functional riboswitch–ligand pairs are identified by growth only in the absence of chloramphenicol, e.g., wells *5* and *11* (color figure online)

5. Divide this culture into two 25 ml aliquots, and add chloramphenicol to a final concentration of 170 μg/ml to one of the aliquots.
6. For each strain, prepare two identical compound library screens in 96-well microplates containing:
   (a) Two wells with 20 μl of DMSO (10 % final concentration) control.
   (b) One well with 20 μl of 5 mM natural ligand (500 μM final ligand concentration and 10 % final DMSO concentration) control.
   (c) *N* wells with 20 μl of different 5 mM ligand stocks (500 μM final ligand concentration and 10 % final DMSO concentration) from the screening compound library (*see* Subheading 3.2).
7. To one of the DMSO wells in each screen add 180 μl of M9 minimal media as a blank (no ligand control).

8. To all other wells in the first screen add 180 μl of the induced culture without chloramphenicol.
9. To all other wells in the second screen add 180 μl of the induced culture supplemented with chloramphenicol.
10. Incubate at 37 °C for 4 h with shaking.
11. Measure the absorbance at 620 nm of each well using a plate reader, and subtract the reading from the media-only well from all other wells to assess the extent of bacterial growth.

3.4 In Vivo Characterization of Functional Mutant Riboswitch–Synthetic Ligand Pairs

The unambiguous phenotype (bacterial growth) of the chloramphenicol reporter is ideal for a qualitative initial screen or selection. However, it is less suitable for quantitatively characterizing the dynamic range and dose-dependent response of riboswitch–ligand pairs. Therefore, we use an eGFP-based screen to further characterize candidates that are identified in the original chloramphenicol-based screen (*see* **Note 16**). The following procedure should be carried out in triplicate.

1. Using established methods, construct a parental eGFP reporter plasmid (Fig. 2b) by cloning the following elements in order into a suitable high-copy-number vector:
   (a) The IPTG-inducible *lac* promoter and operator derived from the 5′ UTR of the *lacZ* gene from *E. coli* (*see* **Note 2**).
   (b) The natural riboswitch aptamer domain and expression platform up to the native translational start site.
   (c) The open reading frame of the *eGFP* gene.

   The final cassette should be sequenced to ensure that the correct construct has been successfully assembled.
2. Using the parental eGFP reporter plasmid as a template and the QuikChange Site-directed Mutagenesis Kit (Stratagene) with appropriate mutagenic primers perform site-directed mutagenesis to introduce the functional mutations identified in the initial chloramphenicol screen. Final clones should be sequenced to ensure that the desired mutations have been introduced.
3. Using established methods, transform the parental and mutant riboswitch reporter plasmids into *E. coli* TOP10F′ (*see* **Note 5**) cells.
4. Streak each strain to single colonies on LB-agar, supplemented with 100 μg/ml ampicillin, and incubate overnight at 37 °C.
5. Inoculate 2.5 ml of LB, supplemented with 100 μg/ml ampicillin, with individual colonies from each strain and incubate at 37 °C overnight with shaking.

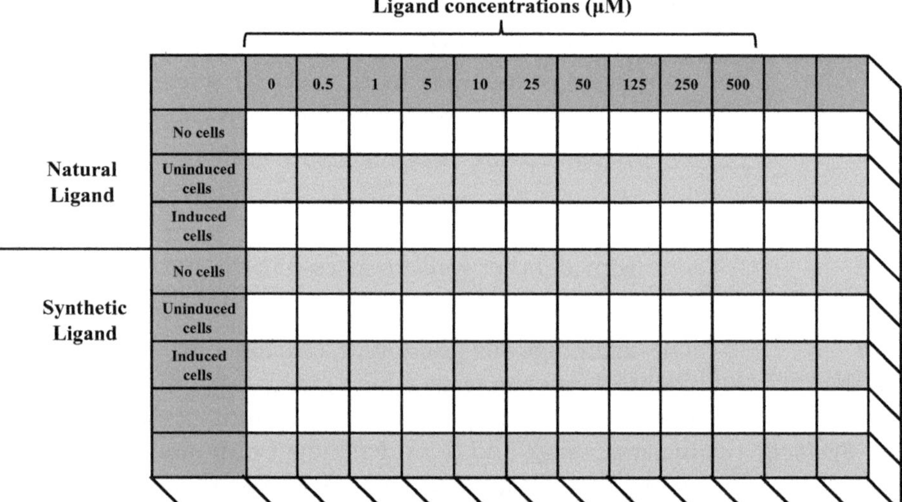

Fig. 4 Example plate layout for quantitative in vivo characterization. Testing different ligand concentrations allows the dose-dependent response to be assessed. Comparison of the natural ligand to synthetic ligands confirms the selectivity of the synthetic ligand

6. Dilute 250 μl of the overnight cultures 20-fold into 5 ml of M9 minimal media, supplemented with 100 μg/ml ampicillin, with or without 1 mM IPTG (required to induce expression of the eGFP reporter cassette), and incubate for 1 h at 37 °C with shaking.
7. Dilute each ligand stock with DMSO to prepare working stocks of, e.g., 0.005, 0.01, 0.05, 0.1, 0.25, 0.5, 1.25, 2.5, and 5 mM (yielding final concentrations of 10 % DMSO and 0.5, 1, 5, 10, 25, 50, 125, 250, and 500 μM ligand) (*see* **Note 17**).
8. For each strain, prepare a ligand concentration screen in a 96-well microplate (*see* Fig. 4):
   (a) Three series of ten concentrations of natural ligand (20 μl of DMSO (0 μM ligand) and each of the working stocks).
   (b) Three series of ten concentrations of synthetic ligand (20 μl of DMSO (0 μM ligand) and each of the working stocks).
9. To one of each series add:
   (a) 180 μl of M9 minimal media (to allow calculation of background absorbance/fluorescence from the ligand).
   (b) 180 μl of uninduced (without IPTG) cells (to allow calculation of background fluorescence from the cells).
   (c) 180 μl of induced (with IPTG) cells (the experiment).
10. Grow cells at 37 °C with shaking in an Anthos Zenyth 3100 plate reader for 5 h measuring the absorbance at 620 nm and eGFP fluorescence (excitation 485 nm, emission 535 nm) at 6-min intervals (50 readings).

11. Calculate normalized fluorescence units as follows:
    (a) Subtract the media-only absorbance/fluorescence readings from the corresponding uninduced and induced cell absorbance/fluorescence readings.
    (b) Divide the adjusted fluorescence values by the adjusted absorbance values to normalize for cell density.
    (c) Subtract the normalized uninduced fluorescence from the normalized induced fluorescence to account for basal expression and background fluorescence.
12. Calculate induction/repression factors by dividing the final normalized fluorescence values by the value for 0 μM ligand (DMSO control).
13. Data can be plotted against time (e.g., Fig. 5, left panels) or, for a single time point, against ligand concentration (e.g., Fig. 5, right panels).

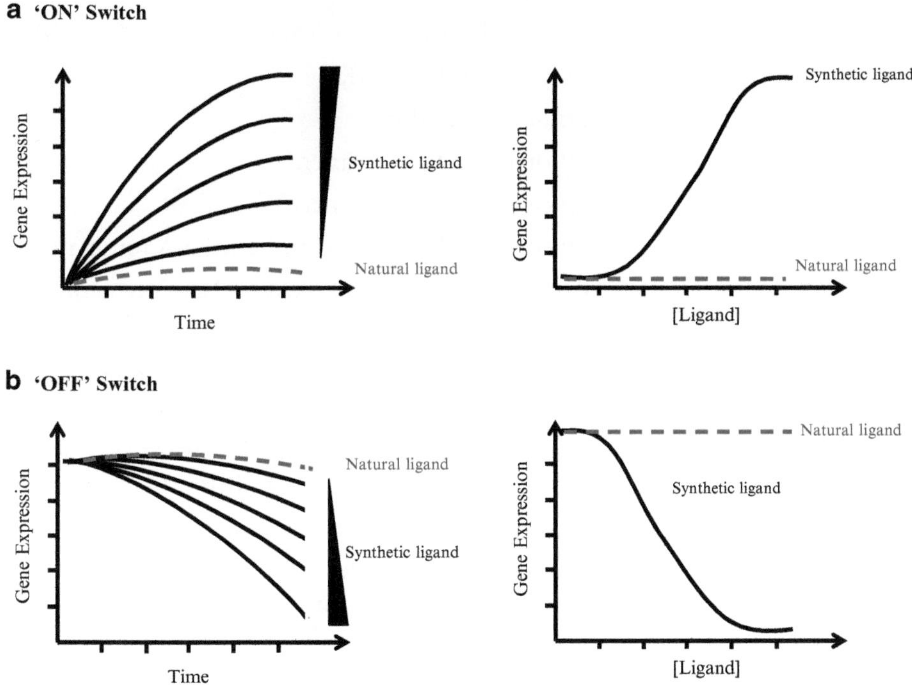

Fig. 5 Schematic representation of expected results from quantitative in vivo characterization. *Left panels* show gene expression plotted against time with different ligand concentrations plotted as individual series. *Right panels* show gene expression plotted against ligand concentration for a single time point. (**a**) "ON" riboswitches. Gene expression increases with increasing time and increasing ligand concentration for the synthetic ligand (dose-dependence). Gene expression is not affected by increasing time or increasing ligand concentration for the natural ligand (orthogonality). (**b**) "OFF" riboswitches. Gene expression decreases with increasing time and increasing ligand concentration (dose-dependence). Gene expression is not affected by increasing time or increasing ligand concentration for the natural ligand (orthogonality)

3.5 Preparation of DNA Template to Produce Aptamer RNA for In Vitro Biophysical Characterization

It may be desirable to characterize the new aptamer domain–synthetic ligand interaction in vitro to demonstrate that the synthetic ligand binds to the aptamer domain directly (i.e., the ligand is not converted into another compound in vivo which subsequently binds to the riboswitch), to confirm the selectivity of binding, and to determine the binding affinity of the aptamer for the ligand. For these studies it is necessary to produce the RNA corresponding to the aptamer domain by in vitro transcription. Since most aptamer domains are less than 100 nucleotides in length, it is possible to generate the required double-stranded transcription template in a PCR reaction using a single-stranded oligonucleotide as the template. This oligonucleotide should begin with the T7 RNA polymerase promoter sequence (TAA TAC GAC TCA CTA TA) immediately followed by two G residues (required for efficient transcription) (*see* **Note 18**), which, in turn, are followed by the desired aptamer sequence.

1. Design 5′ and 3′ oligonucleotide primers 15–20 nucleotides in length with homology to the 5′ end of the template and complementary to the 3′ end of the template, respectively.
2. Prepare the following 1 ml scale (*see* **Note 19**) PCR reaction:
   (a) 200 μl of 5× HF buffer.
   (b) 100 μl of dNTP mix.
   (c) 5 μl of 100 μM 5′ primer.
   (d) 5 μl of 100 μM 3′ primer.
   (e) 1 μl of 100 μM single-stranded DNA template.
   (f) 10 μl of Phusion High Fidelity DNA Polymerase (20 units).
   (g) dH_2O to 1 ml.
3. Using standard PCR conditions, perform 30 cycles with a suitable annealing temperature for the primers.
4. Purify the DNA by phenol:chloroform extraction using MaXtract high-density tubes (*see* **Note 20**).
   (a) Centrifuge two MaXtract tubes for 30 s at 13,000 × *g*.
   (b) Add 500 μl of the PCR reaction and 500 μl of phenol:chloroform:isoamyl alcohol (25:24:1) to each tube, mix by inverting the tube several times (do not vortex), and centrifuge for 5 min at 13,000 × *g*.
   (c) Add 500 μl of chloroform to each tube, mix, and centrifuge for 5 min at 13,000 × *g*.
   (d) Combine the upper aqueous layers from both tubes in a fresh 15 ml tube.
   (e) Add 1/9th volume (110 μl) of 3 M sodium acetate (pH 5.2) and mix thoroughly.

(f) Add 2.5 volumes (2.775 ml) of chilled 100 % ethanol and incubate at −20 °C overnight.

(g) Pellet the DNA by centrifugation at 20,000 × *g* for 20 min at 4 °C.

(h) Decant the ethanol, and wash the pellet with 2 ml chilled 70 % ethanol.

(i) Decant the ethanol, and allow the pellet to air-dry.

(j) Resuspend in 200 μl of nuclease-free H_2O, and determine the concentration by measuring the absorbance at 260 nm.

3.6 Preparation of Aptamer RNA by In Vitro Transcription for In Vitro Biophysical Characterization

1. Prepare the following 5 ml scale (*see* **Note 19**) in vitro transcription reaction:
   (a) 500 μl of 10× transcription buffer.
   (b) 1.2 ml NTP mix.
   (c) 50 μl of 1 M DTT.
   (d) 140 μl of 1 M $MgCl_2$.
   (e) 50–100 μg/ml (final concentration) double-stranded DNA template (*see* Subheading 3.5).
   (f) 0.25 mg/ml (final concentration) T7 RNA polymerase.
   (g) Nuclease-free H_2O to 5 ml.
2. Incubate at 37 °C for 6–8 h (*see* **Note 21**).
3. Add 500 μl of 0.5 M EDTA (pH 8.0) and incubate for a further 30 min at 37 °C.
4. Using a syringe fitted with a needle, transfer the reaction to a Slide-A-Lyzer dialysis cassette (3,000 MWCO, 12 ml capacity), and dialyze overnight at 4 °C in 2 l of freshly prepared riboswitch buffer.
5. Using a syringe fitted with a needle, transfer the dialyzed RNA to a Vivaspin 6 centrifugal concentrator (3,000 MWCO) and concentrate to a final volume of ~2 ml by centrifugation at 4,000 × *g* at 4 °C.
6. Add 200 μl (400 units) of DNase I and incubate for 30 min at 37 °C.
7. Store at 4 °C.

3.7 Purification of Aptamer RNA for In Vitro Biophysical Characterization

We purify our RNA transcripts (from T7 RNA polymerase, DNase I, and prematurely terminated transcripts) by size-exclusion chromatography. This intentionally avoids denaturation or precipitation of the RNA which could trap the RNA in inactive, nonnative conformations [23]. The following purification protocol is designed to produce RNA of suitable quality and quantity for carrying out ITC and X-ray crystallography and assumes that the user is familiar with operating an FPLC system since FPLC training is beyond the scope of this chapter.

1. Attach a HiLoad 26/600 Superdex 200 size-exclusion column (320 ml column volume) to the FPLC system and equilibrate with three column volumes of riboswitch buffer.
2. Attach a 5 ml sample loop, flush with 15 ml of buffer, and load ~2.2 ml RNA (*see* Subheading 3.6) into the loop.
3. Set the FPLC system to run isocratically at 1 ml/min, with UV monitoring at 260 and 280 nm.
4. Inject the sample onto the column and elute with one column volume riboswitch buffer, collecting 1.5 ml fractions.
5. Pool fractions containing full-length RNA transcript (*see* **Note 22**), transfer the RNA to a Vivaspin 6 centrifugal concentrator (3,000 MWCO), and concentrate it to a final volume of ~2 ml by centrifugation at 4,000 × *g* at 4 °C.
6. Transfer the RNA to a Slide-A-Lyzer dialysis cassette (3,000 MWCO, 3 ml capacity) using a syringe fitted with a needle and dialyze overnight at 4 °C in 2 l of riboswitch buffer.
7. Recover the RNA, and determine the concentration by measuring the absorbance at 260 nm.
8. The RNA is ready for ITC. Filter and degas the dialysis buffer and store at 4 °C for cleaning the ITC sample cell, diluting the RNA sample, and for making up ligand solutions.
9. For crystallography trials:
   (a) Mix 1 ml of 100 μM RNA aptamer with 9 ml of 550 μM syringe-filtered ligand solution, prepared in crystallography buffer, and incubate at 4 °C for 30 min.
   (b) Concentrate the aptamer–ligand complex to 200 μl (~500 μM) using Vivaspin 6 and 0.5 ml Amicon Ultra centrifugal concentrators.
   (c) Use immediately or flash freeze in liquid nitrogen, and store at −80 °C until needed.

3.8 *In Vitro* Biophysical Characterization of Aptamer–Ligand Interactions: Isothermal Titration Calorimetry

We typically characterize the aptamer–ligand interactions using ITC and X-ray crystallography (*see* **Note 23**). Detailed protocols for both of these techniques, with respect to riboswitches/RNA [24–27], have been reported previously; therefore, we will not go into detail here. As an example, we present a brief outline of our ITC method, focusing on how we design our experiments to obtain useful binding data for our mutant aptamer–ligand pairs. The following method assumes that the user is familiar with ITC, as ITC training is beyond the scope of this chapter.

1. Before the first use, clean the sample cell as follows (*see* **Note 24**):
   (a) Fill the sample cell with 12 % NaOH and incubate at 65 °C for 1 h.
   (b) Wash the sample cell with DEPC-treated dH_2O.
   (c) Fill the sample cell with 5 % RNaseZap and incubate at 25 °C for 1 h.
   (d) Wash the sample cell with DEPC-treated dH_2O.
   (e) Wash the sample cell thoroughly with dialysis buffer.
2. Before the first use, clean the titrant syringe and the Hamilton syringe used to load the sample cell as follows:
   (a) Fill with 5 % RNaseZap, and leave for 1 h.
   (b) Wash with DEPC-treated dH_2O.
   (c) Wash with dialysis buffer.
3. Degas the aptamer RNA and ligand solutions using a degassing unit.
4. Add 2 ml of dialysis buffer to the sample cell using a Hamilton syringe (the blank).
5. Purge the titrant syringe twice and fill with 100 μM (*see* **Note 25**) ligand solution using a syringe fitted with tubing.
6. Dry the titrant syringe needle gently using a clean KimWipe, taking care not to touch the outlet pore at the base of the needle, and carefully lower into the sample cell until it clicks into place.
7. Set the cell temperature to 25 °C, reference power to 5 μcal/s, and stirrer speed to 310 rpm.
8. After a 60-s pre-titration delay, inject 2 μl of ligand over 4.8 s followed by 24 × 12 μl injections of ligand, each over 28.8 s with a 300-s delay between injections (ligand concentration range of ~0.5 to ~12.5 μM when starting from a 100 μM ligand stock solution).
9. Thoroughly wash the sample cell with dialysis buffer and then load the sample cell with 2 ml of 10 μM (*see* **Note 26**) aptamer RNA in dialysis buffer using a Hamilton syringe (the experiment).
10. Carry out the same injection protocol as for ligand into buffer (*see* **step 8**).
11. Fit the data using Origin software (*see* **Note 27**). A good data set will have a clear sigmoidal curve (with $c = 20–100$) and an appropriate stoichiometry ($n = 1$ for most riboswitch aptamer–ligand interactions), and the first few injections should cover at least 10 μcal each, with an average of 5 μcal per injection over the course of the experiment (*see* **Note 28**).

4 Notes

1. The chosen vector should not require the *CAT* gene for plasmid maintenance.
2. We use the inducible *lac* promoter/operator to provide tighter gene expression control when developing novel switches, but alternative inducible or constitutive promoters could be substituted.
3. The *CAT* gene could be substituted for alternative reporter genes, e.g., *tetA* (allowing for selection or screening with tetracycline and/or $NiCl_2$ [5, 28–30]), *lacZ* (allowing enzymatic screening for β-galactosidase activity [6, 16]), *GFP* (allowing for flow cytometry-based screening for fluorescence [8, 9]), or *cheZ* (allowing for screening for motility [7, 12]).
4. CAS SciFinder and ChemFinder require a site licence.
5. The F′ episome carries the *lacI*^{q} gene required for inducible expression from the *lac* promoter.
6. Residual DEPC can inhibit the in vitro transcription reaction.
7. We purify T7 RNA polymerase in-house from an *E. coli* expression strain [31], since the quantities required for the preparative-scale (5–20 ml) in vitro transcription reactions that we routinely carry out can become costly if commercial sources are used.
8. In our laboratory we keep a dedicated column for RNA purification.
9. In our laboratory we have set aside an RNA bench space, which includes a Labcare PCR cabinet and a dedicated set of pipettes. Apparatus is frequently treated with RNaseZap (Invitrogen) and washed with copious amounts of DEPC-treated dH_2O.
10. Riboswitch aptamer domains that bind small, uncharged, planar ligands with four or fewer conserved ligand-interacting nucleotides are most suitable for mutagenesis. Riboswitches that respond to more complex coenzymes (e.g., adenosylcobalamin and flavin mononucleotide) are not suitable as alternative synthetic ligands are unavailable due to synthetic inaccessibility.
11. In the case of riboswitches for which there is no X-ray crystal structure available, in vitro mutagenesis analysis and/or structural probing methods (e.g., in-line probing and enzymatic footprinting) could provide information about ligand-binding nucleotides.
12. This approach has been validated and is feasible for small ligands (e.g., purines), where initially up to two ligand-binding

nucleotides need to be mutated ($4^n - 1 = 15$ mutants in total if mutated to all other possibilities) to alter the ligand specificity of the aptamer. However, for binding pockets requiring the mutation of more than two nucleotides simultaneously, a library of mutants can be generated by site-saturated mutagenesis PCR using the parental reporter plasmid as a template and degenerate primers [32]. The resultant library would then need to be subjected to a selection or a screen for functional riboswitches and individual clones sequenced to identify the mutations.

13. We typically filter out compounds with three or fewer hydrogen bond acceptors/donors.
14. The synthetic accessibility of the compounds should also be considered in the eventuality that further derivatization and optimization are necessary.
15. We typically set a maximum cost of £100 for 50 mg, as this is sufficient for both in vivo and in vitro characterization, but this can be revised up or down depending on the compounds available.
16. Alternative reporters include *lacZ* [6, 16] and *cheZ* [7, 12].
17. At least ten ligand concentrations should be tested. The concentrations tested will depend upon the affinity of the aptamer for the ligand, the kinetics of ligand binding, and the bioavailability of the ligand.
18. The two G residues will be transcribed; therefore, it is important to consider how these should be incorporated. Assuming that the aptamer sequence does not begin GG, the required G residues can be added to the 5′ end of the sequence. However, this extension may prevent later crystallization of the RNA transcript. Alternatively, the 5′ sequence can be modified, along with any compensatory mutations to maintain secondary structure, to include the G residues.
19. The scale of the PCR/in vitro transcription reaction depends upon how much RNA will be needed for biophysical characterization. The scales suggested here are suitable for ITC and X-ray crystallography, both of which typically require milligram quantities of RNA.
20. See the manufacturer's instructions for a more detailed protocol.
21. The appearance of a white precipitate (magnesium pyrophosphate) indicates that the reaction is complete.
22. We avoid pooling shoulders or tails around the central RNA peak, sacrificing recovery for purity, to ensure that the sample is homogenous.

23. ITC has the advantage over many other in vitro methods in that the interaction is monitored directly, in solution, without immobilization or modification of either binding partner, and it provides a complete biophysical characterization of the interaction including binding affinity, binding stoichiometry, and thermodynamic parameters (enthalpy change (ΔH_{obs}), entropy change (ΔS), and Gibbs free energy change (ΔG)). However, it does require significant quantities of both aptamer RNA and ligand (10–50 nmol in the sample cell, 2.5–250 nmol in the titrant syringe) for each experimental run. Other biochemical techniques can also be used, e.g., in-line probing [33, 34], equilibrium dialysis [35], and selective 2′-hydroxyl acylation and primer extension (SHAPE) [34].
24. Samples can become adsorbed onto the surface of the sample cell, interfering with heat transfer and causing erratic, noisy baselines.
25. The concentration of the ligand in the titrant syringe should be 10–20 times the concentration of the aptamer RNA in the sample cell.
26. To determine the lower and upper limits for the concentration of the aptamer in the sample cell (*Mcell*), we use the equation

$$c = (n \times Mcell) / K_d$$

where c describes the sigmoidicity of the binding curve (should be between 20 and 100 [36]), n is the stoichiometry of the interaction (1 for most riboswitch aptamer–ligand interactions), and K_d is the expected dissociation constant. If the expected K_d is unknown, 10 μM aptamer is a reasonable concentration for pilot studies.

27. The baseline and integration limits for each injection should be checked manually before attempting to fit the data since noise can perturb the automatic fitting performed by the software.
28. The aptamer/ligand concentrations or injection settings can be adjusted in subsequent experiments until these criteria are satisfied.

References

1. Breaker RR (2011) Prospects for riboswitch discovery and analysis. Mol Cell 43:867–879
2. Serganov A, Nudler E (2012) A decade of riboswitches. Cell 152:17–24
3. Barrick JE, Breaker RR (2007) The distributions, mechanisms, and structures of metabolite-binding riboswitches. Genome Biol 8:R239
4. Codrea V, Hayner M, Hall B et al (2010) In vitro selection of RNA aptamers to a small molecule target. Curr Protoc Nucleic Acid Chem Chapter 9:Unit 9.5.1–5.23
5. Sharma V, Nomura Y, Yokobayashi Y (2008) Engineering complex riboswitch regulation by dual genetic selection. J Am Chem Soc 130:16310–16315
6. Lynch SA, Desai SK, Sajja HK et al (2007) A high-throughput screen for synthetic riboswitches reveals mechanistic insights into their function. Chem Biol 14:173–184

7. Topp S, Gallivan JP (2008) Random walks to synthetic riboswitches-a high-throughput selection based on cell motility. Chembiochem 9:210–213
8. Fowler CC, Brown ED, Li Y (2008) A FACS-based approach to engineering artificial riboswitches. Chembiochem 9:1906–1911
9. Lynch SA, Gallivan JP (2009) A flow cytometry-based screen for synthetic riboswitches. Nucleic Acids Res 37:184–192
10. Dixon N, Duncan JN, Geerlings T et al (2010) Reengineering orthogonally selective riboswitches. Proc Natl Acad Sci U S A 107:2830–2835
11. Desai SK, Gallivan JP (2004) Genetic screens and selections for small molecules based on a synthetic riboswitch that activates protein translation. J Am Chem Soc 126:13247–13254
12. Sinha J, Reyes SJ, Gallivan JP (2010) Reprogramming bacteria to seek and destroy an herbicide. Nat Chem Biol 6:464–470
13. Topp S, Reynoso CMK, Seeliger JC et al (2010) Synthetic riboswitches that induce gene expression in diverse bacterial species. Appl Environ Microbiol 76:7881–7884
14. Reynoso CMK, Miller MA, Bina JE et al (2012) Riboswitches for intracellular study of genes involved in *Francisella* pathogenesis. MBio 3:e00253–00212
15. Dixon N, Robinson CJ, Geerlings T et al (2012) Orthogonal riboswitches for tuneable coexpression in bacteria. Angew Chem Int Ed 51:3620–3624
16. Suess B, Fink B, Berens C et al (2004) A theophylline responsive riboswitch based on helix slipping controls gene expression in vivo. Nucleic Acids Res 32:1610–1614
17. Wachsmuth M, Findeiß S, Weissheimer N et al (2013) De novo design of a synthetic riboswitch that regulates transcription termination. Nucleic Acids Res 41:2541–2551
18. Link KH, Breaker RR (2009) Engineering ligand-responsive gene-control elements: lessons learned from natural riboswitches. Gene Ther 16:1189–1201
19. Carothers JM, Oestreich SC, Szostak JW (2006) Aptamers selected for higher-affinity binding are not more specific for the target ligand. J Am Chem Soc 128:7929–7937
20. Johnson JE Jr, Reyes FE, Polaski JT et al (2012) B12 cofactors directly stabilize an mRNA regulatory switch. Nature 492:133–137
21. Peselis A, Serganov A (2012) Structural insights into ligand binding and gene expression control by an adenosylcobalamin riboswitch. Nat Struct Mol Biol 19:1182–1184
22. Serganov A, Patel DJ (2012) Molecular recognition and function of riboswitches. Curr Opin Struct Biol 22:279–286
23. Uhlenbeck OC (1995) Keeping RNA happy. RNA 1:4–6
24. Gilbert SD, Batey RT (2009) Monitoring RNA-ligand interactions using isothermal titration calorimetry. Methods Mol Biol 540:97–114
25. Salim NN, Feig AL (2009) Isothermal titration calorimetry of RNA. Methods 47:198–205
26. Pikovskaya O, Serganov AA, Polonskaia A et al (2009) Preparation and crystallization of riboswitch-ligand complexes. Methods Mol Biol 540:115–128
27. Edwards AL, Garst AD, Batey RT (2009) Determining structures of RNA aptamers and riboswitches by X-ray crystallography. Methods Mol Biol 535:135–163
28. Nomura Y, Yokobayashi Y (2007) Reengineering a natural riboswitch by dual genetic selection. J Am Chem Soc 129:13814–13815
29. Muranaka N, Sharma V, Nomura Y et al (2009) An efficient platform for genetic selection and screening of gene switches in *Escherichia coli*. Nucleic Acids Res 37:e39
30. Muranaka N, Abe K, Yokobayashi Y (2009) Mechanism-guided library design and dual genetic selection of synthetic OFF riboswitches. Chembiochem 10:2375–2381
31. He B, Rong M, Lyakhov D et al (1997) Rapid mutagenesis and purification of phage RNA polymerases. Protein Expr Purif 9:142–151
32. Chronopoulou EG, Labrou NE (2011) Site-saturation mutagenesis: a powerful tool for structure-based design of combinatorial mutation libraries. Curr Protoc Protein Sci Chapter 26:Unit 26.6
33. Regulski EE, Breaker RR (2008) In-line probing analysis of riboswitches. Methods Mol Biol 419:53–67
34. Wakeman CA, Winkler WC (2009) Analysis of the RNA backbone: structural analysis of riboswitches by in-line probing and selective 2′-hydroxyl acylation and primer extension. Methods Mol Biol 540:173–191
35. Mandal M, Breaker RR (2004) Adenine riboswitches and gene activation by disruption of a transcription terminator. Nat Struct Mol Biol 11:29–35
36. Myszka DG, Abdiche YN, Arisaka F et al (2003) The ABRF-MIRG'02 study: assembly state, thermodynamic, and kinetic analysis of an enzyme/inhibitor interaction. J Biomol Tech 14:247–269

Chapter 9

Dual Genetic Selection of Synthetic Riboswitches in *Escherichia coli*

Yoko Nomura and Yohei Yokobayashi

Abstract

This chapter describes a genetic selection strategy to engineer synthetic riboswitches that can chemically regulate gene expression in *Escherichia coli*. Riboswitch libraries are constructed by randomizing the nucleotides that potentially comprise an expression platform and fused to the hybrid selection/screening marker *tetA–gfpuv*. Iterative ON and OFF selections are performed under appropriate conditions that favor the survival or the growth of the cells harboring the desired riboswitches. After the selection, rapid screening of individual riboswitch clones is performed by measuring GFPuv fluorescence without subcloning. This optimized dual genetic selection strategy can be used to rapidly develop synthetic riboswitches without detailed computational design or structural knowledge.

Key words Riboswitch, Aptamer, Gene regulation, RNA engineering, Translational regulation

1 Introduction

As synthetic biology progresses toward more practical applications, we will inevitably face challenges to interface synthetic gene circuits with the complex chemical environment. For example, it may be desirable to negatively modulate an enzyme level in response to accumulation of an intermediate in metabolic engineering or activate a chemical degradation pathway in response to an environmental pollutant [1]. Riboswitches are used extensively by many bacteria to regulate endogenous gene expression [2], and they have inspired researchers to engineer synthetic riboswitches [3].

In this chapter, we describe a strategy to rapidly engineer synthetic riboswitches based on synthetic or natural RNA aptamers using genetic selection in *Escherichia coli* (Fig. 1). The strategy starts with the design and construction of a plasmid library of partially randomized riboswitch mutants (>10^5) fused to the selection/screening marker *tetA–gfpuv* fusion [4]. The *E. coli* cells transformed with the riboswitch mutants are serially cultured in media that favor the growth of either ON or OFF state in the presence or

Atsushi Ogawa (ed.), *Artificial Riboswitches: Methods and Protocols*, Methods in Molecular Biology, vol. 1111,
DOI 10.1007/978-1-62703-755-6_9,

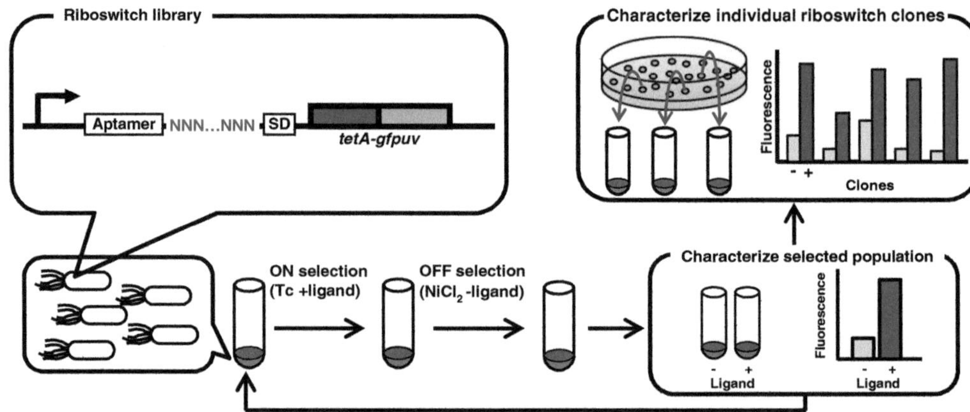

Fig. 1 Dual genetic selection strategy. A riboswitch library is constructed in *E. coli* by partially randomizing an appropriate region within the 5′ UTR. The library is subjected to one or more cycles of ON (using Tc: tetracycline) and OFF (using $NiCl_2$) selections to enrich the functional riboswitches. The selected population of *E. coli* cells can be characterized as a population or as individual clones by measuring the fluorescence reporter gene expression

the absence of the aptamer ligand. The tetracycline antiporter (*tetA*) functions as a dual selection marker by enabling ON cells to grow in the presence of tetracycline. Conversely, the OFF selection is performed in the presence of $NiCl_2$ because *E. coli* cells overexpressing *tetA* are more sensitive to Ni^{2+} compared to the cells that do not [5, 6]. After the selection, the surviving cells can be characterized for riboswitch activity as a population or cloned and individually characterized using the fluorescence reporter (*gfpuv*) without subcloning or retransformation. The salient feature of our selection strategy is the use of a single selection marker (*tetA*) to perform both ON and OFF selections as well as the fused fluorescence reporter (*gfpuv*) which enables rapid quantitative screening of the selected cell populations and individual clones. This greatly streamlines the selection as well as the subsequent screening processes while minimizing the emergence of false positives that often complicate genetic selection experiments.

2 Materials

2.1 Cell Culture Media and Strains

1. LB: 1 % bacto tryptone, 0.5 % yeast extract, 1 % NaCl in diH_2O supplemented with an appropriate antibiotic for plasmid maintenance. LB agar plates are prepared with 1.5 % w/v agar.
2. M9 minimal medium: Mix separately sterilized reagents in the following order, thoroughly mixing after adding each reagent to prepare 1 L (*see* **Note 1**): 777 mL of diH_2O, 10 mL of 40 % v/v glycerol, 0.1 mL of 1 M $CaCl_2$, 2 mL of 1 M $MgSO_4$, 10 mL of 10 % w/v casamino acids, 200 mL of 5× M9 salts (64 g/L $Na_2HPO_4{\cdot}7H_2O$, 15 g/L KH_2PO_4, 2.5 g/L NaCl,

5 g/L NH_4Cl), 1 mL of 1,000× filter-sterilized antibiotic for plasmid maintenance.

3. Host *E. coli* strain for genetic selection: TOP10 (Life Technologies) was used in our previous work [4]. In principle, any strain that is compatible with the riboswitch's aptamer ligand and lacks tetracycline resistance should work. Competent cells with a sufficient transformation efficiency to construct a riboswitch library (>10^5 independent transformants) must be prepared or purchased.

2.2 Equipment

1. UV transilluminator (320 and 360 nm).
2. Thermal cycler.
3. Incubator (37 °C for *E. coli* incubation).
4. Incubator shaker (37 °C for *E. coli* culture).
5. Microcentrifuge (plasmid miniprep and DNA purification).
6. Agarose gel electrophoresis apparatus (chamber and power supply).
7. Microplate reader (for cell fluorescence and OD_{600} measurements).

2.3 Plasmids and Primers

1. A template plasmid with an aptamer ligand and *tetA–gfpuv* selection marker (*see* **Note 2**).
2. PCR primers for library construction (sequence design considerations are given in Subheading 3).
3. An appropriate sequencing primer for sequencing selected clones.
4. (Optional) 5′ phosphorylated primers p-GFP-f (5′ATTGA GTAAA GGAGA AGAAC TTTTC AC 3′) and p-ORF-r (5′ GGATC CAGCA GGTCG ACTTG CAT 3′) for *tetA* removal.

2.4 Reagents

1. High-fidelity DNA polymerase, e.g., Phusion DNA polymerase (Thermo Scientific).
2. dNTPs.
3. Tetracycline (*see* **Note 3**).
4. Nickel (II) chloride ($NiCl_2$) (*see* **Note 4**).
5. Glycerol.
6. Restriction enzyme Dpn I (New England Biolabs).
7. Agarose gel electrophoresis reagents (buffer, agarose, dye, etc.).
8. DNA Clean and Concentrator-5 (Zymo) or other silica column DNA purification kit.
9. Zyppy Plasmid Miniprep Kit (Zymo) or other plasmid miniprep kit.

10. Quick Ligation Kit (New England Biolabs) or other T4 DNA ligase.
11. Phosphate-buffered saline (PBS).

3 Methods

3.1 Cell Culture Conditions

All liquid cultures are grown in M9 minimal medium or LB medium supplemented with an appropriate antibiotic to maintain the plasmids and kept at 37 °C in an incubator shaker (~275 rpm). All agar plates are incubated at 37 °C overnight after plating.

3.2 Design and Construction ON and OFF Controls

1. Successful genetic selection critically depends on the use of optimal selection pressures that favor the growth or the survival of the desired phenotypes. The optimal selection conditions depend on various parameters such as promoter strength, plasmid copy number, and host genotype. Therefore, it is important to carefully optimize the selection conditions *prior* to the actual riboswitch selection. In order to optimize the selection conditions, we recommend the construction of two controls, one that represents an ON state and another that represents an OFF state of the presumed riboswitches (Fig. 2a). The controls should be constructed using the same plasmid backbone and the promoter with which the researcher desires to use the functional riboswitches. The ON control could be any canonical 5′ UTR that allows efficient translation initiation fused to the *tetA–gfpuv* marker gene (*see* **Note 5**). The OFF control should be carefully designed so that it expresses a sufficiently low level of TetA–GFPuv that is acceptable for the desired riboswitches at their OFF state. One might consider using a "blank" plasmid (no encoded *tetA–gfpuv*) as an OFF control. In reality, however, most functional riboswitches (and other protein-based switches) produce low levels of basal expression even at OFF states. Using a blank plasmid as the OFF control could result in the optimized selection condition being too stringent such that no reasonably functional riboswitches would survive the selection. To obtain a "realistic" OFF control, a stable hairpin loop may be engineered to sequester the Shine–Dalgarno (SD) sequence to suppress translation (Fig. 2a). Such sequestration of an SD region is often observed in natural and engineered riboswitches.
2. After construction of the ON and OFF controls, transform the plasmids into an appropriate selection host such as TOP10 (*see* **Note 6**).
3. Start overnight cultures of the control strains in 1.0 mL of M9 medium.

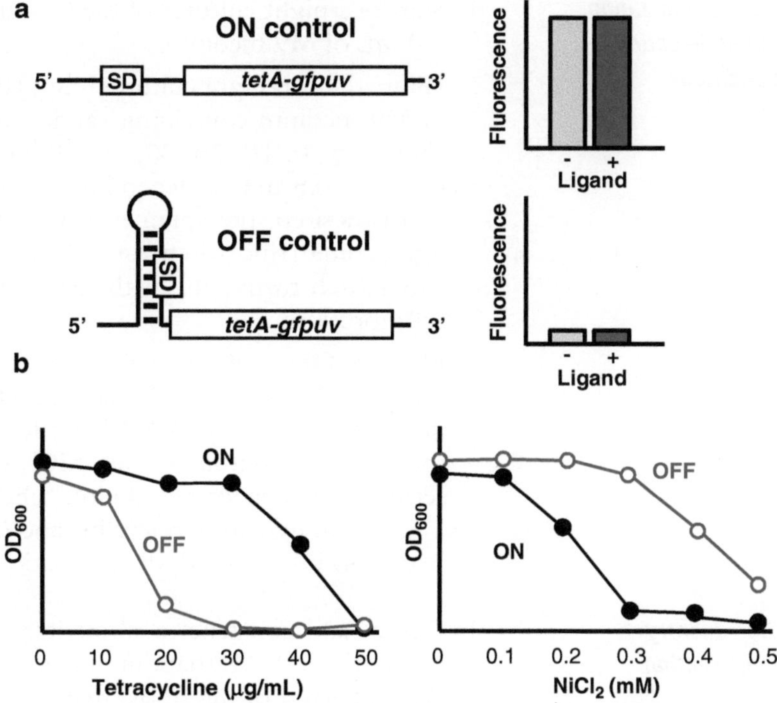

Fig. 2 Suggested ON and OFF control constructs to be used for optimization of the selection conditions. (**a**) The ON and OFF controls should constitutively express TetA–GFPuv marker at high and low levels, respectively, to mimic presumed outputs of the desired riboswitches. The ON control should possess a canonical SD sequence without strong secondary structures. The OFF control could contain a rationally designed stem–loop that sequesters the SD sequence. Both constructs should be evaluated for cellular fluorescence to confirm the desired expression levels. (**b**) An example of the survival profiles of the ON (*filled circles, black*) and OFF (*open circles, red*) controls under varying concentrations of the selective reagents (tetracycline or $NiCl_2$). The desirable selection conditions are where the growth difference between the two phenotypes (ON and OFF) is the greatest. In the depicted figure, 30 μg/mL tetracycline and 0.3 mM $NiCl_2$ are likely to result in the best selection (color figure online)

4. Dilute the overnight cultures (e.g., ~100–200-fold) into fresh 1.0 mL of M9 medium (with and without aptamer ligand), and culture the cells to mid to late log phase.
5. Harvest the cells by centrifugation, wash the cells once by 1.0 mL of PBS, and resuspend the cells in an appropriate volume of PBS (e.g., 0.2–1.0 mL).
6. Measure the OD_{600} and GFPuv fluorescence (excitation 395 nm; emission 509 nm) of each culture. Ensure that the growth (OD_{600}) and the GFPuv fluorescence normalized by OD_{600} are not significantly affected by the ligand used. Also verify that the expression levels of the ON and OFF controls are acceptable for the intended application.

3.3 Optimization of the Selection Conditions

1. Start overnight cultures of the ON and OFF control strains in 1.0 mL of M9 medium.
2. Dilute the overnight cultures (e.g., 100-fold) into fresh 1.0 mL of M9 medium containing varying concentrations of tetracycline (e.g., 0, 10, 20, 30, 40, 50 μg/mL) or $NiCl_2$ (0.1, 0.2, 0.3, 0.4, 0.5 mM). Depending on the type of the riboswitch that is desired, the aptamer ligand should be added to either tetracycline (riboswitch turns ON with the ligand) or $NiCl_2$ (riboswitch turns OFF with the ligand) cultures. Culture the cells for 24 h.
3. Measure OD_{600} of each culture using a microplate reader. The ON and OFF controls should exhibit different sensitivity profiles to tetracycline and $NiCl_2$ as illustrated in Fig. 2b. Identify the optimal tetracycline and $NiCl_2$ concentrations that allow selective growth of the desired phenotype. For example, in Fig. 2b, 30 μg/mL tetracycline and 0.3 mM $NiCl_2$ exhibit the best ON/OFF growth discrimination (*see* **Note 7**).

3.4 Library Construction

1. Design and construct a riboswitch library by randomizing a part of the 5′ UTR that can function as an expression platform. Which part(s) of the riboswitch to randomize may need some considerations. An example from our previous work [4] shown in Fig. 3 contains 15 consecutive randomized bases between the thiamine pyrophosphate (TPP) aptamer and a canonical SD sequence. This library yielded a number of riboswitches that respond positively to the aptamer ligand but failed to yield riboswitches that respond negatively [4]. It was necessary to redesign the library to obtain riboswitches that repress gene expression in response to the ligand [7]. Consequently, the library design can influence the characteristics of the riboswitches that can be selected (no riboswitch can be obtained by selection if it does not exist in the library) and thus deserves a careful attention. A typical protocol for a library preparation is described below.
2. Starting with a plasmid that contains an aptamer cloned in the 5′ UTR of *tetA–gfpuv* coding sequence, PCR amplify the whole plasmid using primers that contain degenerate bases as shown in Fig. 3 using a high-fidelity polymerase (e.g., Phusion DNA polymerase). It is recommended that the primers be synthesized with 5′ phosphate. Otherwise, the primers need to be phosphorylated prior to PCR, or have the PCR product phosphorylated using T4 polynucleotide kinase.
3. Add 10 units of Dpn I per 50 μL PCR reaction and incubate at 37 °C for 3 h.
4. Purify the PCR product by column purification (e.g., DNA Clean and Concentrator-5).

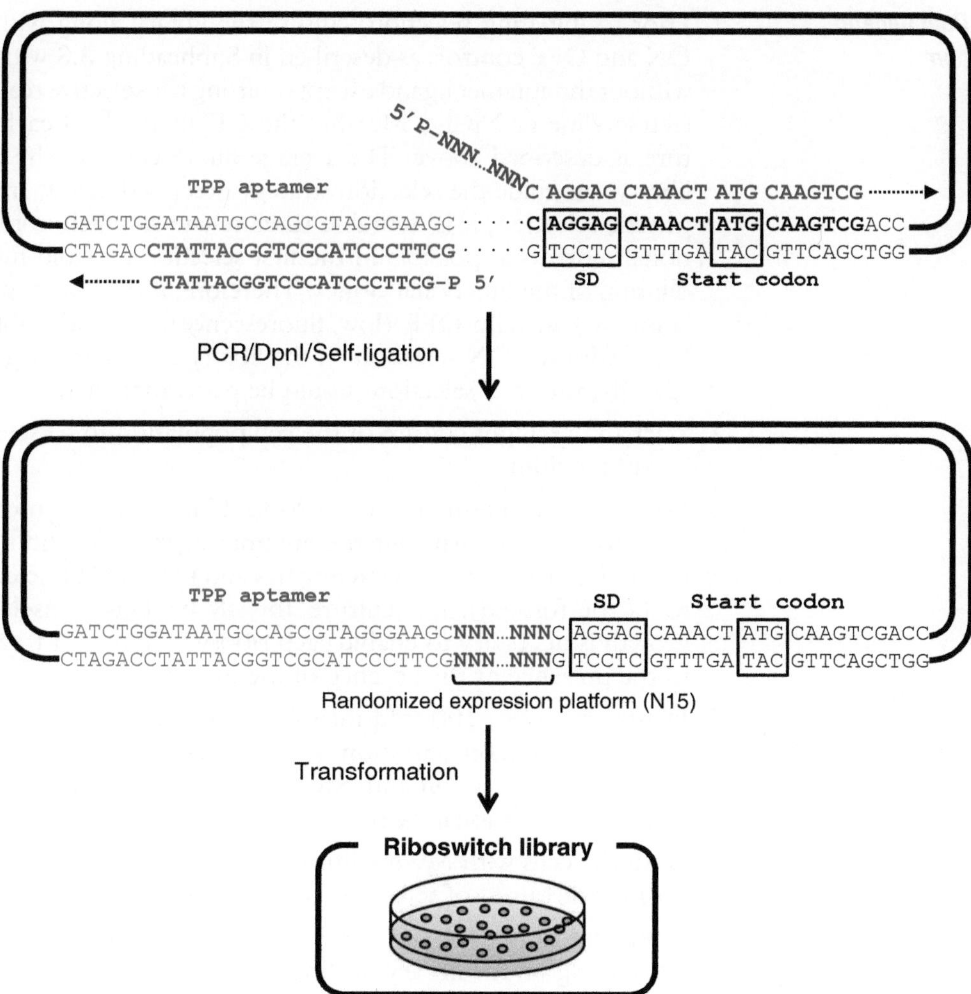

Fig. 3 Library construction strategy. The illustration shows a strategy to randomize a short (15 base) region between an aptamer (TPP) and a canonical SD sequence by whole plasmid PCR. The primers (*blue*) are 5′ phosphorylated, and one of them contains degenerate bases (N) at its 5′ terminus. After PCR, the template plasmid is digested by Dpn I and the PCR product is self-ligated using T4 DNA ligase. Transformation should yield >10^5 colonies (color figure online)

5. Ligate the purified PCR product using T4 DNA ligase following the manufacturer's instructions (*see* **Note 8**).
6. Transform the ligation solution into competent *E. coli* host cells (e.g., TOP10) and spread on LB agar plates. Incubate the plates at 37 °C overnight. Each plate (10 cm) should have no more than 3,000 colonies. Estimate the total number of colonies (library size) which should be as large as possible, preferably >10^5.
7. Recover the library cells by overlaying ~1 mL of liquid M9 medium per plate and gently swirling the plate. Combine the recovered cells, add glycerol to 20 % final volume, and store in aliquots at −80 °C.

3.5 Dual Genetic Selection

1. Prior to initiating selection, culture the library along with the ON and OFF controls as described in Subheading 3.3 with and without the aptamer ligand except omitting the selective reagents (tetracycline or $NiCl_2$). Measure the GFPuv levels of each culture as described above. The average fluorescence of the naive library can guide the selection strategy described below. In dual genetic selection, it is desirable to eliminate as much nonfunctional mutants as possible in the first selection to avoid further dilution of the functional clones. Therefore, if the library population is generally OFF (low fluorescence), it is advisable to begin with the ON selection. If the library population is generally ON, the OFF selection should be performed first.
2. Start an overnight culture of the cell population to be selected in M9 medium.
3. Dilute the overnight culture ~100-fold into fresh M9 medium (1.0 mL) with or without the aptamer ligand (depending on the desired riboswitch characteristics and ON or OFF selection to be performed) and culture for ~8 h. This nonselective growth is necessary to enable each riboswitch mutant to adapt to the presence or the absence of the ligand.
4. Dilute the cells ~100-fold into fresh M9 medium containing an appropriate concentration of the selective reagent (tetracycline for ON selection and $NiCl_2$ for OFF selection) and culture for an additional 24 h.
5. Wash the cells with M9 medium and store in 20 % glycerol for additional rounds of selection and/or characterization.
6. Complete both ON and OFF selections as described above. Depending on the library, additional cycles of ON/OFF selections may need to be performed to enrich the functional riboswitches.
7. After one or more rounds of ON/OFF selection, evaluate the selected cell population's response to the aptamer ligand in liquid culture as described in **step 1** above. If the cells respond to the aptamer ligand as a population, it strongly suggests that the selection successfully enriched the desired riboswitches; therefore, one should proceed to screening of individual clones outlined below (3.6). If no response is observed, selection conditions and/or library design may need to be further optimized or additional cycles of selection may need to be performed (*see* **Note 9**).

3.6 Screening of Individual Clones

1. Plate the selected cells on an LB agar plate to isolate single colonies.
2. Pick and culture appropriate number (>10) of colonies in M9 medium overnight.

3. Dilute each clone ~100-fold into two M9 (1.0 mL) cultures with and without the aptamer ligand. Grow to mid to late log phase.
4. Harvest the cells by centrifugation (*see* **Note 10**), wash the cells once with 1.0 mL of PBS, and resuspend the cells in an appropriate volume of PBS (e.g., 0.2–1.0 mL).
5. Measure the OD_{600} and GFPuv fluorescence (excitation 395 nm, emission 509 nm) of each culture.
6. Identify functional riboswitch clones using the normalized GFPuv fluorescence data.
7. Using a miniprep kit, isolate the plasmids of the functional clones and analyze their sequences.
8. (Optional) Due to some toxicity exhibited by the overexpressed TetA, quantitative responses of the riboswitches to the aptamer ligand concentration may vary when the riboswitches are fused to other downstream genes [4]. If desired, majority of the *tetA* coding sequence can be removed by PCR mutagenesis, leaving GFPuv as the only reporter gene. We use two universal primers p-GFP-f and p-ORF-r to amplify the whole plasmid by PCR followed by self-ligation, similarly as described above for library construction (Fig. 3).

4 Notes

1. The order of mixing is important to avoid precipitation of $CaCl_2$.
2. One of the plasmids described in our previous report [4] is available from the authors upon request.
3. Prepare 20 mg/mL stock solution in 50 % ethanol in diH_2O prior to use.
4. Prepare 2 M stock solution in diH_2O and pass through a 0.2 μm sterile syringe filter. Store in aliquots in the dark at 4 °C.
5. Excessively strong expression of TetA–GFPuv can inhibit cell growth. Some tuning of the translation efficiency may be necessary to moderate the expression level by manipulating the ribosome binding sequence.
6. Due to the use of tetracycline for ON selection, tetracycline-resistant strains cannot be used as host for selection. Other host restrictions may exist depending on the aptamer ligand to be used.
7. The actual optimal concentrations of the selection reagents may vary depending on various factors. Cell culture conditions

such as dilution factor and incubation time may also be adjusted according to each system.

8. **Steps 5** and **6** should be performed in a small scale first to optimize the transformation efficiency and colony counts on each plate. As a negative control, a DNA solution without ligase should be transformed to confirm that the template plasmid is digested.
9. Even if the cell population does not appear to respond to an aptamer ligand, the selected cells may contain sufficiently enriched riboswitches to be discovered by extensive screening of individual clones.
10. For qualitative screening, it may be possible to rapidly screen a large number of clones by visually observing the cell pellets over a UV transilluminator (360 nm). Use appropriate eye protection when using UV.

Acknowledgments

We thank the previous members of our group, in particular, Dr. Norihito Muranaka and Dr. Vandana Sharma who contributed to the development of the dual genetic selection methods. The work was supported by National Science Foundation.

References

1. Sinha J, Reyes SJ, Gallivan JP (2010) Reprogramming bacteria to seek and destroy an herbicide. Nat Chem Biol 6:464–470
2. Winkler WC, Breaker RR (2005) Regulation of bacterial gene expression by riboswitches. Annu Rev Microbiol 59:487–517
3. Wittmann A, Suess B (2012) Engineered riboswitches: expanding researchers' toolbox with synthetic RNA regulators. FEBS Lett 586: 2076–2083
4. Muranaka N, Sharma V, Nomura Y et al (2009) An efficient platform for genetic selection and screening of gene switches in Escherichia coli. Nucleic Acids Res 37:e39
5. Nomura Y, Yokobayashi Y (2007) Dual selection of a genetic switch by a single selection marker. Biosystems 90:115–120
6. Podolsky T, Fong ST, Lee BT (1996) Direct selection of tetracycline-sensitive Escherichia coli cells using nickel salts. Plasmid 36:112–115
7. Muranaka N, Abe K, Yokobayashi Y (2009) Mechanism-guided library design and dual genetic selection of synthetic OFF riboswitches. Chembiochem 10:2375–2381

Chapter 10

Nucleotide Kinase-Based Selection System for Genetic Switches

Kohei Ike and Daisuke Umeno

Abstract

Ever-increasing repertories of RNA-based switching devices are enabling synthetic biologists to construct compact, self-standing, and easy-to-integrate regulatory circuits. However, it is rather rare that the existing RNA-based expression controllers happen to have the exact specification needed for particular applications from the beginning. Evolutionary design of is powerful strategy for quickly tuning functions/specification of genetic switches. Presented here are the steps required for rapid and efficient enrichment of genetic switches with desired specification using recently developed nucleoside kinase-based dual selection system. Here, the library of genetic switches, created by randomizing either the part or the entire sequence coding switching components, is subjected to OFF (negative) selection and ON (positive) selection in various conditions. The entire selection process is completed only by liquid handling, facilitating the parallel and continuous operations of multiple selection projects. This automation-liable platform for genetic selection of functional switches has potential applications for development of RNA-based biosensors, expression controllers, and their integrated forms (genetic circuits).

Key words Thymidine kinase, Nucleoside analog, Genetic selection, Dual selection, Directed evolution, Genetic switch

1 Introduction

Due to the limited prediction capability in sequence/function relationship, rational design of the genetic switches with desired specification remains a daunting challenge. Thus, the evolutionary (or combinatorial) design has been the most successful and realistic approach for the functional tuning of genetic switches. Here, the pool of switch variants is prepared to create a diverse set of switching properties, from which the variants with desired specifications are selected in a high-throughput manner. Because all genetic switches and their assemblies (regulatory circuits) ultimately turn ON or OFF the genetic expression, one can isolate any type of genetic switches, once good selection platform for ON and OFF states of gene expression is established. A variety of positive and negative selection markers are available. However, the use

Atsushi Ogawa (ed.), *Artificial Riboswitches: Methods and Protocols*, Methods in Molecular Biology, vol. 1111,
DOI 10.1007/978-1-62703-755-6_10, © Springer Science+Business Media New York 2014

of independent selection module for the ON/OFF state is in general not recommended because it unnecessarily complicates the selection processes. Another issue is the frequent emergence of false positives. This concern is especially prominent for *cis*-acting regulators such as translation-controlling riboswitches. To circumvent these problems, Yokobayashi and his colleagues have proposed the single-gene dual selecting system for ON/OFF states of the genetic switches [1] or its modified version [2]. This system utilizes the fact that the expression of tetracycline/H+ antiporter TetA confers host cells tetracycline resistance (ON selection) while sensitizing cells to the toxic metal ions such as Ni (II) to host cells (*see* Chapter 9).

Recently, we developed an alternative selection system using herpes simplex virus thymidine kinase (hsvTK) as a dual selector (Fig. 1) [3]. For selecting for the OFF state of the genetic switches, it utilizes the highly efficient kinase activity on artificial nucleoside called dP [3]. dP passes through the cellular membrane, phosphorylated by hsvTK and incorporated into the genomic DNA of host cell. In duplex, it forms base pairs either with A and G, thereby scrambling the genomic information of host cell. Thus, the cells harboring genetic switches in OFF state are enriched by $>10^7$-fold only by 5–15 min of exposure to low concentration (10–100 nM) of dP (Fig. 1). By adding the cocktail of dT and 2′-deoxy-5-fluorouridine (5FdU), one can use hsvTK as efficient selector for ON state of the genetic switches (Fig. 1). When 5FdU is added, it is phosphorylated to yield 5F-dUMP, a potent inhibitor of thymidine synthase (ThyA) [4]. The result is the blockage of the de novo synthesis of thymidine. In this situation, growth of *E. coli* is dependent on exogenously supplemented dT. With *tdk* background, one can enrich the hsvTK-expressing cell, with the efficiency of $>10^6$-fold, simply by culturing overnight in the presence of 5FdU and dT.

Here, we demonstrate the functional selection of quorum sensing switches based on *Vibrio fischeri* LuxR/Plux system (Fig. 2). This system is one of the most popular cell–cell communication devices in synthetic biology community [5–8]. The core component is the LuxR sensory protein that specifically binds to homoserine lactones (HSLs), the signaling molecules used in quorum sensing system shared by diverse gram-negative bacteria, and activates the gene expression under the control of Lux promoter (Plux). Demonstrated here is the creation of LuxR library with error-prone PCR, followed by OFF selection/ON selection to rapidly enrich the functional variants. Although this motif does not includes the RNA-based switching mechanism, the exact procedure shown here should be applicable to the engineering of RNA-based switches including riboswitches, ribozyme switches, and genetic controllers that employ RNA-binding protein.

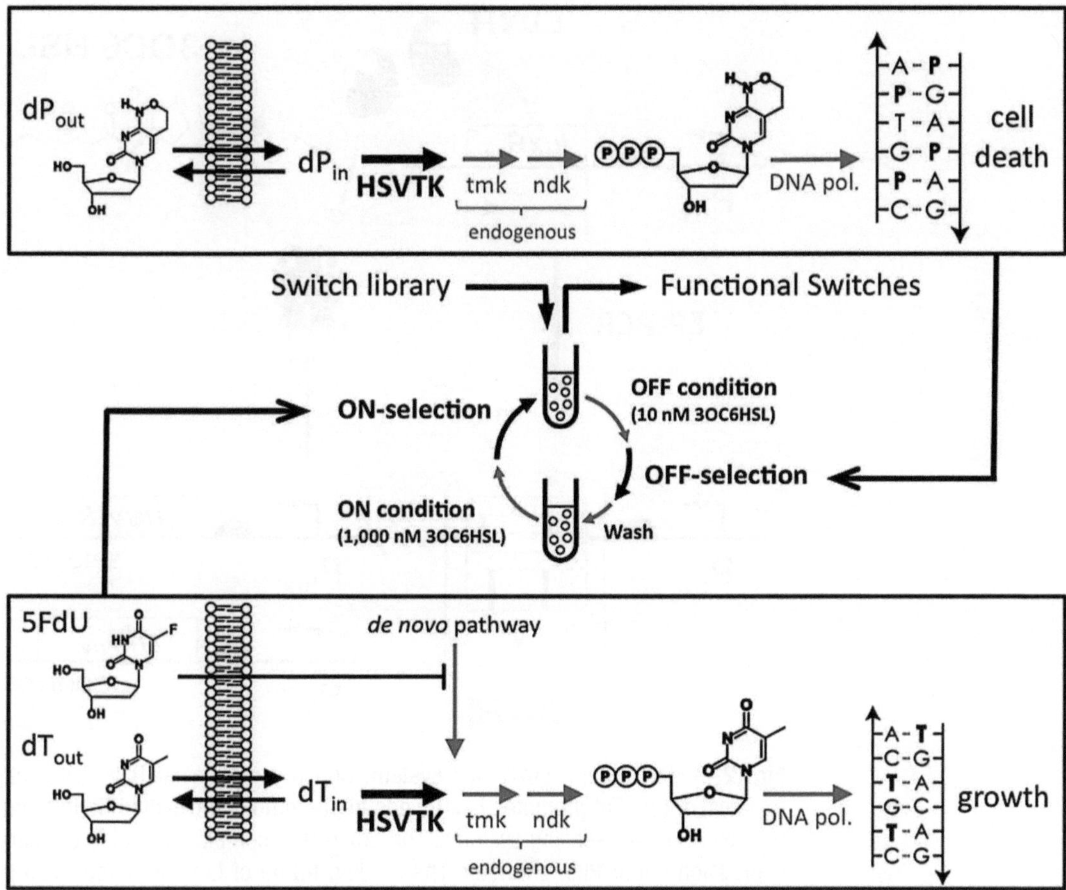

Fig. 1 Dual selection of genetic switches using nucleoside kinase activity. The entire sequence or a part of the genetic switch component is randomized either by error-prone PCR or by oligonucleotide-based randomization. The resultant "switch library" is then placed on the condition to be at OFF state and then subjected to OFF selection by adding nucleoside dP. The survivor pool is then placed on the condition to be ON, subjected to ON selection by the growth in the presence of dT and 5FdU

2 Materials

2.1 Library Creation (PCR Random Mutagenesis)

1. Template plasmid (pTrcHis2-*luxR*) (*see* **Notes 1** and **2**).
2. PCR primers (forward and reverse) (*see* **Note 3**):
   Primer 1: 5′-CAATCTGTGTGGGCACTCGAC-3′.
   Primer 2: 5′-TACTGCCGCCAGGCAAATTC-3′.
3. *Taq* polymerase.
4. 10× *Taq* buffer.
5. 10× dNTP mixture: 2 mM each of dATP, dTTP, dCTP, dGTP. Prepare 50 μL aliquots of this mixture (to avoid excessive freeze/thaw cycles) and store at −20 °C.

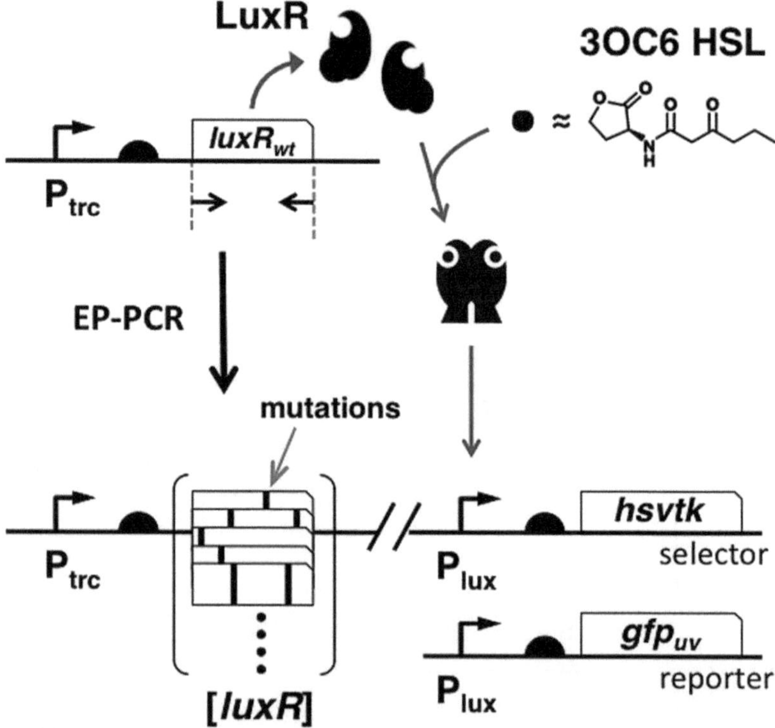

Fig. 2 Model system: LuxR/Plux system. Sensory protein LuxR is constantly expressed from Trc promoter. LuxR binds to pheromone molecule HSL in gram-negative quorum sensing systems. The LuxR–HSL complex activates the gene expression under Plux promoter. The reading frame of LuxR is randomized by error-prone PCR to make LuxR library that is to be subjected to the ON/OFF selection in various HSL concentrations

6. 10× $MnCl_2$ solution: 10 μM prepared in distilled water and stored at room temperature.
7. Nuclease-free water.
8. pTrcHis2-based vector (*see* **Note 2**) [7].
9. DNA clean and concentrator™-5 (Zymo Research Corporation, Orange, CA, USA).
10. Restriction enzymes *Nco*I and *Hind*III.
11. T4 DNA Ligase (Invitrogen, high-conc.)
12. T4 DNA ligase buffer.
13. TAE: 40 mM Tris–acetate, and 1 mM ethylenediaminetetraacetic acid (EDTA).
14. 0.5 μg/mL Ethidium bromide in TAE.
15. Agarose gel: 0.7 % LE agarose in TAE and ethidium bromide.
16. Zymoclean™ Gel DNA Recovery Kit (Zymo Research Corporation, Orange, CA, USA).

2.2 ON/OFF Selection

1. Selector plasmid (for details, *see* **Notes 2** and **4**).
2. *E. coli* JW1226 (KEIO collection, *tdk* strain) (*see* **Note 5**, [9]).
3. 50 mg/mL Carbenicillin (1,000× stock).
4. 30 mg/mL Chloramphenicol (1,000× stock).
5. Luria-Bertani (LB): 20 g of premixed LB medium (Invitrogen) into 1 L of deionized water. Autoclave at 121 °C for 20 min.
6. SOC media: Add 20 g of bacto-tryptone, 5 g of yeast extract, and 0.5 g of NaCl into 950 mL of deionized water. Add 10 mL of a 250 mM solution of KCl. Adjust the volume of the solution to 1 L with deionized water. Autoclave at 121 °C for 20 min. Just before use, add 5 mL of a sterile solution of 2 M $MgCl_2$ and 20 mL of a sterile 1 M solution of glucose [10].
7. dP, 6-(β-D-2-Deoxyribofuranosyl)-3,4-dihydro-8*H*-pyrimido [4,5-c][1,2]oxazin-7-one) [3, 11]: Prepare 1 μM–1 mM (typically 100 nM) dP solution dissolved in DMSO as 1,000× stock. Sample stored at 4 °C.
8. OFF selection medium: LB containing 10–1,000 nM dP [3].
9. 10 mg/mL Thymidine in deionized water (1,000× stock): Stock stored at 4 °C.
10. 1 mg/mL Adenosine in deionized water (1,000× Stock): Stock stored at 4 °C.
11. 20 mg/mL 5FdU in deionized water (1,000× stock): Stored at −20 °C.
12. Tryptone broth: Add 20 g of tryptone and 5 g of NaCl into 1 L of deionized water. Autoclave at 121 °C for 20 min.
13. ON selection medium: 10 μg/mL thymidine, 1 μg/mL adenosine, and 20 μg/mL 5FdU [3, 4] in tryptone broth.
14. 10 mM stock solution of 3OC6-HSL in ethyl acetate acidified with glacial acetic acid (0.001 % (*v*/*v*)): Stored at −20 °C.

2.3 Evaluating the Selected Switches

1. Reporter plasmid (for details, *see* **Note 4**).
2. *E. coli* JW1226 (*see* **Note 5**, [9]).
3. Carbenicillin stock (*see* Subheading 2.2, **item 3**).
4. Chloramphenicol (*see* Subheading 2.2, **item 4**).
5. LB-agar (*see* Subheading 2.2, **item 5**).
6. SOC media (*see* Subheading 2.2, **item 6**).
7. NaCl solution: Add 9 g of NaCl into 1 L of deionized water. Autoclave at 121 °C for 20 min.
8. Flow cytometer: MACS Quant VYB (Miltenyi Biotec).
9. Fluorescence plate reader: Fluoroskan Ascent® (Thermo Fisher Scientific Inc.).
10. Plate reader: SpectraMax Plus384 (molecular devices).

3 Methods

3.1 Library Creation

There are a variety of established methods for the construction of random libraries [12]. Among them, error-prone PCR is the most popular method that could apply for a region or the entire part of the targeted gene. The simplest is the addition of Mn^{2+} into the PCR, where manganese ion lowers the fidelity of the polymerase, inserting random base substitutions into the target sequence [13].

1. For each PCR sample, add to tube 2 fmol of template DNA (pTrcHis2-*luxR*), 5 μL of 10× *Taq* buffer, and 5 μL of 10× dNTP mixture. 5 μL of 5 μM each primers (to be 25 pmol in final concentration), 1 μL of *Taq* polymerase (5 U), 5 μL of 10× $MnCl_2$ solution (to be 10–100 μM in final concentration), and nuclease-free water to a final volume of 50 μL.
2. Mix the sample by pipetting, and confirm all the solution is on the bottom.
3. Place the tubes in a thermal cycler, and run the following PCR program: (1) 5-min initial denature at 94 °C; (2) 30-s denature at 94 °C, 30-s annealing at the temperature designed for primers, and 1-min extension at 72 °C; repeat for 25 cycles; and (3) 10 min at 72 °C for final extension (*see* **Notes 6** and 7).
4. Run the product for gel electrophoresis to estimate the yield of full-length gene (*see* **Notes 6** and **8**).
5. Clean the PCR product by using a Zymo DNA clean and concentrator kit. In parallel, clean up the vector in the same way.
6. Digest both PCR product and the expression vector with appropriate enzymes (*Nco*I/*Hind*III).
7. Gel-purify the digested samples using Zymoclean™ Gel DNA Recovery Kit (*see* **Note 9**).
8. Treat the mixture of the purified insert (100 ng) and vector (100 ng) with ligase buffer and T4 ligase for 2 h (400 units/reaction to total volume of 10 μL) (*see* **Note 10**).
9. Optionally, concentrate/purify the ligation product by using a Zymo DNA clean and concentrator kit (*see* **Note 11**).
10. Electroporate the ligation mixture (1 μL) into the competent cell (40 μL) (*see* **Note 12**).
11. Add 1 mL of SOC media immediately after the electroporation, and shake it for 1 h at 37 °C. After washing the transformants with fresh LB media, resuspend the cells into 10 mL of LB media (*see* **Notes 13** and **14**).
12. After overnight shaking at 37 °C, miniprep the portion (1–2 mL) of the above culture to obtain plasmid library pTrcHis2-[*luxR*].

3.2 Selection of the LuxR Variants with Proper Switching

1. Add 100 ng (1 μL) of plasmid library (pTrcHis2-[*luxR*]) into 40 μL of JW1226 cell harboring pAC-Plux-*hsvtk* (*see* **Note 15**).
2. Add 1 mL of SOC media and gently shake for 1 h at 30 °C.
3. Collect the cells by centrifugation (4,000 rcf, 3 min), resuspend them into the 10 mL of LB media containing chloramphenicol (30 μg/mL) and carbenicillin (50 μg/mL), and allow for overnight shaking at 37 °C.
4. *OFF selection*: Transfer about 10^6 cells into the 1 mL of OFF selection medium containing 0–1,000 nM 3OC6-HSL and leave in the shaking incubator (37 °C) for 2 h (*see* **Notes 16** and **17**).
5. Wash the cell by centrifugation (4,000 rcf, 3 min)/resuspension to fresh LB media containing the same concentration of 3OC6-HSL at OFF selection. Repeat this process twice (*see* **Note 18**).
6. Shake the culture for another 6 h at 37 °C (*see* **Notes 19** and **20**).
7. *ON selection*: From the above culture, take 10 μL (about 10^6 cells) to resuspend into 1 mL of ON selection medium containing desired concentration of inducer (3OC6-HSL).
8. Shake the cell culture for another 20 h at 37 °C (*see* **Note 21**).
9. Miniprep the culture to collect the plasmid containing survivor variants.

3.3 Characterization of the Survivor Pools and Isolation of Switch Variants

1. Electroporate the library plasmids before and after the selection into JW1226 harboring pAC-Plux-*gfpuv*.
2. Add 1 mL of SOC and incubate at 37 °C for 1 h in shaker.
3. After collecting the cells by centrifugation (4,000 rcf, 3 min), resuspend them into the 10 mL of LB media containing chloramphenicol (30 μg/mL) and carbenicillin (50 μg/mL), and allow for overnight shaking at 37 °C.
4. Take 10^6 cells into fresh 500 μL of LB media containing various concentrations (0–1 μM) of 3OC6-HSL and shake them for another 12 h. Then proceed to **step 5** or **6**.
5. *Population analysis* (Fig. 3a): Wash the cell by saline solution. Dilute the cell by about 1,000×. The diluted cell suspension is subjected to flow cytometry (V2 channel, laser 405 nm, filter 525/50 nm) for the analysis of the pool.
6. *Isolation of the switch variants*: A fraction of transformant is plated on LB-agar plate containing chloramphenicol (30 μg/mL) and carbenicillin (50 μg/mL) to form colonies (*see* **Note 20**). Pick several to dozens of colonies from the plates, and inoculate them into LB (2 mL) culture containing chloramphenicol (30 μg/mL) and carbenicillin (50 μg/mL) in deep-well 96 plates, allowing the overnight growth at 37 °C.

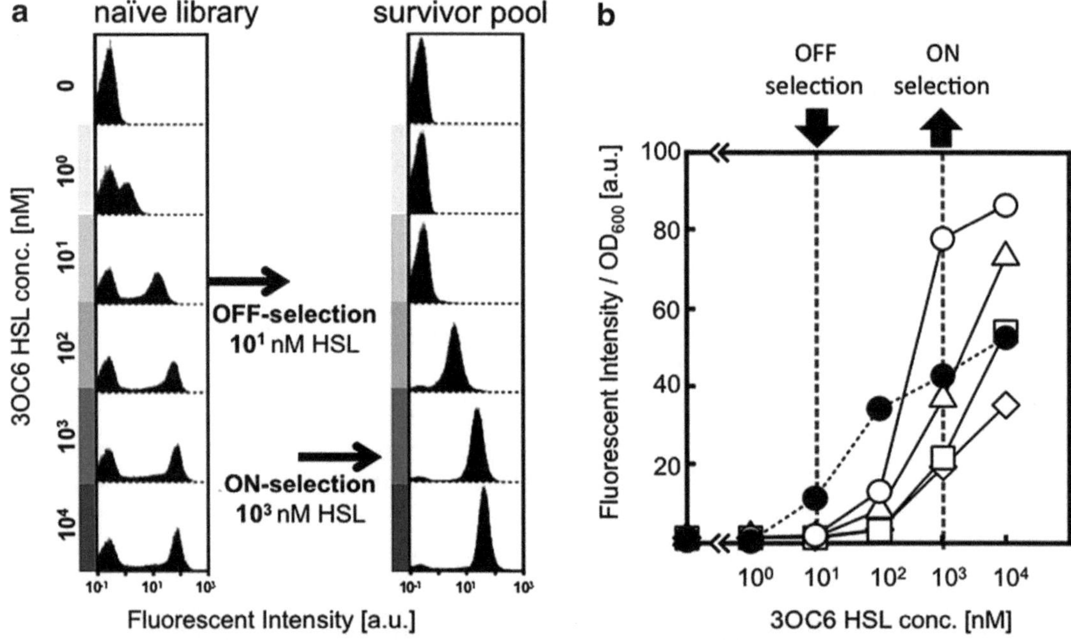

Fig. 3 (**a**) Flow cytometric analysis of the library of HSL-induced genetic switches before and after the OFF and ON selection in the presence of 10 and 1,000 nM 3OC6-HSL. Original variant mixture (library) was two-peaked in any concentration of HSL, but after single round of OFF selection in the presence of 10 nM HSL and OFF selection in the presence of 1,000 nM HSL, the properly responding LuxR/Plux systems were enriched. (**b**) [HSL]-output transfer function of representative LuxR clones isolated from the survivor pool. *Filled circle* represents the switching property of wild-type LuxR, while *open circle*, *open triangle*, and *open squares* represent that of LuxR variants

7. Each of the above cultures is inoculated to the LB plate containing chloramphenicol (30 μg/mL), carbenicillin (50 μg/mL), and various concentrations (0–10 μM) of 3OC6-HSL.
8. After overnight growth at 37 °C, select the variants for further analysis. Analysis includes sequence analysis, input/output transfer function in liquid medium (Fig. 3b), and cytometric analysis of the switch variants. Upon necessity, one can determine the sensitivity (threshold concentration of input molecule needed to turn on the switches), stringency (leakage level at OFF condition), *S*/*N* ratio (maximum output intensity/output intensity at OFF condition), cooperatively (Hill coefficient), signal selectivity, and transition time.

4 Notes

1. It is highly recommended to physically separate the region to be randomized/mutagenized from selector genes. Thus, we always use the double-plasmid system as is shown here. This is mainly to minimize the accidental introduction of mutations

into the selector genes in the experimental processes for mutagenesis and selection. Also, this format enables us to quickly exchange the reporter/selector plasmids. For the functional selection of *cis*-acting regulatory elements such as translational regulator riboswitches, one can think of placing trans-acting expression controllers such as LuxR under the direct control of the riboswitches to be engineered, and the riboswitch region should be mutagenized. This way, the selector/reporter plasmid described here can be used without further modification.

2. The gene coding for protein LuxR from *Vibrio fischeri* was subcloned into pTrcHis2 vector (Invitrogen) to construct pTrcHis2-*luxR*, which serves as template for PCR mutagenesis. *LuxR* gene used in this study has a sequence of *luxR* nearly identical to the originally reported, except the amino acid substitution A4G (K2E) to create *Nco*I site in N terminal start codon. The *luxR* also contains several non-synonymous substitutions A51G (K17K), T174C (I58I), and C624T (G208G).
3. PCR primers are designed to share the same or the similar melting temperature. They are used at 5 μM and stored at −20 °C. The primers for error-prone PCR are designed to anneal to the outside of the reading frame of the gene so that the entire reading frame is randomized. In this case, however, regulator sequence (promoter and ribosome-binding site) is also under mutagenesis. In the situation where the mutations in the regulation sequence could lead to the false positives, design the forward/reverse primer to anneal to the N/C terminus of the reading frame, respectively. In this case, the priming sites of the target genes are virtually "masked" from the mutagenesis.
4. Selector plasmid pAC-Plux-*hsvtk* was constructed by inserting *V. fischeri* Lux promoter (Plux) and the gene for hsvTK into pACYC184-based vector. The reporter plasmid pAC-Plux-*gfpuv* was constructed simply by exchanging reading frame of hsvTK of pAC-Plux-hsvtk with *gfpuv* (Clontech). The sequence of the Lux promoter (Plux) is as follows:

   5′-ACCTGTAGGATCGTACAG**GTTTACG**CAAGAAA ATGGTTTGT**TATAGT**CGAATAAA-3′.

   The sequence underlined is the binding site (operator site) of LuxR–HSL complex, and letters in bold represent the −35/−10 sequences (promoter regions).
5. For selecting ON state of the genetic switches, any tdk strains could be used. In this work, we chose JW1226 (*F^- Δ(araD-araB)567 ΔlacZ4787(::rrnB-3)* λ^- *Δtdk-747::kan rph-1 Δ(rhaD-rhaB)568, hsdR514*).
6. With a high amplification yield of error-prone PCR (>1,000-fold), one can keep the library free from contamination by the unamplified wild-type sequence.

7. Mutation rates of the library can be controlled by the concentration of Mn^{2+} in a PCR reaction. Typical concentration range for Mn^{2+} is from 0 to 100 μM. The higher concentration results in the higher mutation rate of the resultant library. In the high side (>200 μM), polymerase reaction is inhibited, resulting in the low amount or no product.
8. In the given (fixed) concentration of Mn^{2+}, the average mutation rate of the library is proportional to the effective cycle numbers of the PCR. The most convenient way to control the cycle number is to change the amount of target DNA in the PCR. In theory, tenfold decrease in template in error-prone PCR results in $\log_2 10$ ~ca. 3.3-fold elevation in mutation rate, given that final yield of PCR stays the same.
9. This step is important to remove the template plasmid. In addition to this size fractionation, one can further eliminate the template plasmid by the treatment of PCR product with *Dpn*I, which selectively digests plasmid-borne methylated DNA, prior to the gel purification.
10. Addition of 10 mM ATP in the reaction mixture would increase the ligation efficiency.
11. After purification, ligation can be stored at −20 °C without losing the transforming efficiency (library size). In case where the larger library is necessary, transform the (concentrated) ligation into commercial super-competent cells and propagate the transformant in a single mixed culture for 12 h. Miniprep this mixed culture to prepare plasmid library ready for transforming to screening strains.
12. To maximize the library size, use the ultrahigh competent cells commercially available.
13. In case the density of competent cell is high, especially when electroporation is applied, the subsequent culture should be of relatively large volume (ca. 10 mL). Presence of too much non-transformed cell (dead cell) hampers the normal propagation of transformant cell, resulting in the poor yield in plasmid prep and in decreased library size.
14. Plate a portion (0.1–1 %) of the transformant mixture for colony counting to evaluate the library size, which is an important measure of genetic diversity.
15. Here again, plate a portion (1 %) of the transformant mixture for colony counting in order to evaluate the effective selection size (the number of variants subjected to the selection). For the determination of selection size, the number of colonies appeared on the plate is multiplied with 100.
16. *E. coli* cells (with and without *tdk* genotype) not expressing hsvTK can tolerate up to 100 μM of dP without any detectable

change in cell viability and growth speed. However, we observed ca. tenfold increase in mutation rate on the genomic DNA, which was determined by rifampicin assay (unpublished results).

17. In principle, one can start from ON selection and then move on to OFF selection. We prefer to conduct OFF selection first and then proceed to ON selection. This is firstly because OFF selection is completed in a very short period of time and ON selection requires overnight growth. Another and probably the most important reason for this is that one cannot immediately move on to the OFF selection after the ON selection; the transition from ON state to OFF state requires hours in proteomic level due to the remainder of hsvTK (selector gene) when expressed in the previous "ON" selection. Consequently, one needs to wait for 4–6 h until the level of selector enzyme hsvTK decreases to undetectable level before setting the media condition for OFF state. On the other hand, the transition from OFF state to ON state is quick in proteomic level and free from wrongly eliminating the proper switches (false negatives).

18. Washing should be repeated at least twice to minimize the dP remaining in the media. Remaining dP could swipe off the properly behaving switch variants by inducing the hsvTK at the subsequent "ON" selection process.

19. To check whether OFF selection is properly conducted, you can miniprep the rest of the OFF-selected culture at this point and proceed for flow cytometric analysis. This also gives you the backups for the possible failure in subsequent ON selection. With this "OFF-selected pool," you can re-try the ON selection in various different conditions.

20. In case you do not observe the elevation in cell density, this could be due to the improper setup of the construct. Frequently, genetic switches have the weak but certain level of the leaky expression. Note that dP kinase activity is very powerful in killing hsvTK-expressing cell, and even the minimal level of leaky expression could be enough for cell death. In such cases, basal expression construct should be downward adjusted by changing either the promoter sequence or ribosome-binding sequences.

21. In case you do not observe the increase in cell density, leave the culture shaken for an additional 6–12 h. If the cell density increases, then it means that the survivors (the cell harboring switch variants properly in ON state) consist of the very minor fraction of the library. If you do not observe the cell growth at all, even after 24 h of shaking, it probably means that the initial library did not contain the variants with desired specificity. You should redesign the library.

References

1. Nomura Y, Yokobayashi Y (2007) Dual selection of a genetic switch by a single selection marker. Biosystems 90:115–120
2. Muranaka N, Sharma V, Nomura Y et al (2009) An efficient platform for genetic selection and screening of gene switches in *Escherichia coli*. Nucleic Acids Res 37:e39
3. Tashiro Y, Fukutomi H, Terakubo K et al (2011) A nucleoside kinase as a dual selector for genetic switches and circuits. Nucleic Acids Res 39:e12
4. Yagila E, Rosnerb A (1971) Phosphorolysis of 5-fluoro-2′-deoxyuridine in *Escherichia coli* and its inhibition by nucleosides. J Bacteriol 108: 760–764
5. Weiss R, Knight TF (2001) Engineered communications for microbial robotics, vol 2054, Lecture Notes in Computer Science., pp 1–16
6. Basu S, Gerchman Y, Collins CH et al (2005) A synthetic multicellular system for programmed pattern formation. Nature 434: 1130–1134
7. Tabor JJ, Salis HM, Simpson ZB et al (2009) A synthetic genetic edge detection program. Cell 137:1272–1281
8. Danino T, Mondragón-Palomino O, Tsimring L et al (2010) A synchronized quorum of genetic clocks. Nature 463:326–330
9. Baba T, Ara T, Hasegawa M et al (2006) Construction of *Escherichia coli* K-12 in-frame, single-gene knockout mutants: the Keio collection. Mol Syst Biol 2:1–11
10. Sambrook J, Russell DW (2001) Molecular cloning: a laboratory manual, 3rd edn. Cold Spring Harbor Laboratory, Cold Spring Harbor, NY
11. Negishi K, Loakes D, Schaaper RM (2002) Saturation of DNA mismatch repair and error catastrophe by a base analogue in *Escherichia coli*. Genetics 161:1363–1371
12. Arnold FH, Georgiou G (2003) Directed evolution library creation: methods and protocols. Humana, Totowa, NJ
13. Cirino PC, Mayer KM, Umeno D (2003) Generating mutant libraries using error-prone PCR. Methods Mol Biol 231:3–9

Chapter 11

Measuring Riboswitch Activity In Vitro and in Artificial Cells with Purified Transcription–Translation Machinery

Laura Martini and Sheref S. Mansy

Abstract

We present a simple method to measure the real-time activity of riboswitches with purified components in vitro and inside of artificial cells. Typically, riboswitch activity is measured in vivo by exploiting β-galactosidase encoding constructs with a putative riboswitch sequence in the untranslated region. Additional in vitro characterization often makes use of in-line probing to explore conformational changes induced by ligand binding to the mRNA or analyses of transcript lengths in the presence and absence of ligand. However, riboswitches ultimately control protein levels and often times require accessory factors. Therefore, an in vitro system capable of monitoring protein production with fully defined components that can be supplemented with accessory factors would greatly aid riboswitch studies. Herein we present a system that is amenable to such analyses. Further, since the described system can be easily reconstituted within compartments to build artificial, cellular mimics with sensing capability, protocols are provided for building sense-response systems within water-in-oil emulsion compartments and lipid vesicles. Only standard laboratory equipment and commercially available material are exploited for the described assays, including DNA, purified transcription–translation machinery, i.e., the PURE system, and a spectrofluorometer.

Key words Riboswitch, Transcription–translation, In vitro compartmentalization, Liposome, Emulsion, Cell-free synthetic biology

1 Introduction

Riboswitches are genetically encoded control elements that respond to small molecules through direct binding. Sensing is mediated by an aptamer [1–4] sequence within the mRNA that controls the conformation of the expression platform. Usually ligand binding turns off gene expression; however, natural on-riboswitches exist [5]. The induced conformational changes either regulate transcription through terminator–anti-terminator activity, translation by modulating the accessibility of the ribosome binding

Atsushi Ogawa (ed.), *Artificial Riboswitches: Methods and Protocols*, Methods in Molecular Biology, vol. 1111,
DOI 10.1007/978-1-62703-755-6_11,

site, mRNA processing, or splicing [6]. In addition to natural riboswitches, many riboswitches have been engineered by modifying previously selected aptamer sequences [7] or by mutating natural riboswitches to display new functionality [8]. Most of the characterized natural riboswitches control transcription, whereas synthetic riboswitches typically control translation.

Monitoring transcription in vitro is straight forward [9, 10] thereby allowing for the characterization of riboswitches that alter transcript length in a manner dependent upon the presence or absence of ligand. However, riboswitches that control ribosome binding site accessibility produce transcripts of the same length regardless of the presence or absence of ligand, making methods that quantify differences at the RNA level less insightful. Moreover, riboswitches ultimately control protein synthesis, regardless of the specific mechanism exploited. Therefore, more direct methods that probe the influence of riboswitch activity on protein synthesis are desirable. This is most often achieved by placing the riboswitch in question within a genetic construct that encodes β-galactosidase, a fluorescent protein [7, 11] or more recently, a protein involved in motility [12] and monitoring the activity of the reporter protein in *Escherichia coli*. In other words, the assay is carried out within the cell and absorbance or fluorescence are quantified.

The advantage of such methods is that the activity of the riboswitch within a living cell is monitored, meaning that the measured activity is not a result of imperfect in vitro approximations of in vivo conditions. However, there are several limitations of such in-cell assays. First, the influence of accessory proteins could easily be missed, since their participation in sensing or transducing chemical messages is largely uncontrolled in such experiments. Second, the putative ligand either must be capable of crossing the membrane (to allow for exogenous delivery) or easy to manipulate in terms of concentration. For example, the activity of the flavin mononucleotide (FMN) riboswitch was characterized at the transcriptional level in vitro [13, 14], but the influence of FMN on protein synthesis was not investigated, presumably due to the difficulty in quantifying and modulating intracellular FMN concentrations.

Herein we present a simple method to characterize the influence of riboswitch activity on protein synthesis in vitro. Guidelines for the design and assembly of the genetic construct, and the evaluation of in vitro riboswitch activity by monitoring the synthesis of fluorescent protein with fully defined components are described (Fig. 1). Importantly, this real-time fluorescence assay is amenable to the screening of protein accessory factors and ligands, including ligands that are metabolites, impermeable, or toxic. It should be noted that the described protein synthesis assay does not replace current methods that characterize transcriptional activity. The investigation of both transcription and translation is needed in order to fully define the mechanistic details of riboswitch activity.

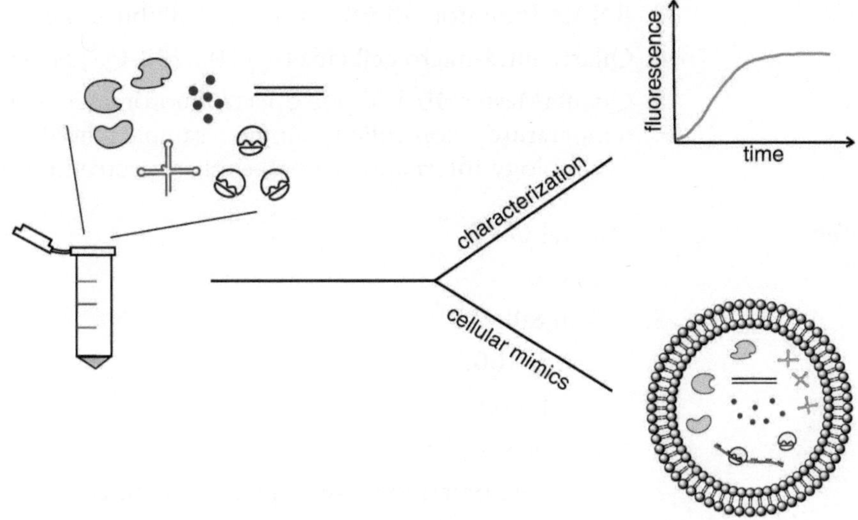

Fig. 1 Cell-free systems for in vitro riboswitch characterization and the construction of artificial, cellular mimics. The PURE system is used to characterize in real-time riboswitch activity through the expression of a reporter protein either in vitro or inside of a compartment with dimensions similar to living cells

We additionally describe how this riboswitch controlled in vitro transcription–translation system can be encapsulated within compartments to build cellular mimics (Fig. 1). As opposed to the majority of artificial cell studies that focus on self-replication, riboswitch sensing-based cellular mimics integrate more fully with the environment and thus could potentially serve as a platform for future technologies. The example described here uses water-in-oil (w/o) emulsion droplets [15], vesicles [16], and a previously reported theophylline riboswitch [7, 17, 18].

2 Materials

All solutions should be prepared using diethyl pyrocarbonate (DEPC) treated water. All reagents are nuclease-free, molecular biology grade. The theophylline riboswitch sequence used here is available from the Registry of Standard Biological Parts (BBa_J89000).

2.1 Template Preparation

1. *E. coli* DH5α or similar laboratory, cloning strain.
2. Commercial plasmid miniprep kit.
3. 25:24:1 Phenol–chloroform–isoamyl alcohol mixture (*see* **Note 1**).

2.2 In Vitro Transcription and Translation

1. PURExpress in vitro protein synthesis kit (New England Biolabs).
2. Riboswitch ligand molecule (e.g., theophylline).

3. RNAse Inhibitor (RiboLock RNase Inhibitor, Fermentas).
4. Quartz ultra-micro cell cuvette (105.252-QS, Hellma).
5. QuantaMaster 40 UV–Vis Spectrofluorometer with a Peltier temperature controlled single sample holder (Photon Technology International) or a similar spectrofluorometer.

2.3 Emulsion Preparation

1. Mineral Oil.
2. Span 80.
3. Tween 80.
4. Triton X-100.
5. 9 mm Teflon stir bar.
6. Magnetic stir plate.
7. Gilson Microman Positive displacement pipettes.

2.4 Liposome Preparation

1. 1-palmitoyl-2-oleoyl-*sn*-glycero-3-phosphocholine (POPC).
2. Cholesterol.
3. *N*-(carbonyl-methoxypolyethyleneglycol5000)-1,2-distearoyl-*sn*-glycero-3-phosphoethanolamine (DSPE-PEG 5000, NOF-Europe).
4. Rotary evaporator, e.g., Rotavapor R-210 with Vacuum Pump V-700 (Buchi).
5. IKA T 10 basic ULTRA-TURRAX disperser with a 5 mm diameter dispersing tool.
6. Mini-Extruder (Avanti Polar Lipids, Inc.).
7. Nucleopore Track-Etch Membrane 0.4 μm (Whatman).
8. Centrifugal evaporator, e.g., CentriVap Centrifugal Vacuum Concentrator (Labconco).
9. Tris saline buffer; 50 mM Tris–HCl, 50 mM NaCl, pH 7.4 supplemented with 10 mg/mL Proteinase K (Fermentas).

2.5 Imaging

1. Zeiss Observer Z1 microscope (Carl Zeiss S.p.A.) or similar fluorescence microscope.

3 Methods

3.1 DNA Template Preparation

The DNA template can be either a circular plasmid or a linear PCR product that contains a series of modular elements to allow for riboswitch controlled protein synthesis (Fig. 2). The sequence should contain a transcriptional promoter, a sequence encoding the riboswitch that contains a ribosome binding site (preferably the natural ribosome binding site sequence, if possible), a gene coding for a fluorescent protein to act as a reporter, and a transcriptional terminator. The transcriptional promoters for T7 and *E. coli* RNA polymerases are typically used. However, since the activity of a riboswitch can depend

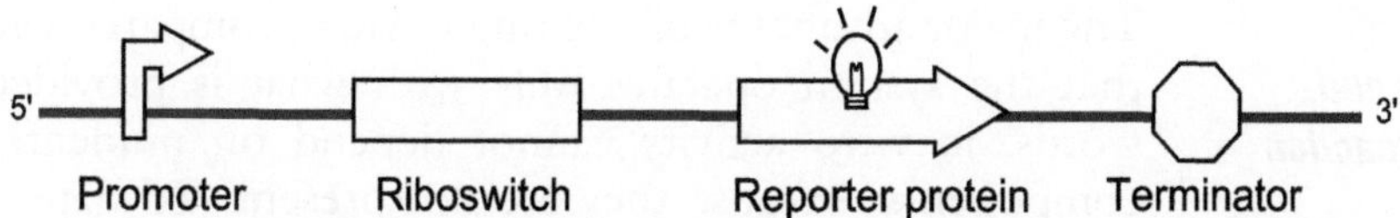

5' TAATACGACTCACTATAG **GGTGATACCAGCATCGTCTTGATGCCCTTGGCAGCACCCTGCTAAGG**
TAACAACAAG *ATGmYPetTAA* AGAGAATATAAAAAGCCAGATTATTAATCCGGCTTTTTTGTTATTT 3'

Fig. 2 The composition of a theophylline riboswitch for the real-time observation of riboswitch activity. The construct shown here contains a standard T7 transcriptional promoter, a theophylline riboswitch that contains a ribosome binding site, a gene that codes for A206K YPet (mYPet), and a T7 transcriptional terminator. Only the start and stop codons of mYPet are shown. The riboswitch sequence is shown in *bold*

on the RNA polymerase [19], particularly for riboswitches that use a terminator–anti-terminator mechanism, the choice of which promoter to use can significantly impact in vitro riboswitch activity. Similarly, it is advisable that the sequence chosen for the riboswitch portion of the construct contains the ribosome binding site. Riboswitches are often associated with inefficient ribosome binding sites, either due to a lack of potential base-pairing interactions with the ribosome or because of structural features of the riboswitch that obstruct ribosome binding site—ribosome interaction. Further, since riboswitches typically do not fully block protein synthesis in the off-state nor mediate robust expression in the on-state, i.e., riboswitch control is leaky and generally mediates more subtle changes in expression [7], a ribosome binding site not tuned to the activity of the riboswitch could complicate analyses. The reporter should be a protein that expresses well in vitro and is easily detectable. We find the green fluorescent proteins super folder GFP (sfGFP) and GFPmut3b and the yellow fluorescent proteins YPet and Venus to be particularly good choices [20]. Finally, incorporating a hairpin transcriptional terminator is advisable, even if not absolutely required when using a linear PCR product as a template, because structured RNA termini increase RNA stability and thus protein yield [21].

1. The DNA template should contain from 5′ to 3′ a T7 promoter followed by two GG nucleotides to enhance transcription, a sequence encoding the riboswitch [17] and the RBS, and the gene coding for the reporter protein followed by a transcriptional terminator (Fig. 2).
2. The template should be amplified either by PCR or by transforming a typical laboratory cloning strain of *E. coli*, such as DH5α, and purified with a commercially available kit according to the manufacturer's instructions.
3. Subsequently, the DNA is phenol–chloroform extracted [22] with an equal volume of Tris-buffered Phenol–Chloroform (*see* **Note 1**).
4. The DNA is ethanol precipitated [23], resuspended in sterile water, and stored at −20 °C.

3.2 In Vitro Transcription and Translation Reaction

The major advantage of working in vitro, compared with in vivo, is that the system operates only with what is provided. In other words, in vitro activity cannot depend on unidentified cellular components, because they are not present. To date, only *E. coli* [24] and *Thermus thermophilus* [25] translation machinery have been reconstituted in vitro from purified components. Of these two, only the *E. coli* system, i.e., the PURE system, is commercially available. It should be noted that in contrast to in vivo or cell-extract conditions, reactions with purified transcription–translation machinery do not contain nucleases. The lack of nucleases decreases the amount of DNA template needed. Whether protein production is more or less efficient with the PURE system in comparison with cell-extract based systems depends on the specific folding properties of the expressed protein.

1. The PURE system components should be aliquoted on ice and stored in 0.2 mL microcentrifuge tubes. Convenient volumes are 10 μL aliquots of solution A and 7.5 μL aliquots of solution B. Aliquots are stored at −80 °C.
2. Assemble the reaction components, except for the DNA template, on ice following the manufacturer's instructions. Supplementary reagents can also be added, such as RNase inhibitor (e.g., 20 U RiboLock RNase Inhibitor) or the riboswitch ligand (e.g., 0.5 mM theophylline).
3. The assembled reaction is transferred to a quartz cuvette and incubated at 37 °C.
4. The reaction is initiated by the addition of the DNA template and monitored by fluorescence spectroscopy for 6 h (Fig. 3) (*see* **Note 2**). The DNA template concentration should be screened. We used 250 ng of plasmid template in 25.5 μL total reaction volume. If YPet is used as the reporter protein, the excitation and emission wavelengths are 517 nm and 530 nm, respectively.

3.3 In Vitro Compartmentalization

Since cellular life is chemically distinct from the environment, efforts to mimic cells in the laboratory often times exploit w/o emulsion droplets or vesicles to approximate the compartment of the living, chemical system [26, 27]. However, even if cellular life is distinct from the environment, life cannot exist in isolation and must in some manner interface with the environment to survive [28]. Since lipid vesicles are semipermeable and more similar to the types of barriers found in biology, vesicles are better suited than w/o emulsion droplets for the construction of cellular mimics. Nevertheless, the encapsulation efficiency of w/o emulsions is nearly 100 %, whereas encapsulation efficiency in vesicles is at best 30 % [29]. It is for this reason that the screening of compartmentalized reactions is carried out with w/o emulsions. Once optimal conditions are identified, similar vesicle systems are setup.

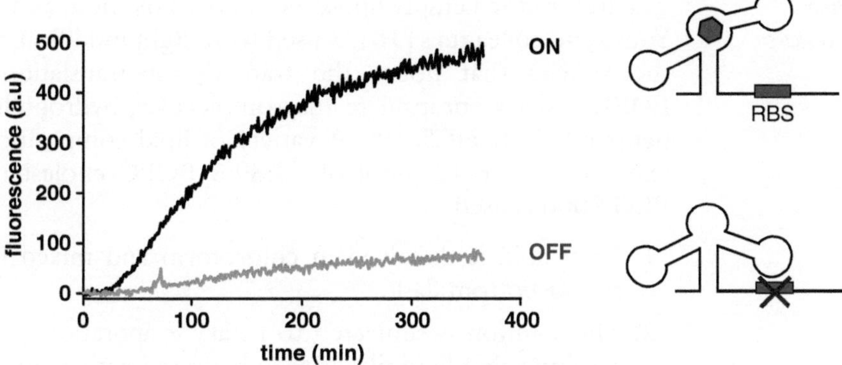

Fig. 3 In vitro theophylline riboswitch activity observed by measuring the expression of the reporter protein mYPet. The presence of the ligand activates protein expression. A schematic representation of the mRNA in the ligand-bound ON state and the uncomplexed OFF state are shown. Protein production is inhibited in this case in the OFF state because the RBS is not available for base-pairing with the ribosome. The data were taken from a previous in vitro theophylline riboswitch study [18]

3.3.1 Cell-Free Expression in w/o Emulsion

The method described here is based on that of Davidson et al. [30] but has been scaled down to be compatible with small volume PURE system reactions. Positive displacement pipettes are used to handle oil samples.

1. The oil phase that will be used for the w/o emulsion is assembled in a 15 mL Falcon tube (*see* **Note 3**), in the following order: 474.75 μL of mineral oil, 22.5 μL of span 80, 2.5 μL of tween 80, and 0.25 μL of triton X-100.
2. The aqueous phase is first assembled in a microcentrifuge tube by mixing the PURE system components on ice as described above for the in vitro reactions. Note that the optimal DNA template concentration may be different when transcription–translation is performed in a compartment versus in vitro. We used 500 ng of plasmid DNA in 25.5 μL of total aqueous volume (*see* **Note 4**) for expression in w/o emulsion droplets. The theophylline concentration was also increased to 5 mM to compensate for partitioning into the oil phase (*see* **Note 5**).
3. The 15 mL tube containing the oil phase is placed in a 250 mL beaker filled with ice water on a magnetic stir plate. A teflon stir bar is inserted in the oil phase and the oil is mixed by stirring at maximum speed for 1 min.
4. The emulsion is formed by the drop-wise addition of the aqueous phase containing the PURE system reaction to the oil phase over 1 min with continuous stirring. The emulsion is then stirred for an additional 3 min (*see* **Note 6**).
5. Finally, the emulsion is transferred to a 2 mL microcentrifuge tube and incubated at 37 °C for 6 h. 5 μL aliquots are removed every hour for observation by fluorescence microscopy.

3.3.2 Cell-Free Expression in Vesicles

The freeze-dried empty liposome (FDEL) method, as described by Yomo and colleagues [16], is used with slight modification to build the vesicles that house the transcription–translation reaction. FDEL vesicles encapsulate macromolecular, hydrophilic components relatively efficiently. A variety of lipid compositions can be exploited. Here 12 μmol of 58:39:3 POPC–cholesterol–DSPE-PEG 5000 is used.

1. Each lipid is dissolved in chloroform and mixed in a 5 mL round-bottom flask.
2. The solution is subjected to rotary evaporation for 1 h. The resulting thin lipid film is then hydrated with 1 mL of DEPC-treated water and vortexed for 20 s or until a homogeneous opaque solution is formed.
3. The vesicle solution is then transferred to a 2 mL microcentrifuge tube and disrupted with an IKA T 10 basic homogenizer at high speed (level 4 setting) for 1 min (*see* **Note 7**).
4. Samples are extruded through 400 nm polycarbonate filters 11 times with an Avanti mini-extruder. 40 μL aliquots of the vesicles are placed in 1.5 mL microcentrifuge tubes, frozen in liquid nitrogen (*see* **Note 8**), and lyophilized with a centrifugal evaporator overnight at 30 °C. A thin opaque lipid layer can be observed at the bottom of the microcentrifuge tube after lyophilization. At this stage the samples can be stored at −20 °C.
5. A PURE system reaction is assembled on ice in a total volume of 22.1 μL, including 500 ng of the template plasmid (*see* **Note 4**). 20 U of RiboLock RNase inhibitor is added to the solution.
6. 10 μL of the assembled PURE system reaction is added to an aliquot of FDEL vesicles on ice and incubated without agitation for 2.5 h (*see* **Note 9**). The unused portion of the PURE system reaction can be stored at −80 °C.
7. The hydrated liposomes are then diluted 20-fold in Tris saline buffer supplemented with proteinase K in a 0.2 mL microcentrifuge tube and incubated at 37 °C. The inclusion of proteinase K is to degrade extravesicular proteins.
8. At this point the ligand to be sensed, e.g., 5 mM theophylline (*see* **Note 5**) is added. Theophylline is capable of diffusing across the membrane, binding directly to the mRNA, and activating translation thereby resulting in fluorescence. Control reactions in the absence of ligand should result in no or significantly reduced fluorescence.
9. The reactions are incubated at 37 °C for 6 h. 5 μL aliquots are removed from the reaction every 1.5 h and visualized by fluorescence microscopy.

3.4 Microscope Sample Preparation

Detecting activity inside of vesicles by fluorescence microscopy is more difficult than in vitro measurements with a spectrofluorometer. First, encapsulation efficiency is low, particularly when over 80 different components need to be encapsulated within one vesicle in order for protein synthesis to proceed [31] (*see* **Notes 4** and **9**). Second, riboswitches typically have weak ribosome binding sites and thus produce less protein than constructs typically exploited for recombinant expression. Finally, fluorescent proteins photobleach and require time to mature. We use a monomeric version of the yellow fluorescent protein YPet as a reporter because the characteristics of YPet are more amenable to investigation by microscopy. YPet is one of the brightest fluorescent proteins and is more photostable than the majority of available fluorescent proteins [32]. The expression of monomeric YPet with the PURE system requires approximately 2 h to reach half maximal fluorescence [20].

1. 5 μL aliquots are removed from the reaction and spotted on a clean glass slide. A cover slip is added.
2. The slide is then left for 2 min to rest on the bench. This step helps decrease the number of rapidly moving vesicles.
3. The sample is then observed by bright field and epifluorescence with 100× magnification (Fig. 4). Care should be taken to decrease photobleaching by decreasing exposure time.

4 Notes

1. It is preferable to avoid Phenol–Chloroform solutions that contain ethylenediaminetetraacetic acid (EDTA), since the chelation of metals by EDTA can interfere with enzyme activity.
2. A plate reader or a real-time PCR machine can be used in place of a spectrofluorometer.
3. We also made w/o emulsions in 13 mL Sarstedt tubes with stirring with x-shaped spinplus stir bars, as described by Davison et al. [30]. The resulting emulsions were more homogeneous and more stable than the emulsions we obtained with Falcon tubes and linear stir bars. However, we found the Davidson et al. emulsion droplets to be smaller and thus more difficult to observe by microscopy.
4. The PURE system instruction manual suggests screening between 25 to 250 ng of template DNA for 25 μL reactions. For the in vitro characterization of the theophylline riboswitch, 250 ng of plasmid template was found to be optimal. However, better results were obtained with 500 ng DNA template for compartmentalized reactions.

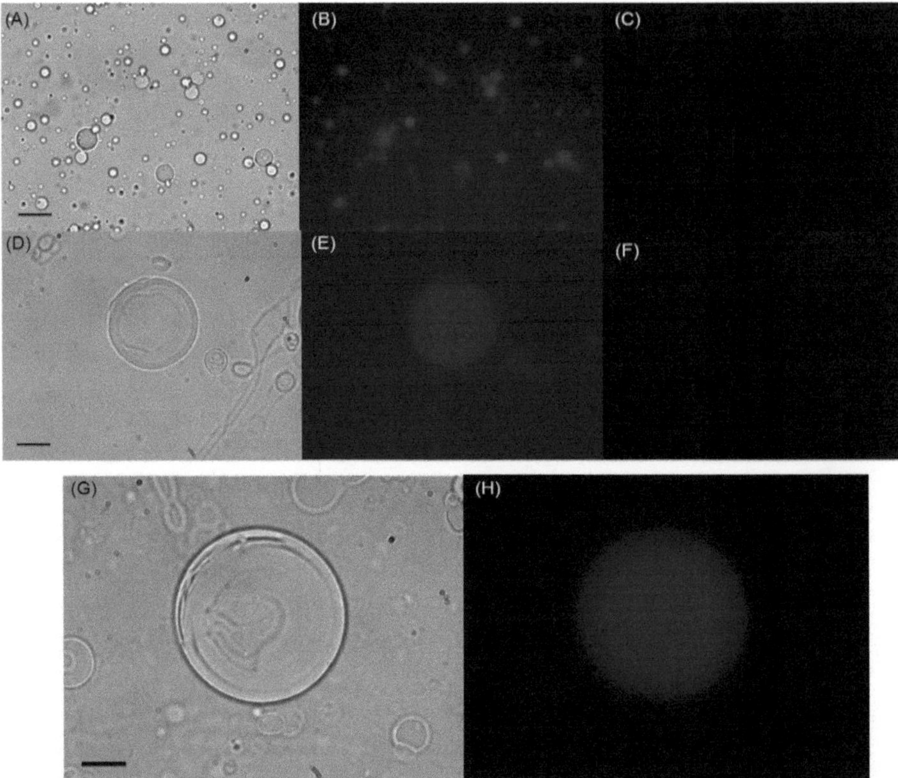

Fig. 4 Fluorescence microscopy images of riboswitch activity in compartments. Upper panels (**a**–**c**): activity recorded in emulsion compartments. Middle panels (**d**–**f**): activity recorded in liposomes. Lower panels (**g**, **h**): activity due to riboswitch sensing of the environment. Images are all epifluorescence, except for panels **a**, **d**, **g**, which are bright-field. Panel **b**, **e** show the fluorescence resulting in compartments when the ligand is present. The corresponding controls (i.e., in the absence of ligand) are also shown (**c**, **f**). In panel **h**, the activity of a riboswitch as a sensor element is shown. The data are from the characterization of a cell-free theophylline riboswitch system and panels **g** and **h** are reproduced with permission from the Royal Society of Chemistry [18]

5. Since the oil–water partition coefficient of theophylline is low [33], much more theophylline is required to activate the riboswitch in the presence of oil than in aqueous solution.
6. Stirring is an important parameter to consider when generating an emulsion. A constant stir force should be used and the stir bar must be compatible with the tube holding the aqueous-oil mixture. For example, the conical shape of a Falcon tube is not compatible with x-shaped stir bars. The efficiency of mixing can be qualitatively assed by eye by including in the aqueous phase 1 mM HPTS (8-hydroxypyrene-1,3,6-trisulfonic acid) and observing the distribution of green color throughout the tube during stirring.
7. We also tested sonication as a possible disruption method. Sonication at 70 A of amplitude for 5 min resulted in higher dispersion. However, the number of liposomes observed by

microscopy was lower than that observed with the IKA T homogenizer.

8. Freezing can also be performed with dry ice without any appreciable difference in liposomes formation.
9. The slow vesicle formation process mediated by the natural swelling method described herein results in fewer, but larger vesicles that are easier to observe by microscopy than by other methods that exploit vortexing. Additionally, compartment size impacts protein synthesis efficiency with larger vesicles being more compatible with protein synthesis [34].

References

1. Soukup GA, Breaker RR (1999) Relationship between internucleotide linkage geometry and the stability of RNA. RNA 5:1308–1325
2. Nahvi A, Sudarsan N, Ebert MS, Zou X, Brown KL, Braker RR (2002) Genetic control by a metabolite-binding mRNA. Chem Biol 9:1043
3. Wakeman CA, Winkler WC (2009) Analysis of the RNA backbone: structural analysis of riboswitches by in-line probing and selective 2′-hydroxyl acylation and primer extension. In: Serganov A (ed) Riboswitches. Methods Mol Biol, vol 540, pp 173–191
4. Ellington AD, Szostak JW (1990) *In vitro* selection of RNA molecules that bind specific ligands. Nature 346:818–822
5. Serganov A, Yuan Y-R, Pikovskaya O, Polonskaia A, Malinina L, Phan AT, Hobartner C, Micura R, Breaker RR, Patel DJ (2004) Structural basis for discriminative regulation of gene expression by adenine- and guanine-sensing mRNAs. Chem Biol 11:1729–1741
6. Serganov A, Nudler E (2013) A decade of riboswitches. Cell 152:17–24
7. Desai SK, Gallivan JP (2004) Genetic screens and selections for small molecules based on a synthetic riboswitch that activates protein translation. J Am Chem Soc 126:13247–13254
8. Nomura Y, Yokobayashi Y (2007) Reengineering a natural riboswitch by dual genetic selection. J Am Chem Soc 129:13814–13815
9. Artsimovitch I, Henkin TM (2009) In vitro approaches to analysis of transcription termination. Methods 47:37–43
10. Mironov A, Epshtein V, Nudler E (2009) Transcriptional approaches to riboswitch studies. IN: Serganov A (Ed) Riboswitches. Methods Mol Biol, vol 540: 39-51, Humana Press, NY, USA
11. Muranaka N, Yokobayashi Y (2010) Posttranscriptional signal integration of engineered riboswitches yields band-pass output. Angew Chem Int Ed 49:4653–4655
12. Topp S, Gallivan JP (2007) Guiding bacteria with small molecules and RNA. J Am Chem Soc 129:6807–6811
13. Soukup GA, Breaker RR (1999) Engineering precision RNA molecular switches. Proc Natl Acad Sci U S A 96:3584–3589
14. Wickiser JK, Winkler WC, Breaker RR, Crothers DM (2005) The speed of RNA transcription and metabolite binding kinetics operate an FMN riboswitch. Mol Cell 18:49–60
15. Tawfik DS, Griffiths AD (1998) Man-made cell-like compartments for molecular evolution. Nat Biotechnol 16:652–656
16. Sunami T, Matsuura T, Suzuki H, Yomo T (2010) Synthesis of functional proteins within liposomes. Methods Mol Biol 607:243–256
17. Lynch SA, Gallivan JP (2009) A flow cytometry-based screen for synthetic riboswitches. Nucleic Acids Res 37:184–192
18. Martini L, Mansy SS (2011) Cell-like systems with riboswitch controlled gene expression. Chem Commun 47:10734–10736
19. Lemay J-F, Desnoyers G, Blouin S, Heppell B, Bastet L, St-Pierre P, Massé E, Lafontaine DA (2011) Comparative study between transcriptionally- and translationally-acting adenine riboswitches reveals key differences in riboswitch regulatory mechanisms. PLoS Genet 7:e1001278
20. Lentini R, Forlin M, Martini L, Del Bianco C, Spencer AC, Torino D, Mansy SS (2013) Fluorescent proteins and in vitro genetic organization for cell-free synthetic biology. ACS Synth Biol 2:482–489
21. Aiba H, Hanamura A, Yamano H (1991) Transcriptional terminator is a positive regulatory element in the expression of the *Escherichia coli crp* gene. J Biol Chem 266:1721–1727
22. Sambrook J, Russel DW (2001) Commonly used techniques in molecular cloning. In: Molecular cloning. (Ed) Jan Argentine Cold Spring Harbor Laboratory Press, NY, USA

23. Sambrook J, Russell DW (2006) Standard ethanol precipitation of DNA in microcentrifuge tubes. Cold Spring Harb Protoc. doi:10.1101/pdb.prot4456
24. Shimizu Y, Inoue A, Tomari Y, Suzuki T, Yokogawa T, Nishikawa K, Ueda T (2001) Cell-free translation reconstituted with purified components. Nat Biotechnol 19:752–755
25. Zhou Y, Asahara H, Gaucher EA, Chong S (2012) Reconstitution of translation from *Thermus thermophilus* reveals a minimal set of components sufficient for protein synthesis at high temperatures and functional conservation of modern and ancient translation components. Nucleic Acids Res 40:7932–7945
26. Forlin M, Lentini R, Mansy SS (2012) Cellular imitations. Curr Opin Chem Biol 16:586–592
27. Ichihashi N, Matsuura T, Kita H, Sunami T, Suzuki H, Yomo T (2010) Constructing partial models of cells. Cold Spring Harb Perspect Biol. doi:10.1101/cshperspect.a004945
28. Bianco CD, Mansy SS (2012) Non replicating protocells. Acc Chem Res 45:2125–2130
29. Colletier J-P, Chaize B, Winterhalter M, Fournier D (2002) Protein encapsulation in liposomes: efficiency depends on interaction between protein and phospholipid bilayer. BMC Biotechnol 2:9
30. Davidson EA, Dlugosz PJ, Levy M, Ellington AD (2009) Directed evolution of proteins in vitro using compartmentalization in emulsions. Curr Protoc Mol Biol. doi:10.1002/0471142727.mb2406s87
31. Lazzerini-Ospri L, Stano P, Luisi P, Marangoni R (2012) Characterization of the emergent properties of a synthetic quasi-cellular system. BMC Bioinformat 13(Suppl 4):S9
32. Nguyen AW, Daugherty PS (2005) Evolutionary optimization of fluorescent proteins for intracellular FRET. Nat Biotechnol 23:355–360
33. Donoso P, O'Neill SC, Dilly KW, Negretti N, Eisner DA (1994) Comparison of the effects of caffeine and other methylxanthines on $[Ca^{2+}]_i$ in rat ventricular myocytes. Br J Pharmacol 111:455–458
34. Pereira de Souza T, Stano P, Luisi PL (2009) The minimal size of liposome-based model cells brings about a remarkably enhanced entrapment and protein synthesis. Chembiochem 10:1056–1063

Chapter 12

Rational Design of Artificial ON-Riboswitches

Atsushi Ogawa

Abstract

Riboswitches are composed of two regions: one for binding to the ligand (the aptamer domain) and the other for regulating the expression of the gene (the expression platform). In most riboswitches (both natural and artificial), a part of the aptamer domain required for ligand binding is directly involved in the regulation of expression, so that it is difficult to design other ligand-responsive riboswitches based on these riboswitches even by using artificial aptamers obtained through in vitro selection. This chapter describes a method for rationally constructing a foundational ON-riboswitch, which is easily available for the design of other ligand-dependent riboswitches, by introducing a new region (a modulator sequence: MS) in addition to the two basic regions. A facile method for preparing arbitrary molecule-dependent riboswitches based on the foundational riboswitch is also presented.

Key words Riboswitch, IRES, Aptamer, Gene regulation, Biosensor, Cell-free translation

1 Introduction

Riboswitches are ligand-dependent and cis-regulatory RNAs, discovered in nature in 2002 [1, 2]. Many types of natural riboswitches have been identified in the untranslated regions (UTRs) of bacterial mRNAs, and some are present in plants and fungi [3–5]. The riboswitch consists of two regions: "an aptamer domain" which specifically binds to its ligand, and "an expression platform" whose conformational change induced by aptamer–ligand complex formation directly switches ON or OFF transcriptional termination or translational initiation (in some cases, mRNA cleavage or splicing). There is however no definite borderline between these two regions because the aptamer domain generally includes a part of the expression platform: a part of a transcriptional terminator (e.g., a stem-loop and poly U tail) and/or the complementary sequence thereof in transcription-regulating riboswitches; an anti-sequence to a translational initiator (TI: the ribosome binding site (RBS) and start codon in the case of bacteria) and/or the complementary sequence thereof in translation-regulating ones [3–5].

Atsushi Ogawa (ed.), *Artificial Riboswitches: Methods and Protocols*, Methods in Molecular Biology, vol. 1111,
DOI 10.1007/978-1-62703-755-6_12, © Springer Science+Business Media New York 2014

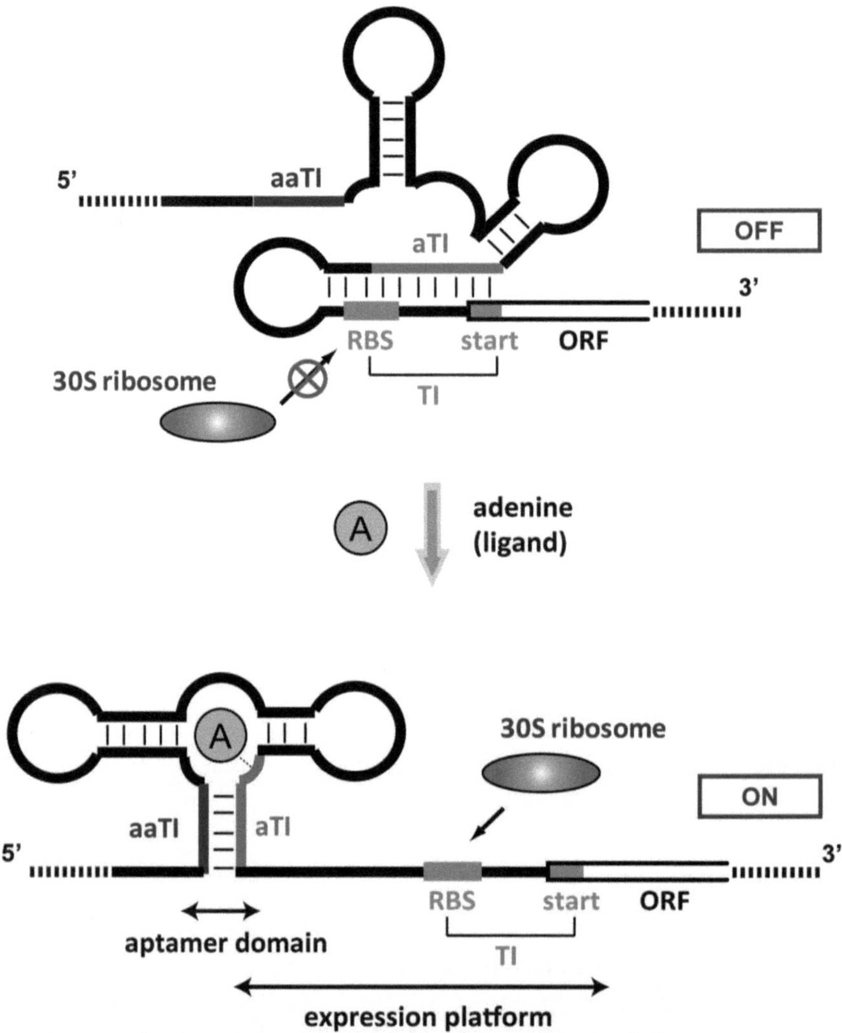

Fig. 1 An example of natural riboswitches: add adenine ON-riboswitch in *Vibrio vulnificus*. In the absence of the ligand (adenine), the aTI hybridizes to the TI (the RBS and the start codon) to prevent the bacterial 30S ribosome from binding to the RBS (OFF state, above). In the presence of the ligand, it leaves the TI to bind to the aaTI and the ligand (*dashed thin line*), so that the ribosome can access the RBS to start translation (ON state, *below*). *Bold* and *thin lines* represent mRNA and base-pair interactions, respectively

Although this part of the aptamer domain is involved in expression regulation in the absence of the ligand to promote or inhibit the expression, it is required for ligand binding, so that aptamer–ligand complex formation makes it leave the expression platform core to inversely inhibit or promote the expression, respectively (in the case of OFF-riboswitches or ON-riboswitches, respectively). Taking a translation-regulating bacterial add ON-riboswitch as an example (Fig. 1) [6], the anti-TI sequence (aTI) corresponding to the part described above hybridizes to the TI to form an aTI–TI duplex in the absence of the ligand (OFF state), while it releases the TI and then binds to another partially complementary sequence (anti-aTI: aaTI) in the presence of the ligand (ON state), because aTI–aaTI duplex formation is required for ligand–aptamer complex formation, and also because a part of aTI that does not hybridize to the aaTI directly contributes to ligand binding. Briefly, the ligand biases the equilibrium between these two states toward the ON state to upregulate gene expression.

Although the ligands of natural riboswitches are metabolites (thiamine pyrophosphate, s-adenosyl methionine, purine, etc.) related to the regulated gene, an arbitrary ligand-dependent riboswitch can be artificially constructed by using an in vitro-selected aptamer (*see* **Note 1**). However, artificial riboswitches cannot be obtained only by converting an aptamer in a natural riboswitch into an in vitro-selected one, because a part of the natural aptamer required for ligand binding plays a role as a part of the expression platform in the absence of the ligand, as described above. The most powerful methods for obtaining artificial riboswitches are in vivo screening methods, with which some groups have actually obtained bacterial artificial riboswitches in response to ligands that bacteria do not originally have (*see* **Note 2**). In typical screening methods, an in vitro-selected aptamer and some randomized nucleotides are inserted before the TI, and then only mRNAs with the riboswitch activity are selected based on their expression efficiencies with and without the ligand. Thus, these methods allow us to obtain efficient translation-regulating riboswitches. Nonetheless, the artificial riboswitches selected through these methods generally have the same feature as the natural riboswitches: a part of the region used for ligand binding is also used for directly regulating the expression in the absence of the ligand. Therefore, these artificial riboswitches, just like natural ones, are not available as foundations for rationally constructing other ligand-dependent riboswitches.

On the other hand, I have recently established a method for rationally designing translation-regulating riboswitches [7, 8]. In the design strategy in this method, a "variable" modulator sequence (MS) is used, separately from the aptamer domain and the expression platform, for regulating aTI–aaTI duplex formation in response to the ligand. Thanks to this new element, aTI and aaTI can be designed so as not be involved directly with ligand binding, meaning that they can be fixed for any aptamer to be shared with other ligand-dependent riboswitches. Thus, although optimization of these sequences is required for obtaining the first riboswitch, it is not necessary for the designs of subsequent riboswitches. Even the design of the MS variable for each aptamer sequence needs only a simple calculation with a computer. In other words, based on the first riboswitch, we can easily design arbitrary artificial riboswitches with only sequence information of the corresponding in vitro-selected aptamer.

For the optimization of the first riboswitch (i.e., the foundational riboswitch), cell-free translation systems are very useful because we can easily synthesize many types of mRNAs in vitro, whereas much time and work is required to prepare various transformants, especially in eukaryotic cells, wherein we also need to consider the maturation of the mRNAs (*see* **Note 3**). Although there are some differences in translation conditions between in vitro and in vivo, artificial riboswitches optimized in vitro are also expected to work well in vivo, with minor corrections. In addition, artificial riboswitches are very useful even in vitro: for example, they can be used as label-free biosensors and logic gate elements [7, 8].

This chapter describes, as an example of how to use the method for constructing artificial riboswitches, the rational design of eukaryotic ON-riboswitches modulating internal ribosome entry site (IRES)-mediated translation (i.e., TI = IRES) in wheat germ extract (eukaryotic cell-free translation system) [9] (*see* **Note 4**).

2 Materials

2.1 General

1. Nuclease-free water (*see* **Note 5**).
2. Various sizes of nuclease-free microtubes and filter tips.

2.2 Preparation of DNA Templates

2.2.1 Amplification of DNA Templates

1. DNA template with the IRES sequence chosen as described in Subheading 3.1.1 (*see* **Note 6**).
2. Plasmid encoding the luciferase gene (*see* **Note 7**).
3. PrimeSTAR MAX DNA polymerase including dNTPs (Takara Bio, Shiga, Japan) or other high-fidelity DNA polymerase with dNTPs.
4. Primers for the upstream part: various forward primers including a variety of regulator sequences (*see* **Note 8**) and a reverse primer with a unique restriction enzyme site (RES).
5. Primers for the downstream part: a forward primer with the same RES as in the reverse primer for the upstream part and a reverse primer (*see* **Note 9**).
6. Forward primer including the T7 promoter (5′-TAATACGA CTCACTATA-3′) for the final PCR.

2.2.2 Digestion and Ligation

1. The restriction enzyme whose recognition site is included in the primers described above (*see* **Note 10**).
2. Ligation high Ver.2 (TOYOBO, Osaka, Japan) or other ligation kit.

2.2.3 Agarose Gel Purification

1. 1.0 % agarose gel containing 1 mg/L ethidium bromide (*see* **Note 11**).
2. Running buffer: 1× Tris/acetate/EDTA (TAE) prepared from 50× solution.

3. 1-kb DNA ladder marker.
4. Gel loading dye for agarose gel electrophoresis.
5. Disposable knives for cutting the gel.
6. illustra GFX PCR DNA and Gel Band Purification Kit (GE Healthcare) or a similar spin column kit for DNA purification.

2.3 In Vitro Transcription

1. MEGAscript T7 Kit (Invitrogen) (*see* **Note 12**).
2. RNeasy MinElute Cleanup Kit (QIAGEN) (*see* **Note 13**).

2.4 In Vitro Translation

1. WEPRO1240 Expression Kit (CellFree Sciences, Yokohama, Japan) (*see* **Note 14**). The following preparation should be performed beforehand by using materials attached to the kit: Add 125 μL each of S1, S2, S3, and S4 one by one to 500 μL of nuclease-free water to prepare 1 mL of 5× SUB-AMIX, and aliquot it into 30-μL aliquots. Dilute 20 μL of 20 mg/mL creatine kinase with 380 μL of nuclease-free water to 1 mg/mL, and aliquot it into 6-μL aliquots. Aliquot 1 mL of WEPRO (wheat germ extract) into 30-μL aliquots. Store them at less than −80 °C until use.
2. 12-strip PCR tubes.
3. Theophylline and other ligands.

2.5 Luciferase Assay

1. Luciferase Assay System (Promega). It is better to aliquot 10 mL of a Luciferase Assay Reagent (prepared with the system) into 1-mL aliquots and store them at less than −80 °C beforehand.
2. 96-well microplates (black).
3. 12-channel pipette.

2.6 Calculation of Delta G Values

1. RNAstructure software (http://rna.urmc.rochester.edu/RNAstructure.html).

2.7 Equipment

1. Thermal cycler for PCR, digestion, ligation, transcription, and translation.
2. Agarose gel electrophoresis apparatus.
3. UV transilluminator for excitation of ethidium bromide bound to DNA.
4. Incubator for dissolving of the gel slices at 50–60 °C.
5. Deep freezer.
6. UV spectrometer.
7. Microplate luminometer.
8. Centrifuge.

3 Methods

This methods section is divided into two subsections: one is for the design of riboswitches, and the other is for the actual experiments, such as the construction of DNA templates, transcription, translation, and reporter protein assays. Experimental procedures are common in any design step, and thus see the experimental subsection (Subheading 3.2) for preparing mRNAs and measuring their riboswitch activities in each design step.

3.1 Design Methods

3.1.1 Choice of an IRES and a Reporter Gene Downstream of the IRES

An IRES that has been much studied (especially regarding its secondary and tertiary structure) and that works with higher translation efficiency is a better choice than a less-studied IRES. One of the best candidates is the intergenic IRES of the *Plautia stali* intestine virus (PSIV), which meets these conditions and functions in many types of eukaryotic translation systems [10–12] (*see* **Note 15**). Other IRESs could be used depending on the user's circumstances. For a reporter gene, the firefly luciferase gene is recommended due to its low background, low detection limit, and linear response over a wide range (*see* **Note 16**).

3.1.2 General Design of mRNAs

The following parts are required for mRNAs (in this order): a stable stem-loop (5′-SL), a mimic start codon (*optional*), regulators (*variable*), the IRES (*see* **Note 17**), the reporter gene (*see* **Note 18**), and a 3′ untranslated region (UTR) (Fig. 2). The stable stem-loop structure in the 5′-terminus is required for inhibiting 5′ terminus-mediated translation of the reporter gene [13] (*see* **Note 19**). An 8-base stem with an ACAAC pentaloop (e.g., 5′-GGGAGAC CACAACGGUUUCCC-3′) is sufficient for inhibition, but it should start from 5′ G for efficient transcription by T7 RNA polymerase. A mimic gene in the 5′ UTR is also usable for the inhibition, because the eukaryotic ribosome tends to start translation at the start codon that it encounters first (*see* **Note 20**). Regulators are a set of several sequences involved in regulation of gene expression: a modulator sequence (MS), an anti-anti-IRES (aaIRES), an aptamer, and an anti-IRES (aIRES), the details of which are described below. More than 250 nucleotides (nt) of the 3′ UTR are needed to suppress mRNA degradation from the 3′ end during translation.

3.1.3 Optimization of the IRES Length

The first step in rationally designing artificial riboswitches is to optimize the IRES length toward higher translation efficiency by using mRNAs without regulators. This is an optional but important step because the translation efficiency of riboswitches in the ON state should be less than that of their original mRNA. If the sequence has been already optimized for translation, this step can be skipped. Otherwise, it is recommended that some regulator-free

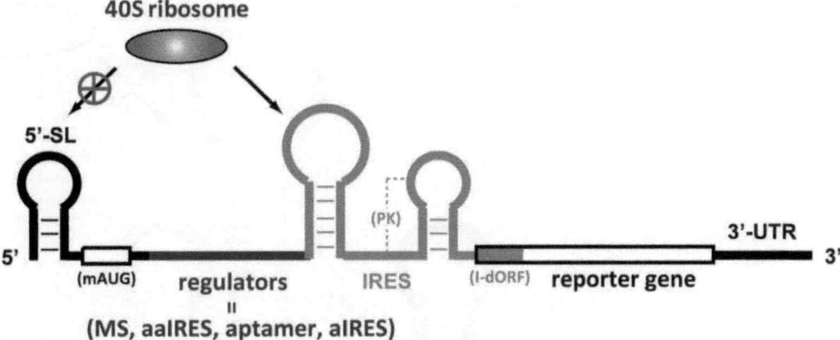

Fig. 2 General design of mRNAs. *SL* stem-loop, *mAUG* mimic start codon, *UTR* untranslated region, *PK* pseudoknot, *I-dORF* a 5′ part of the original downstream ORF of the chosen IRES. The existence or nonexistence of parts in *parentheses* depends on the circumstances of the experiment. The eukaryotic 40S ribosome preferably binds to the IRES (not to the 5′ terminus) due to the 5′-SL

mRNAs with various lengths of the IRES (including or excluding a terminal part of the original upstream and/or downstream ORF: I-uORF and/or I-dORF, respectively) be prepared, and these mRNAs should be translated to determine the optimal IRES length. It is also advisable to prepare mRNAs without IRES and/or 5′-SL to verify whether the 5′-SL actually inhibits 5′ terminus-mediated translation (*see* **Note 21**).

3.1.4 Inhibition of IRES-Mediated Translation by an aIRES

The next step is to inhibit IRES-mediated translation by inserting an aIRES, an anti-sequence complementary to a part of the IRES, into the regulator region (Fig. 3a). This is a very crucial step, because lower false expression of protein in the OFF state leads to higher switching efficiencies. The procedure for determining an effective aIRES is as follows:

1. Choose some parts of the IRES as aIRES targets. The best candidates for aIRES targets are important sequences for the IRES function or structure such as stem-loops and pseudoknots. If the structure of the chosen IRES is unknown, pick up as many aIRES targets as possible. Regarding the aIRES length (equal to the target length), 8 nt is empirically long enough to hybridize to its target when they are sterically close in the antiparallel direction.
2. Prepare mRNAs with various aIRES sequences and check their translation efficiencies to determine the most effective aIRES (*see* **Note 22**). If the translation efficiency of the champion is much higher than that of the IRES-free mRNA (*see* Subheading 3.1.3), try other aIRES targets or other IRES.
3. Prepare mRNAs with various lengths of the best aIRES to investigate the minimum length for efficiently inhibiting IRES-mediated translation as if there are no IRES and the

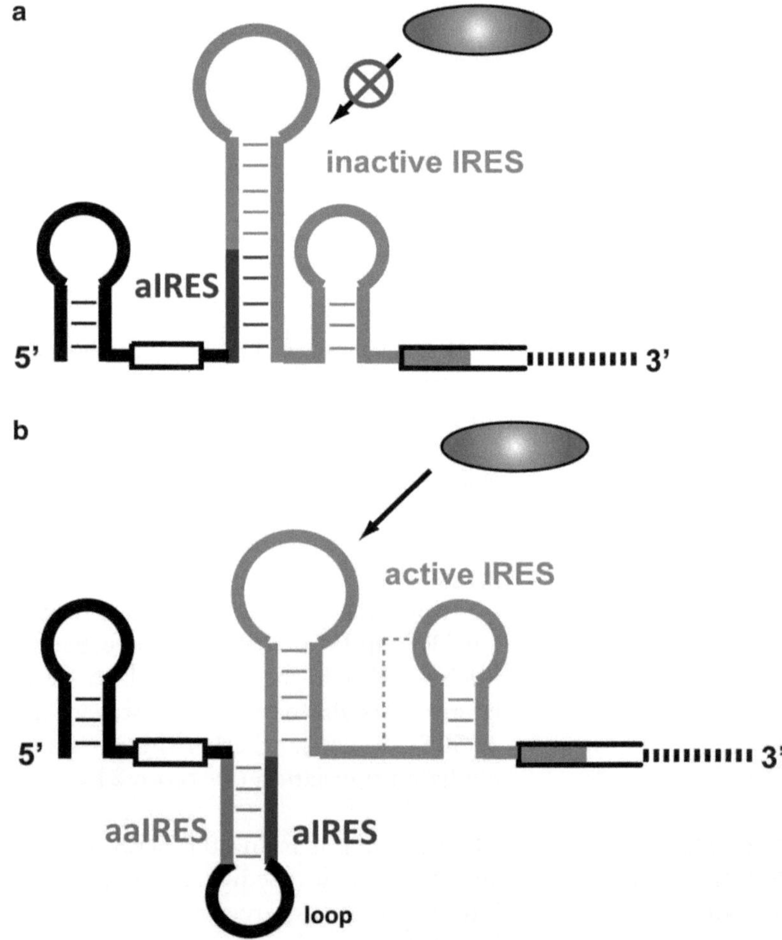

Fig. 3 Regulation of IRES-mediated translation by antisense sequences. The aIRES disrupts the IRES conformation by hybridizing to a part of the IRES (**a**). The aaIRES binds to the aIRES to restore the disrupted conformation to the active state (**b**)

maximum length for hardly affecting IRES-mediated translation (*see* **Note 23**). The latter information is required for designing riboswitches later.

3.1.5 Restoration of IRES-Mediated Translation by an aaIRES

To rationally construct ON-riboswitches, a part for restoring IRES-mediated translation inhibited by the aIRES is required independently from an aptamer (Fig. 3b). It is much easier to identify an effective part: an aaIRES, an anti-sequence complementary to aIRES. Nonetheless, it is recommended to confirm whether IRES-mediated translation is restored by actually inserting the aaIRES and some extra bases as a loop (e.g., 5′-ACCAC-3′) before the aIRES. The same length of the aaIRES as that of the aIRES should sufficiently restore translation (*see* **Notes 24** and **25**).

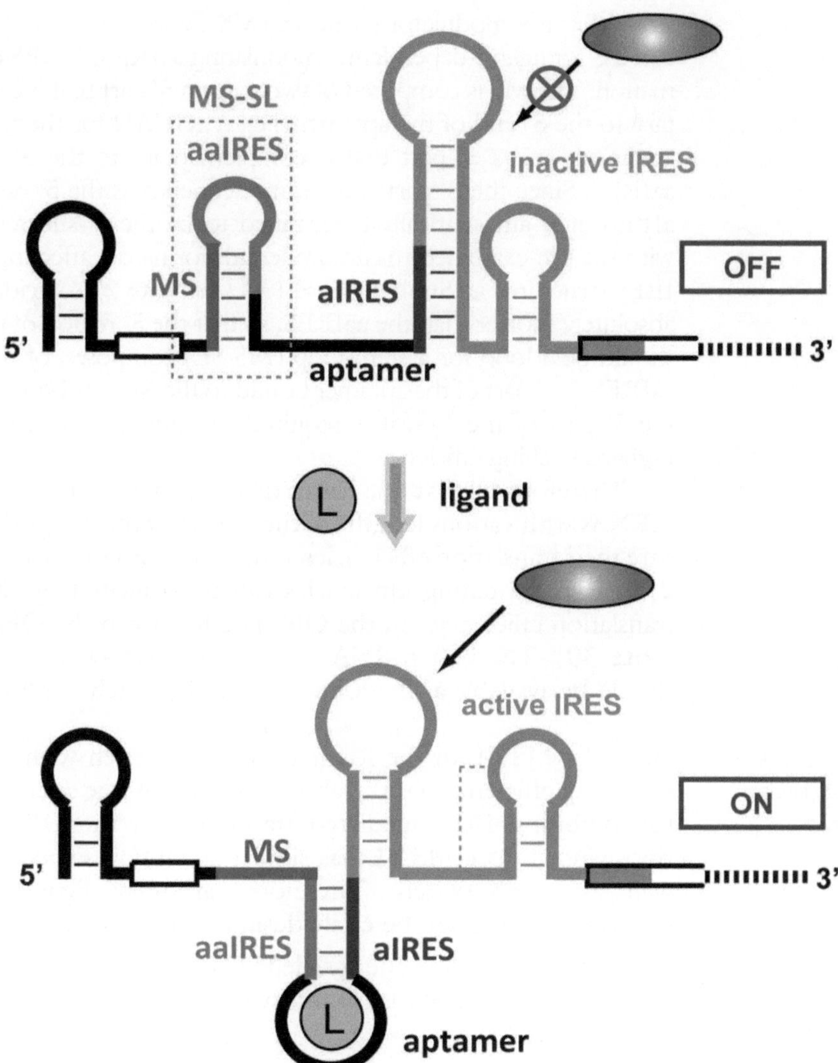

Fig. 4 Design of IRES-based ON-riboswitches. In the absence of the ligand, since the MS hybridizes to the 3' end of the aaIRES (and the 5' end of the aptamer), the aIRES binds to the IRES to inhibit IRES-mediated translation (OFF state, *above*). In the presence of the ligand, it releases the aptamer (due to aptamer–ligand complex formation) and then leaves the aaIRES (due to destabilization of the MS-SL), so that the aIRES binds not to the IRES but to the aaIRES to promote IRES-mediated translation (ON state, *below*)

3.1.6 Design and Optimization of the Foundational IRES-Based ON-Riboswitch

To construct the foundational IRES-based ON-riboswitch, which is necessary and sufficient for designing various ligand–riboswitch pairs, another two regulator parts in addition to the aIRES and the aaIRES are required (Fig. 4). One is of course an aptamer, which is inserted between the aaIRES and the aIRES instead of the intervening loop described above. A theophylline aptamer (5′-AUACCAGCCGAAAGGCCCUUGGCAG-3′) is recommended for the foundational riboswitch because it is well minimized and strongly binds to the ligand (theophylline, K_d=0.1 μM) (*see* **Note 26**) [14].

The other is a modulator sequence (MS), which is inserted before the aaIRES for ligand-dependently modulating aaIRES/aIRES duplex formation. The MS is composed of two parts: a 5′ part that is complementary to the 5′ end of the aptamer (i.e., ...GUAU for the theophylline aptamer), and a 3′ part that is complementary to the 3′ end of the aaIRES. Since the 3′ part is the same sequence as the 5′ region of the aIRES, it is automatically determined to be the 3'-shortened aIRES with the pre-examined maximum length for hardly affecting the active IRES structure (*see* Subheading 3.1.4) (*see* **Note 27**). Incidentally, it is absolutely shorter than the aaIRES, so that the 5′ region of the aaIRES should be a loop for a stem-loop (MS-SL) composed of the MS, the aaIRES, and part of the aptamer bound to the MS (*see* **Note 28**). As for the 5′ part of the MS, it is required to optimize the length toward higher switching efficiency.

Therefore, all that one has to do in actual experiments is prepare mRNAs with various lengths of the 5′ part of the MS and then measure their translation efficiencies without or with the ligand (*see* **Note 29**). Their switching efficiencies can be evaluated by the ratio of translation efficiencies in the ON state to that in the OFF state (*see* **Note 30**). The best mRNA with the highest switching efficiency should be available as the foundational riboswitch (*see* **Note 31**).

3.1.7 Rational Design of Other Ligand-Dependent ON-Riboswitches

Since the MS-SL in the foundational riboswitch with the highest switching efficiency has the almost-minimum free energy necessary for inhibiting IRES-mediated translation in the OFF state, any riboswitch whose MS-SL has similar stability is expected to work well as a riboswitch. Therefore, arbitrary ligand-dependent ON-riboswitches can be easily designed as follows:

1. Calculate the free energy (delta G value) of the cropped MS-SL part in the foundational riboswitch by using the RNAstructure software [15].
2. On a computer, replace the aptamer and the 5′ part of the MS in the foundational riboswitch with an aptamer for the ligand of interest and various lengths of the sequence complementary to the 5′ end of the chosen aptamer, respectively (*see* **Note 26**).
3. Calculate free energies of their MS-SL part and determine which has the nearest value to the free energy of the MS-SL in the foundational riboswitch (*see* **Note 32**).
4. Prepare actual mRNA with the most promising MS-SL (and mRNAs with the second and third most promising MS-SL if necessary), and confirm its switching efficiency in the presence or absence of the ligand (*see* **Notes 29** and **31**).

3.2 Experimental Methods

3.2.1 Construction of DNA Templates for mRNA

DNA templates are constructed by preparing the upstream part including the IRES and the downstream part including the luciferase gene separately and then joining them together (Fig. 5) (*see* **Note 33**).

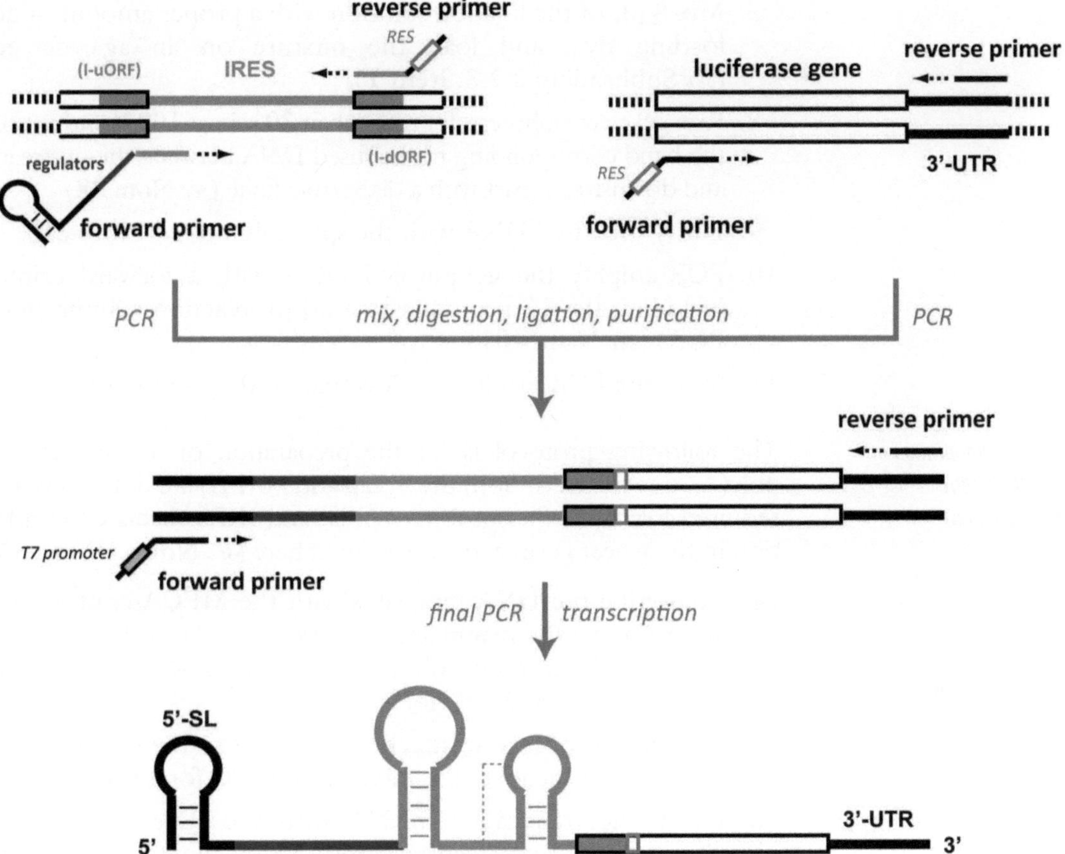

Fig. 5 General procedures of the construction of DNA templates and their transcripts (mRNA)

1. PCR-amplify the upstream part including a 5′-SL, a mimic start codon (optional), regulators, the IRES, and a unique restriction enzyme site (RES) in a 20-μL reaction volume (*see* **Notes 8** and **34**).
2. PCR-amplify the downstream part including the RES, the luciferase gene (without the original start codon), and a 3′ UTR (*see* **Note 35**) in a 20-μL reaction volume (*see* **Note 34**).
3. Mix both parts and purify the mixture with a commercially available spin column kit (*see* **Notes 36** and **37**).
4. Digest the purified DNAs with the restriction enzyme in a 20-μL reaction volume (*see* **Note 34**).
5. Purify the digested DNAs with the spin column (*see* **Note 36**) or denaturation of the restriction enzyme with incubation at a slightly high temperature (*see* **Note 34**).
6. Incubate a mixture of 20 μL of the resultant product and 10 μL of Ligation high Ver.2 at 16 °C for 30 min.

7. Mix 8 µL of the ligation solution with a proper amount of gel loading dye, and load the mixture on an agarose gel (*see* Subheading 2.2.3, **item 1**).
8. Run gel electrophoresis for more than 30 min at 100 V and cut off the band corresponding to the fused DNA between the upstream and downstream part with a disposable knife (*see* **Note 38**)
9. Purify the fused DNA with the spin column (*see* **Note 39**).
10. PCR-amplify the gel-purified DNA with a forward primer including the T7 promoter in a 20-µL reaction volume (final PCR) (*see* **Note 40**).
11. Store the PCR products at less than −20 °C until use.

3.2.2 Preparation of mRNA (In Vitro Transcription)

The following protocol is for the preparation of 5′ cap- and 3′ poly(A)-free mRNAs. Both the 5′ cap and poyl(A) are not necessarily required for both efficient 5′ terminus- and IRES-mediated translation in the wheat germ extract described here (*see* **Notes 19** and **35**)

1. Transcribe the DNA templates with the MEGAscript T7 Kit according to the manufacturer's protocol. Typically, incubate 10 µL of the reaction solution containing 2–4 µL of the final PCR products at 37 °C for 2–4 h.
2. Add 0.5 µL of DNase attached to the kit to the solution, and then further incubate the mixture at 37 °C for 15 min.
3. Purify the transcribed mRNA with the RNeasy MinElute Cleanup Kit. To concentrate mRNA sufficiently, use 14 µL of nuclease-free water to elute mRNA in the final step. Typically, approx. 50 µg of mRNA can be obtained.
4. Quantify the purified mRNA with a UV spectrometer and adjust the concentration to 3 pmol/µL with nuclease-free water.
5. Aliquot the mRNA into smaller volumes (e.g., 3.4 µL for three times' translation) and store them at less than −80 °C until use.

3.2.3 Cell-Free Translation

As many translation reactions as possible should be performed simultaneously to compare many types of mRNAs with the regulator-free optimal mRNA (*see* Subheading 3.1.3), to measure error bars, and to effectively use the materials in the cell-free translation kit. The following protocol is for 12 reactions. Note that the protocol is different from that of the manufacturer.

1. Thaw 30 µL of 5× SUB-AMIX, 6 µL of 1 mg/mL creatine kinase, and 30 µL of WEPRO prepared as described in Subheading 2.4 and mRNA(s) prepared as described in Subheading 3.2.2 on ice.
2. Put 1 µL of mRNA on the inner side surface of each PCR tube (in a total of 12 tubes).

3. Put 1 μL of a ligand or nuclease-free water on a site next to the mRNA spot so as not to mix them.
4. Mix 52.2 μL of nuclease-free water, 29 μL of WEPRO, 29 μL of 5× SUB-AMIX, and 5.8 μL of 1 mg/mL creatine kinase.
5. Put 8 μL of the mixture on the site next to the ligand (or water) spot so as not to mix them.
6. Centrifuge (lightly) the tubes simultaneously and mix them thoroughly, and then centrifuge (lightly) them again.
7. Incubate the tubes at 26 °C for 1 h and then at 4 °C to stop the translation.

3.2.4 Luciferase Assay

To use the Luciferase Assay Reagent prepared as described in Subheading 2.5 immediately after translation, start to thaw the aliquot on ice while protecting it from light right after starting the translation in Subheading 3.2.3, **step** 7.

1. Add 100 μL of water to 10 μL of each translation solution to dilute them 11-fold.
2. Put 5 μL of each diluted solution in a row (12 wells) of a 96-well microplate.
3. Aliquot 1 mL of the Luciferase Assay Reagent into 12 aliquots (~80 μL) in new PCR tubes.
4. Add 75 μL of the reagent to each diluted translation solution in the plate by using a 12-channel pipette, and mix them well by pipetting three times.
5. Immediately after mixing, measure the luminescence values of the mixtures with a luminometer.

4 Notes

1. See previous chapters for descriptions of the in vitro selection of aptamers.
2. See previous chapters for descriptions of the in vivo screening of artificial riboswitches.
3. Nonetheless, the design strategy described here (i.e., optimization of the foundational riboswitch and construction of other riboswitches in reference to the foundational riboswitch) should also be available in vivo.
4. An IRES functions as a receiver of the ribosome in eukaryotes as well as an RBS in bacteria, so that bacterial riboswitches can be constructed by reading "IRES" (precisely, a part targeted by the aIRES) as "RBS (and the start codon)". For the rational design of OFF-riboswitches, refer also to my previous report [8].

5. Since a slight amount of contaminated nucleases and proteases can degrade nucleic acids and proteins, respectively, nuclease-free water should be used in all reactions (PCR, digestion, ligation, transcription, and translation). In addition, the experimenter should wear clean gloves and a mask to prevent nuclease and protease contamination.
6. Commercial gene synthesis services are available for obtaining the template.
7. Some plasmids encoding the luciferase gene are obtainable from Promega.
8. If the forward primer is too long for PCR (more than 100 nt), split it and perform subsequent PCRs.
9. This reverse primer is also used as the reverse primer in the final PCR.
10. Various types of restriction enzymes are available for purchase from New England Biolabs or Takara Bio.
11. Care should be taken when handling ethidium bromide, which is thought to act as a mutagen.
12. A similar transcription kit or a hand-made transcription reaction solution can be used instead.
13. Although mRNA can be also purified by phenol-chloroform extraction and ethanol (or isopropanol) precipitation at low cost, the use of the purification kit should give mRNAs with higher purities.
14. For a eukaryotic cell-free translation system, the WEPRO1240 Expression Kit is recommended because the translation system is very stable, like in vivo, and because the 5′ capping of mRNAs is not necessary for efficient 5′ terminus-mediated translation, although another eukaryotic translation system or hand-made extract can be used instead.
15. The foundational riboswitch sequence for PSIV-IRES is obtainable from my previous report [7].
16. Other reporter genes such as GFP and beta-galactosidase may be available, but the assay methods for these proteins are described elsewhere.
17. The IRES should include an I-dORF (a 5′ part of the original downstream open reading frame), which generally increases the IRES-mediated translation efficiency (*see* Subheading 3.1.3).
18. The start codon of the reporter gene can be removed because the I-dORF is connected to the reporter gene as a part of the ORF.
19. Even 5′ cap-free mRNA can be translated via 5′ terminus-mediated translation in the wheat germ extract presented here, depending on the 5′ terminal sequence of mRNA [13].

20. It should be noted that the ribosome tends to skip the start codon that is too near to the 5′ terminus. To avoid this problem, a mimic start codon can also be inserted between regulators or just before the IRES. The chosen IRES may originally have a start codon(s) available as the mimic codon(s). Incidentally, the PSIV-IRES has two mimic genes in itself. Moreover, IRES-mediated translation via the PSIV-IRES can start at a non-AUG codon, so that 5′ terminus-mediated translation of the reporter gene should be completely suppressed even if the ribosome skips the mimic start codons or remains on mRNA after translation of the mimic genes. In this case, the start codon of reporter gene must absolutely be deleted (*see* **Note 18**).
21. If the inhibition effect is inadequate, strengthen the stem of the 5′-SL by extending the length, or insert a mimic start codon(s).
22. In the case of the PSIV-IRES, an aIRES to a 5′ side of the pseudoknot III (PK-III) is the best [7].
23. In the latter case, aIRESs with the 3′ side shortened (or preferably replaced with the same length of nucleotides uninvolved with the aIRES/IRES duplex) should be investigated. Incidentally, for the PSIV-IRES, the former and the latter are 8 nt and 4 nt, respectively [7].
24. If the restoration efficiency is much lower, longer aaIRES may work well, but a too-long aaIRES would lead to lower ON/OFF switching efficiencies.
25. In terms of the PSIV-IRES, an 8-mer aaIRES for the complementary 8-mer aIRES shows an 80 % recovery of IRES-mediated translation.
26. If an aptamer other than the theophylline aptamer is used, it must be minimized beforehand, because it is required that the ligand–aptamer interaction induces a release of the MS from the aptamer.
27. Longer ones inhibit IRES-mediated translation even when the aaIRES releases the MS and hybridizes to the aIRES in the ON state, whereas shorter ones cannot prevent aaIRES/aIRES duplex formation even in the OFF state.
28. If this loop length is very short (less than 3 or 4 nt), some bases can be inserted between the MS and the aaIRES to stabilize the stem-loop structure.
29. Use the maximum concentration of the ligand that does not affect IRES-mediated translation, which should be investigated beforehand with the regulator-free optimal mRNA (*see* Subheading 3.1.3). To examine the ligand dependency, use various concentrations of the ligand.

30. Generally, the optimal length of the 5′ part of MS is the maximum length for satisfactorily inhibiting IRES-mediated translation in the OFF state.

31. If the highest switching efficiency is lower than 5 at the maximum concentration of the ligand (*see* **Note 29**), it is likely that the aptamer sequence inhibits IRES-mediated translation or the ligand–aptamer interaction is too weak at the maximum ligand concentration. In this case, it is recommended that the ligand–aptamer pair or the IRES be changed. Incidentally, the switching efficiency of the foundational PSIV-IRES-based riboswitch is approx. 10 at 1 mM theophylline [7].

32. To precisely adjust the free energy, it might be possible to include a G–U wobble pair in the MS–aptamer duplex.

33. A "unique" restriction enzyme site (RES) is required for joining these two parts. Since it remains in the fused ORF, it must be inserted in order to keep the translational frame.

34. Refer to the manufacturers' protocol for reaction conditions.

35. Although the 3′ UTR that the luciferase-encoding plasmid originally has could be used as-is, the length should be more than 250 nt for efficient translation.

36. Elute the PCR products with 20 μL of nuclease-free water.

37. In cases in which the luciferase-encoding plasmid has the T7 promoter, DpnI treatment or agarose gel purification is needed to remove the plasmid. Otherwise, the plasmid sequence may be amplified in the final PCR.

38. In cases in which the restriction enzyme recognizes a palindromic sequence, five bands should be observed: two reactants (the upstream part and the downstream part) and three ligated products between two upstream parts, two downstream parts, or each part. To cut off only the band corresponding to the ligated product between the upstream and downstream part without contamination of others, separate these bands completely with long electrophoresis. If this operation is difficult, use restriction enzymes that recognize asymmetric sequences.

39. Elute the ligated product with 50 μL of nuclease-free water and then dilute it 10- to 100-fold with nuclease-free water, to be used as the template in the final PCR.

40. Make sure that only a single clear band is satisfactorily amplified on an agarose gel. Incidentally, purification after the final PCR is not necessary.

Acknowledgments

This work was supported by 'Special Coordination Funds for Promoting Science and Technology' and 'Grant-in-Aid for Scientific Research on Innovative Areas' from the Ministry of Education, Culture, Sports, Science and Technology, Japan.

References

1. Winkler W, Nahvi A, Breaker RR (2002) Thiamine derivatives bind messenger RNAs directly to regulate bacterial gene expression. Nature 419:952–956
2. Mironov AS, Gusarov I, Rafikov R et al (2002) Sensing small molecules by nascent RNA: a mechanism to control transcription in bacteria. Cell 111:747–756
3. Blount KF, Breaker RR (2006) Riboswitches as antibacterial drug targets. Nat Biotechnol 24:1558–1564
4. Nudler E, Mironov AS (2004) The riboswitch control of bacterial metabolism. Trends Biochem Sci 29:11–17
5. Schwalbe H, Buck J, Fürtig B et al (2007) Structures of RNA switches: insight into molecular recognition and tertiary structure. Angew Chem Int Ed 46:1212–1219
6. Serganov A, Yuan Y-R, Pikovskaya O et al (2004) Structural basis for discriminative regulation of gene expression by adenine- and guanine-sensing mRNAs. Chem Biol 11:1729–1741
7. Ogawa A (2011) Rational design of artificial riboswitches based on ligand-dependent modulation of internal ribosome entry in wheat germ extract and their applications as label-free biosensors. RNA 17:478–488
8. Ogawa A (2011) Rational construction of eukaryotic OFF-riboswitches that downregulate internal ribosome entry site-mediated translation in response to their ligands. Bioorg Med Chem Lett 22:1639–1642
9. Madin K, Sawasaki T, Ogasawara T, Endo Y (2000) A highly efficient and robust cell-free protein synthesis system prepared from wheat embryos: plants apparently contain a suicide system directed at ribosomes. Proc Natl Acad Sci U S A 97:559–564
10. Sasaki J, Nakashima N (1999) Translation initiation at the CUU codon in mediated by the internal ribosome entry site of an insect picorna-like virus in vitro. J Virol 73:1219–1226
11. Shibuya N, Nishiyama T, Kanamori Y et al (2003) Conditional rather than absolute requirements of the capsid coding sequence for initiation of methionine-independent translation in *Plautia stali* intestine virus. J Virol 77:12002–12010
12. Pfingsten JS, Costantino DA, Kieft JS (2006) Structural basis for ribosome recruitment and manipulation by a viral IRES RNA. Science 314:1450–1454
13. Ogawa A (2009) Biofunction-assisted sensors based on a new method for converting aptazyme activity into reporter protein expression with high efficiency in wheat germ extract. ChemBioChem 10:2465–2468
14. Jenison RD, Gill SC, Pardi A, Polisky B (1994) High-resolution molecular discrimination by RNA. Science 263:1425–1429
15. Mathews DH, Disney MD, Childs JL et al (2004) Incorporating chemical modification constraints into a dynamic programming algorithm for prediction of RNA secondary structure. Proc Natl Acad Sci U S A 101:7287–7292

Chapter 13

Engineering Protein-Responsive mRNA Switch in Mammalian Cells

Kei Endo and Hirohide Saito

Abstract

Engineering of translation provides an alternative regulatory layer for controlling transgene expression in addition to transcriptional regulation. Synthetic mRNA switches that modulate translation of a target gene of interest in response to an intracellular protein could be a key regulator to construct a genetic circuit. Insertion of a protein binding RNA sequence in the 5′ UTR of mRNA would allow for the generation of a protein-responsive RNA switch. Here we describe the design principle of the switch and methods for tuning and analyzing its translational activity in mammalian cells.

Key words Riboswitch, Mammalian cell, Transfection, Reporter assay, RNA–protein interaction, RNP, Synthetic biology, Translation, RNA module

1 Introduction

The function of riboswitches, monitoring intracellular conditions and controlling synthesis of a specific protein, has attracted increasing attention for their potential as synthetic regulators of various biological processes [1–3]. In particular, synthetic switches in animals could be applied to various fields, such as medicine, foodstuffs, and the environment [4–6]. Protein-responsive mRNA switches control translation of an output protein in response to an input protein expressed in the target cells. They provide an alternative regulatory layer of transgene expression in addition to transcriptional regulation, and combination of these two layers has successfully produced sophisticated gene circuits [7]. Ideally, parameters of switches should be adjusted to desired levels. Thus, the design principle of a protein-responsive RNA switch that enables the level of the output protein to be tuned is increasingly important.

In principle, translation in eukaryotic cells is initiated at the 5′ terminus of an mRNA. The 5′-cap-binding complex recruits the

Atsushi Ogawa (ed.), *Artificial Riboswitches: Methods and Protocols*, Methods in Molecular Biology, vol. 1111,
DOI 10.1007/978-1-62703-755-6_13, © Springer Science+Business Media New York 2014

pre-initiation complex containing the small ribosomal subunit, followed by scanning to the first AUG codon and assembly of the large ribosomal subunit [8]. This complicated, multistep process is one of the most common targets of translational control [9]. Iron regulatory protein (IRP) is a well-studied example of a translational regulatory protein [10]. It binds to the specific sequence (iron-responsive element, IRE) in the 5′ UTR of mRNA encoding the iron-storage protein, Ferritin, and represses synthesis of the protein to maintain a concentration of Fe(III) in a cell. It is proposed that IRP does not interfere with a specific function of translational apparatus nor does it recruit another regulatory protein, but blocks translation based on steric hindrance [10]. Accordingly, replacement of a specific RNA–protein (RNP) complex of IRP and IRE with another pair produced synthetic protein-responsive switches [11–14]. In addition to RNA–protein complexes, small molecules that bind an artificial motif inserted into the 5′ UTR can also block protein synthesis [15]. This is reminiscent of natural riboswitches found in bacteria.

This chapter describes details in constructing, tuning, and evaluating a protein-responsive RNA switch in mammalian cultured cells. Furthermore, short introductions to the uses of the switches are also provided.

2 Materials

All reagents are obtained from Life Technologies or otherwise indicated in the text.

2.1 Plasmid Construction

1. Expression vectors for mammalian cells, such as pIRES Bicistronic Expression Vector (Clontech) and pcDNA Mammalian Expression Vector.
2. Oligodeoxynucleotides as primers to amplify DNA fragments or a DNA fragment to be inserted into a plasmid vector.
3. Annealing buffer: 20 mM Tris–HCl, pH 7.6, 100 mM NaCl.
4. A thermal cycler and appropriate test tubes.
5. Enzymes for plasmid construction (ligase, alkaline phosphatase and appropriate restriction enzymes). We use Ligation High ver.2 (TOYOBO) and rAPid Alkaline Phosphatase (Roche).
6. Reagents for ethanol precipitation: Ethanol, 70 % ethanol, and 3 M potassium acetate, pH 5.2.
7. Competent *E. coli* cells, LB medium, and appropriate antibiotics for transformation and prepping plasmids.
8. A kit for purification of PCR products. We use MinElute PCR purification kit (QIAGEN).

2.2 Cell Culture and DNA Transfection

1. A kit for purification of transfection-grade plasmid DNA. In our case, we use Plasmid Midi Kit (QIAGEN).
2. Mammalian cultured cell lines. We use HeLa cells in this protocol.
3. Dulbecco's modified Eagle medium (DMEM) supplemented with 10 % fetal bovine serum and 1 % antibiotic antimycotic solution (Sigma-Aldrich).
4. Phosphate buffered saline (PBS).
5. 0.25 % Trypsin–EDTA.
6. Equipment for cell counting. We use Countess Automated Cell Counter.
7. Opti-MEM.
8. Lipofectamine 2000.

2.3 Flow Cytometry Analysis

1. Polystyrene tube with 35 μm cell strainer cap (BD Biosciences).
2. A flow cytometer, such as BD Accuri C6 (BD Biosciences) and FACSAria (BD Biosciences).

2.4 Apoptosis Assay

Following reagents are also provided as a package (Pacific Blue AnnexinV/SYTOX AADvanced Apoptosis kit, for flow cytometry).

1. Pacific Blue-conjugated Annexin V.
2. Annexin V binding buffer.
3. SYTOX AADvance.

3 Methods

3.1 Modular Expression Vector for Protein-Responsive OFF Switch

A protein-responsive mRNA OFF switch consists on an aptamer domain, which binds to a specific RNA-binding protein as an input, and the open reading frame of an output protein. A newly synthesized switch needs to be constructed in a mammalian expression vector to assess its activity in a cell. Changes not only in the sequence of the aptamer but in configuration of the 5′ UTR modulate the activity of the switch (*see* Subheading 3.2). Thus, the modular expression vector could help efficient construction, comparison and screening of new synthetic switches. This section provides information on the modular design of an expression vector for switches (switch vector), and on a protocol for plasmid construction when using it. An expression vector for an input protein (trigger vector) is also described because it will be required to evaluate the activity of the switch (*see* **Note 1**).

3.1.1 Switch Vector Expressing OFF Switch mRNA

Figure 1b shows the modular design of a switch vector. The 5′ UTR of the switch has four specific sites of restriction enzymes. *Nhe* I is placed just downstream of the CMV promoter and *Age* I

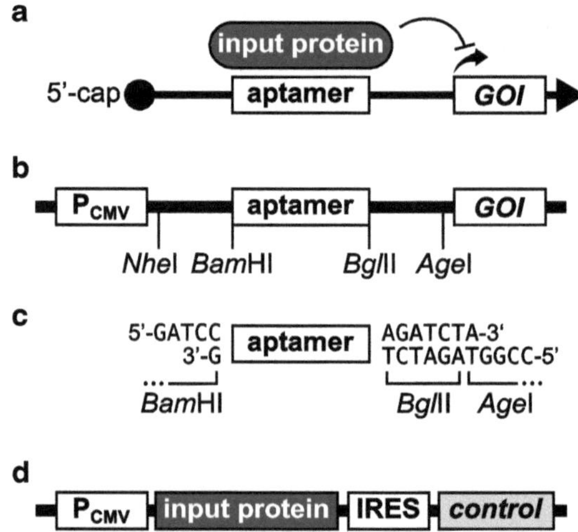

Fig. 1 Genetic configuration of plasmids expressing a protein-responsive RNA switch and its trigger protein. (**a**) Schematic illustration of a protein-responsive RNA switch. An input protein binds to an aptamer sequence placed in the 5′ UTR of the switch and blocks translation of a gene of interest (GOI), which is the output of the switch. (**b**) Configuration of a switch vector. (**c**) A fragment of an aptamer to be inserted into the switch vector. *Bam*H I and *Age* I are used to replace an aptamer, and *Bgl* II and *Age* I are used to add an aptamer. (**d**) Configuration of a trigger vector

is located upstream of the ORF. A sequence of an aptamer is inserted between them with the *Bam*H I and *Bgl* II sites at the 5′ end and 3′ end, respectively. As a result, the 5′ UTR is divided into three parts; upstream of an aptamer (*Nhe* I/*Bam*H I), the aptamer (*Bam*H I/*Bgl* II), and downstream of the aptamer (*Bgl* II/*Age* I).

In most cases, the length of an aptamer domain is relatively short (<100 nucleotides), so here we show a protocol to insert a short dsDNA derived from two synthetic oligodeoxynucleotides (oligo DNAs) into the aptamer part in the switch vector (*see* **Note 2**). In principle, plasmids are constructed according to general methods.

1. Design two oligo DNAs forming dsDNA as shown in Fig. 1c. After annealing two oligo DNAs, the resulting dsDNA should contain 5′-GATC and 5′-CCGG overhangs up- and downstream of the aptamer. This configuration is the same as a product of *Bam*H I and *Age* I digestion.
2. Mix 1 μL each of two oligo DNAs (100 μM) with 38 μL of annealing buffer (*see* **Note 3**) in a PCR tube (f.c. 2.5 μM each).
3. Denature oligo DNAs by heating at 96 °C for 30 s and anneal them by gradual cooling to 10 °C under the melting temperature of the annealing sequence using a thermal cycler.
4. Dilute the resulting dsDNA by 100-fold in DW (f.c. 25 nM).

5. Digest 250 ng of the switch vector (plasmid) with *Bam*H I and *Age* I. Do not add alkaline phosphatase because synthesized oligo DNAs do not have 5′ phosphate groups.
6. Purify the digested plasmid by ethanol precipitation.
7. Resolve/resuspend the pellet in 15 μL of DW.
8. Mix 1 μL aliquot of the diluted dsDNA and 1.5 μL aliquot of the resolved plasmid.
9. Incubate a mixture at 42 °C for 3 min to promote interaction between the dsDNA and the plasmid, and cool it on ice.
10. Add 2.5 μL of Ligation High, which is a premix of ligation reaction by T4 DNA ligase.
11. Incubate at 16 °C for 45 min reaction.
12. Transform *E. coli* with the product.

3.1.2 Trigger Vector Expressing an Input Protein

The level of a transgene transiently expressed from a plasmid varies among the transfected cells. Therefore, an indicator for transfection efficiency must be required to evaluate the activity of a synthetic switch. To monitor the efficiency, a reporter protein expressed by IRES in the same mRNA as the input protein can be used as a control (Fig. 1d), because the ratio of gene expression from two co-transfected plasmids is highly correlated in a population (*see* **Note 4**). The trigger vector is derived from pIRES Bicistronic Expression Vector.

3.2 Design and Tuning of OFF Switch

Both the position and the number of the aptamer motifs on the 5′ UTR of mRNA are critical elements to control the activity of a protein-responsive RNA switch in mammalian cells. Shortening the distance from the 5′ end of mRNA to the aptamer and repeating the aptamer sequences in the 5′ UTR enhance the activity of the OFF switch by the input protein [16]. In other words, engineering mRNAs by modifying these features can tune the activity of the switch composed of the same RNP pair (Fig. 2). This section provides a strategy for tuning protein-responsive switches.

3.2.1 Spacing Between the RNA Motif and the 5′ End

An aptamer located closer to the 5′ terminus of mRNA has more impact on repressing translation. The activity of the OFF switch decreases as the distance of the aptamer from the 5′ end increases [16]. Thus, spacing them could modify the activity in a continuous manner. It is notable that a spacer sequence should not include ATG triplets, because it is going to be inserted upstream of the ORF. The following protocol describes a procedure for insertion of a spacer DNA fragment upstream of an aptamer (*Nhe* I/*Bam*H I) in the switch vector (Fig. 1b). It is also possible to reduce the distance via PCR-based deletion methods [17, 18].

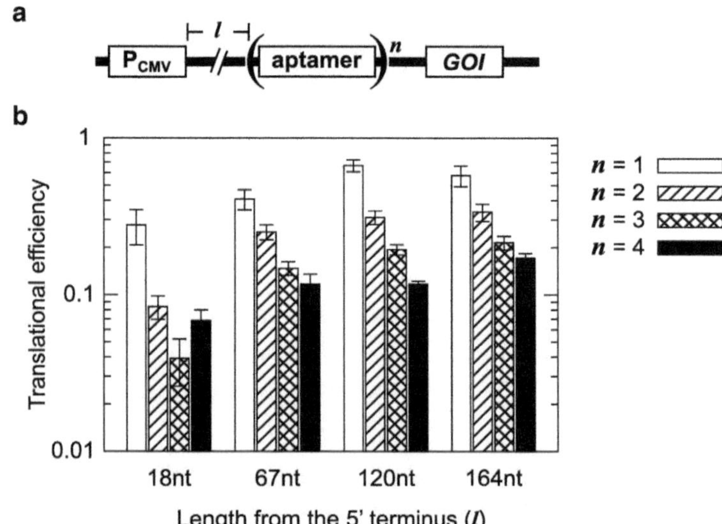

Fig. 2 Tuning the activity of the switch by modification of the copy number and the position of the aptamer. (**a**) Schematic illustration of tuning strategy. Increasing the distance of the aptamer from the 5′ end (*l*) and the number of the aptamer (*n*) in the switch continuously decrease and discretely increase the activity of the switch, respectively. (**b**) Tuning of the L7Ae-responsive switch composed of a K-loop RNA motif (Kl) as an aptamer. Shown is the translational efficiency (*see* also Subheading 3.3.4) of switches that contains one to four times repeated Kl, the first one of which was placed at 18th, 67th, 120th, and 164th nucleotides (nt; *l*) from the 5′ end

1. Design a spacer sequence absent of ATG triplets (*see* **Note 5**).
2. Design a primer set to amplify the spacer. The 5′ end of the forward and reverse primer should contain *Nhe* I and *Bam*H I restriction sites, respectively.
3. Amplify the spacer via PCR.
4. Add *Dpn* I to remove methylated plasmid DNA.
5. Incubate at 37 °C for 30 min.
6. Purify the PCR product (e.g., QIAGEN, MinElute PCR purification kit).
7. Digest 200 ng of the DNA fragment with *Nhe* I and *Bam*H I.
8. Digest 250 ng of the switch vector with *Nhe* I and *Bam*H I, together with alkaline phosphatase.
9. Purify the digested PCR fragment and plasmid, ligate them, and transform *E. coli* with it according to commonly used methods.

3.2.2 Multiplying the RNA Motifs

Tandem insertion of multiple RNA motifs into the 5′ UTR of mRNA increases the activity of the OFF switch in a step-wise manner. A DNA fragment, which contains multiple motifs, can be used to replace an aptamer in a switch vector. Alternatively, it is possible to reuse oligo DNAs described earlier (Subheading 3.2, Fig. 1c) if a switch vector is digested with *Bgl* II and *Age* I (Fig. 1b). In this case, the double-stranded oligo DNA is inserted downstream of the slotted aptamer using compatible ends produced by *Bam*H I and *Bgl* II. The resulting plasmid has the same configuration as the parental vector (a repeated aptamer sequence placed between unique *Bam*H I and *Blg* II sites) and enables repeating the same procedure to concatenate aptamers, because neither *Bam*H I nor *Blg* II can cleave the hybrid sequence.

3.3 Assessment of a Behavior of the Switch in Cultured Mammalian Cells

3.3.1 Preparation of the Samples

1. Prepare transfection-grade plasmid DNAs (switch vectors and trigger vectors) by purification on anion-exchange resin according to the manufacturer's protocol (e.g., QIAGEN, Plasmid Midi Kit).
2. Seed HeLa cells to a 24-well plate 24 h prior to transfection (*see* **Note 6**), as follows:
   (a) Aspirate medium from a Petri dish containing HeLa cells at 80 % confluence.
   (b) Add 5 mL of PBS, wash cells and aspirate PBS.
   (c) Add 1 mL of Trypsin–EDTA and incubate at 37 °C for 3 min.
   (d) Add 9 mL of DMEM medium and detach cells by gently pipetting up and down.
   (e) Transfer cells to a 15-mL conical tube.
   (f) Determine cell density using a cell counter.
   (g) Dilute cells to a concentration of 100,000 cells/mL, and seed 500 μL of cell suspension per well of a 24-well format plate.

3.3.2 Cotransfection of Plasmids into HeLa Cells

1. Mix 120 ng of a switch vector and 600 ng of a trigger vector in 60 μL of Opti-MEM.
2. Add 1.2 μL of Lipofectamine 2000 to 58.8 μL of Opti-MEM.
3. Mix the two solutions above and incubate at room temperature for 20 min.
4. Apply 100 μL of the mixture per well and gently shake the plate to mix.
5. Incubate cells at 37 °C for 4 h.
6. Replace the medium by aspiration and add 500 μL of fresh DMEM containing serum and antibiotics.

3.3.3 Quantification of Reporter Fluorescent Proteins Using a Flow Cytometer

1. Aspirate medium from the wells.
2. Add 500 μL of PBS, wash cells and aspirate PBS.
3. Add 100 μL of Trypsin–EDTA and incubate at 37 °C for 3 min.

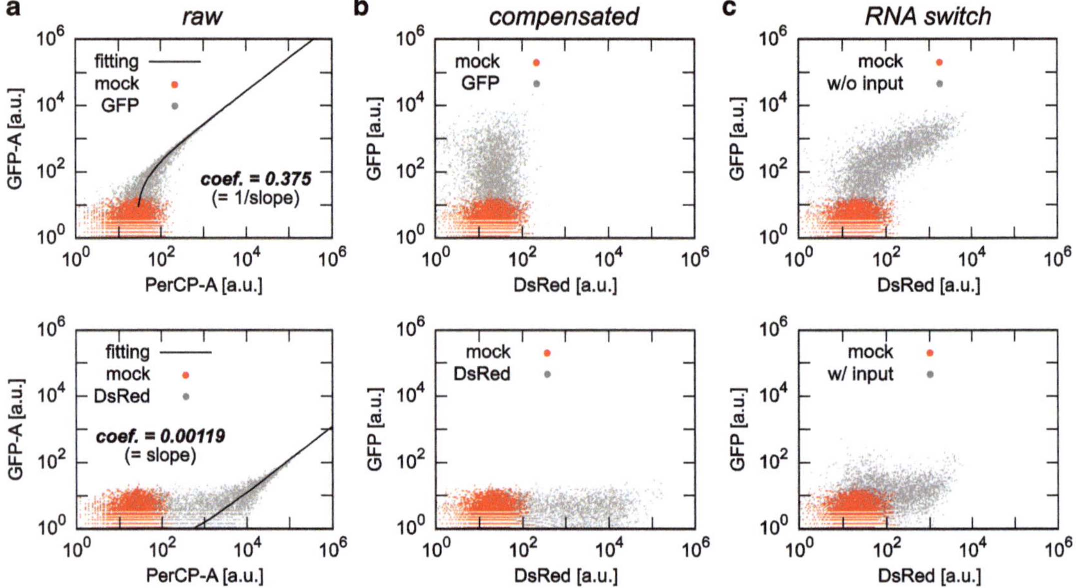

Fig. 3 Plots from flow cytometry analysis. (**a**) Raw data of cells expressing either EGFP (*upper*, reporter) or DsRed (*lower*, control) used for the fluorescence compensation (*gray*). Mock (*red*/*dark gray*) indicates untransfected cells. Fluorescence of EGFP and DsRed is detected in GFP and PerCP channel, respectively. Fitting curve and obtained coefficient are also shown. (**b**) Plots of compensated signals from the same experiments as shown in (**a**). (**c**) Plots of cells expressing L7Ae-responsive RNA switch that outputs EGFP. DsRed was used as a control. The upper and lower panel shows the output in the absence (w/o input) and presence (w/input) of L7Ae, respectively

4. Add 200 μL of medium and gently rock the dish.
5. Immediately before measurement, gently mix cell suspension by pipetting and pass cells through nylon mesh. We use 35 μm cell strainer caps.
6. Measure fluorescence using a flow cytometer.

3.3.4 Post-measurement Compensation of Fluorescent Signals

A broad emission spectrum of one fluorescent protein sometimes results in the signal spilling over into measurement of another (*see* **Note 7**). If necessary, perform the fluorescence compensation as follows.

1. Calculate the mean values for each fluorescence channel (GFP-A and PerCP-A) from untransfected cells to evaluate autofluorescence ($[\text{auto}]_{\text{GFP-A}}$ and $[\text{auto}]_{\text{PerCP-A}}$).
2. Perform linear fitting to obtain a spillover coefficient (coef_1) of the reporter protein (EGFP) into the channel for the control protein (PerCP-A) based on data from cells expressing only EGFP, and the following equation (upper panel, Fig. 3a).

$$[PerCP-\text{A}] = coef_1[EGFP] + [DsRed] + [auto]_{PerCP\text{-A}}$$

3. Perform linear fitting to obtain the spillover coefficient ($coef_2$) of the control protein (DsRed) to the measurement of the reporter protein (GFP-A) based on data from cells expressing only DsRed, and on the following equation (lower panel, Fig. 3a).

$$[GFP-\mathrm{A}]=[EGFP]+coef_2[DsRed]+[auto]_{GFP-\mathrm{A}}$$

4. Invert the spillover matrix (S) to obtain the compensation matrix (C).

   Here, $S=\begin{pmatrix}1 & coef_2\\ coef_1 & 1\end{pmatrix}, C=S^{-1}$

   From the two equations in **steps 2** and **3**, $\begin{pmatrix}[GFP-\mathrm{A}]\\ [PerCP-\mathrm{A}]\end{pmatrix}=S\begin{pmatrix}[EGFP]\\ [DsRed]\end{pmatrix}$

   Therefore, $\begin{pmatrix}[EGFP]\\ [DsRed]\end{pmatrix}=C\begin{pmatrix}[GFP-\mathrm{A}]\\ [PerCP-\mathrm{A}]\end{pmatrix}$

5. Apply the compensation matrix (C) to all the other data to obtain compensated signals from the reporter (EGFP) and the control (DsRed).

3.3.5 Calculation of Translational Efficiency

Cross comparing the activity of the switch with different configurations or in different conditions requires some representative values. It can be considered as the ratio of the translational efficiency of the riboswitch in the presence and in the absence of the input protein. When calculating the ratio, effects of the background fluorescence and the efficiency of transfection should be taken into consideration. Here it is shown one definition that represents translational efficiency of the riboswitch as the average of the output from specific cells determined by the control signal (Fig. 4).

1. Collect cells expressing a specific range of the control fluorescent protein (e.g., 1,000 ± 100 a.u. of DsRed) (Fig. 4a).
2. Calculate the mean level of the reporter protein in the collected cells.
3. Divide the mean value from cells in the presence of the input protein by that in the absence of the input to obtain translational efficiency (Fig. 4b). Confirm that the choice of the specific range selected for the above steps does not significantly change translational efficiency by varying the specific range as shown in Fig. 4b.

3.4 Controlling Apoptotic Genes

Protein-responsive RNA switches can be employed for controlling, in principle, any gene of interest. Induction of a particular gene from a switch has potential to control cellular function, such as apoptosis, in a manner responding to a specific protein. Figure 5 shows an example of a system for switch-mediated

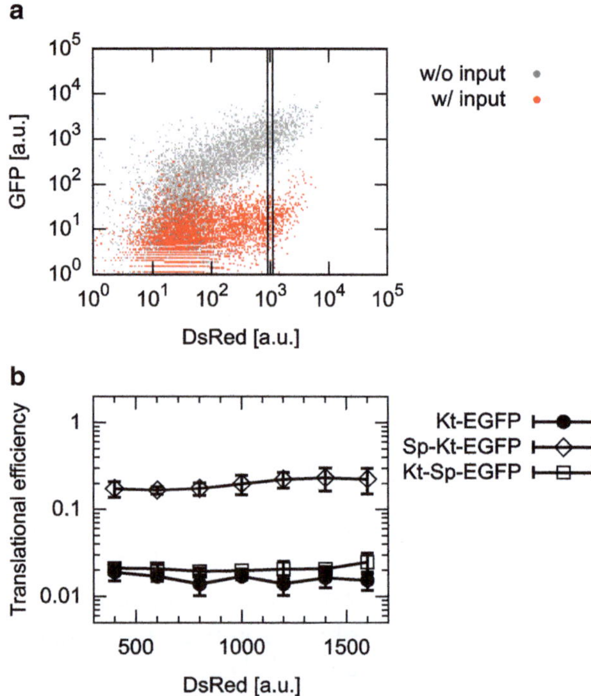

Fig. 4 A definition of translational efficiency. (**a**) An example for a specific window of cells (1,000 ± 100 a.u. of DsRed) used to calculate translation efficiency. (**b**) Comparison of translational efficiencies from ranged different windows (400–1,600 a.u. of DsRed). Those of three different switches are shown. Kt, another aptamer to L7Ae; Sp, a spacer sequence (164 nt). Choice of specific windows does not change translational efficiency significantly

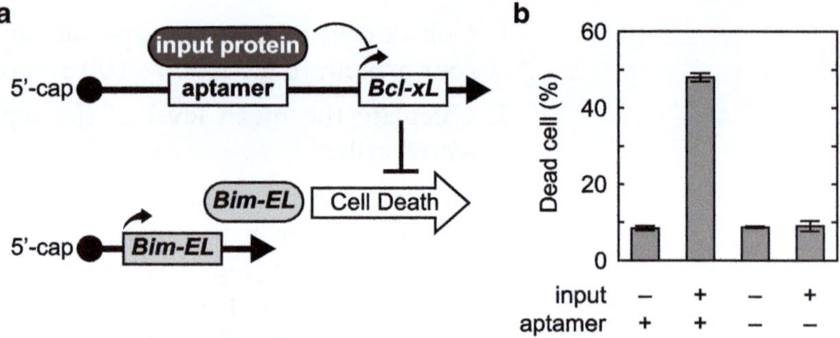

Fig. 5 Cell death induced by a switch. (**a**) Schematic illustration of the system. Bcl-xL protein produced from the switch blocks apoptosis induced by Bim-EL. In the presence of input protein, the switch represses Bcl-xL expression and induces apoptosis. (**b**) Induction of apoptosis was observed in an input protein specific manner

induction of cell death [19]. In this case the switch which outputs to an anti-apoptotic gene, Bcl-xL, is transferred into mammalian cells, together with an apoptosis inducible gene, Bim-EL. Bcl-xL interacts with Bim-EL and therefore blocks the function of Bim-EL and cell death; in the presence of the input protein, the expression of Bcl-xL is repressed and the function of Bim-EL recovers, resulting in the induction of apoptosis.

A procedure for evaluating the induction of apoptosis in transfected cells is as follows.

1. Prepare a switch vector with Bcl-xL as the output instead of the reporter fluorescent protein described previously.
2. Cotransfect both the switch and trigger vectors as well as an additional vector, which expresses Bim-EL (*see* **Note 8**).
3. Change the medium 4 h after the transfection.
4. Transfer media including detached cells from the well into a 1.5 mL-microcentrifuge tube 24 h after the transfection.
5. Add 100 μL of Trypsin–EDTA and incubate at 37 °C for 3 min.
6. Add 200 μL of medium and gently rock the dish.
7. Transfer the cell suspension to the same tube as in **step 4**.
8. Spin the suspension at 200 × *g* for 5 min at room temperature.
9. Aspirate supernatant and add 500 μL of PBS.
10. Spin again at 200 × *g* for 5 min at room temperature and aspirate supernatant again.
11. Suspend cells with 53 μL of premixed staining reagent (Pacific Blue Annexin V, 2.5 μL; SYTOX AADvance, 0.5 μL; binding buffer, 50 μL).
12. Incubate samples in the dark for 30 min at room temperature.
13. Add 200 μL of binding buffer and keep on ice until analysis.
14. Perform the flow cytometry analysis.
15. Gate out untransfected cells based on the fluorescence of the control protein.
16. Evaluate the ratio of Annexin V positive cells in populations (Fig. 5b).

3.5 Dual Switch Regulation

Analogous to transcriptional regulation, multiple transgenes need to be controlled under a single protein in some cases. In addition, it is desirable that expression levels from multiple genes can be tuned simultaneously and specifically. Modification of cis-regulatory elements (Subheading 3.2) in the mRNA is sufficient to produce such switches. In fact we have created such a dual system whereby one input protein can regulate two independent switches simultaneously. We have observed that addition and the configuration of

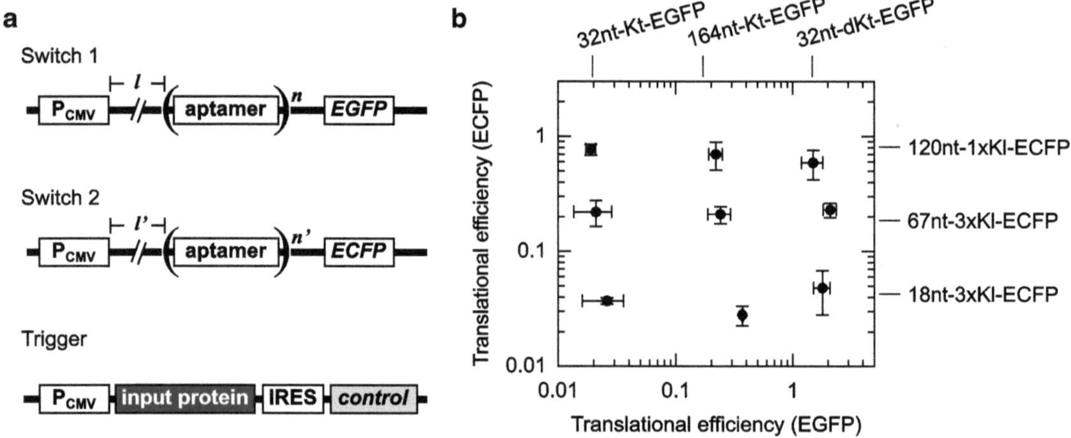

Fig. 6 Simultaneous and specific control of two outputs by a single input. (**a**) Schematic illustration of the system. (**b**) Translational efficiencies of co-transfected two distinct L7Ae-responsive switches. *dKt* defective sequence of Kt

one switch did not affect the translational efficiency of the other switch in the same cell (Fig. 6) [16].

1. Prepare another switch vector composed of the same aptamer but a different reporter protein.
2. Transfect two switch vectors together with a trigger vector (Fig. 6a).
3. Perform flow cytometry analysis with three fluorescent proteins (two reporter and one control).
4. Evaluate translational efficiency of each switch (Fig. 6b).

3.6 Autofeedback Regulation

Metabolite-binding riboswitches in bacteria monitor a physiological condition and control specific genes in the pathway to give feedback on a biosynthetic system [20]. Mammalian IRP also plays a central role in iron homeostasis, controlling iron metabolic genes via iron-dependent binding to a specific sequence in their mRNAs [21]. Similarly, and more simply, a protein-responsive riboswitch itself can form a feedback loop because the same protein can be used for both its input and output.

Figure 7 shows a configuration of translational autoregulation. The genetic construct (Fig. 7a) can be derived from the two basic constructs (Fig. 1b, d) by fusing an input protein in a trigger vector with an output gene (GOI) in a reporter vector. A fusion protein expressed from transcripts binds to its own mRNA and represses further expression of the input–output fusion protein. Modification in the interaction of an RNP pair, using modified RNA motifs and/or variants of the input protein, can also tune the behavior of the system [22]. Furthermore, input proteins under autoregulation can be employed as a regulator for other mRNAs in the same cells [22].

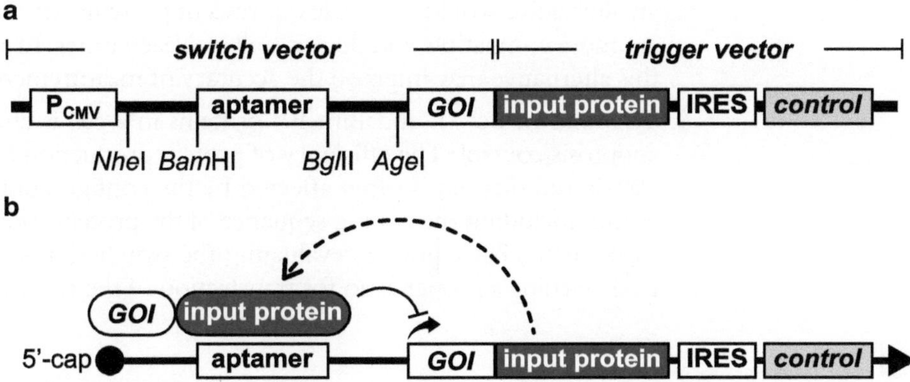

Fig. 7 Autofeedback regulation composed of protein-responsive switch. (**a**) Genetic configuration of the autoregulation. A GOI in a switch vector is fused with an input protein from a trigger vector. (**b**) Schematic illustration of translational negative feedback loop

4 Notes

1. In this protocol, input and output plasmids are transiently transfected into cultured cells. Integration of these constructs into the genome of host cell can produce cells that stably express the switches and/or input proteins [19].
2. Alternatively, DNA fragments digested by *BamH* I (or *Bgl* II) and *Age* I derived from a plasmid or a PCR fragment can be inserted into the switch vector.
3. Reduce the concentration of sodium ions to hybridize the two strands in more strict conditions.
4. It is possible to fuse the input protein with the control protein for some purposes. For example, fluorescent protein (e.g., DsRed) can be fused with the input (e.g., L7Ae) to monitor its expression level [19].
5. The average length of the 5′ UTRs in the human genome is 125 nt [23]. Too long or too short 5′ UTRs may reduce the translation efficiency independent of the switch function.
6. In general, the level of transiently expressed transgene varies widely. However, the ratio of the levels between two genes from two different plasmids seems constant, provided that the two plasmids were mixed with each other before the addition of a transfection reagent. Untransfected cells and cells expressing either of the two fluorescent proteins are needed to examine and to correct for the spillover between the two (*see* Subheading 3.3.4).
7. The fluorescent spillover interferes with the quantification of the signals. In the analysis of reporter fluorescent proteins, it is easy to prepare cells expressing either of the two signals, and so, possible to compensate for the spillover mathematically after the measurement. To circumvent the necessity for compensation,

an alternative would be to use fluorescent proteins whose signals are more compatible and do not spill into each other. In addition this alternative may improve the accuracy of measurement.

8. The ratio of Bcl-xL and Bim-EL proteins in a cell is critical for apoptosis control. The efficiency of protein production from the transferred plasmid is often affected by the configuration of the vector, including the coding sequence of the protein, and is hard to estimate. Thus, prior to evaluating the switches, it is required to determine a proper ratio for transfection of the two plasmids.

References

1. Isaacs FJ, Dwyer DJ, Collins JJ (2006) RNA synthetic biology. Nat Biotechnol 24: 545–554
2. Wieland M, Fussenegger M (2010) Ligand-dependent regulatory RNA parts for synthetic biology in eukaryotes. Curr Opin Biotechnol 21:760–765
3. Chang AL, Wolf JJ, Smolke CD (2012) Synthetic RNA switches as a tool for temporal and spatial control over gene expression. Curr Opin Biotechnol 23:679–688
4. Nandagopal N, Elowitz MB (2011) Synthetic biology: integrated gene circuits. Science 333:1244–1248
5. Ruder WC, Lu T, Collins JJ (2011) Synthetic biology moving into the clinic. Science 333:1248–1252
6. Wieland M, Fussenegger M (2012) Engineering molecular circuits using synthetic biology in mammalian cells. Annu Rev Chem Biomol Eng 3:209–234
7. Ausländer S, Ausländer D, Müller M et al (2012) Programmable single-cell mammalian biocomputers. Nature 487:123–127
8. Jackson RJ, Hellen CU, Pestova TV (2010) The mechanism of eukaryotic translation initiation and principles of its regulation. Nat Rev Mol Cell Biol 11:113–127
9. Sonenberg N, Hinnebusch AG (2009) Regulation of translation initiation in eukaryotes: mechanisms and biological targets. Cell 136:731–745
10. Hentze MW, Kühn LC (1996) Molecular control of vertebrate iron metabolism: mRNA-based regulatory circuits operated by iron, nitric oxide, and oxidative stress. Proc Natl Acad Sci U S A 93:8175–8182
11. Stripecke R, Oliveira CC, McCarthy JE et al (1994) Proteins binding to 5′ untranslated region sites: a general mechanism for translational regulation of mRNAs in human and yeast cells. Mol Cell Biol 14:5898–5909
12. Paraskeva E, Atzberger A, Hentze MW (1998) A translational repression assay procedure (TRAP) for RNA-protein interactions in vivo. Proc Natl Acad Sci U S A 95:951–956
13. Nie M, Htun H (2006) Different modes and potencies of translational repression by sequence-specific RNA-protein interaction at the 5′-UTR. Nucleic Acids Res 34:5528–5540
14. Saito H, Kobayashi T, Hara T et al (2010) Synthetic translational regulation by an L7Ae-kink-turn RNP switch. Nat Chem Biol 6:71–78
15. Werstuck G, Green MR (1998) Controlling gene expression in living cells through small molecule-RNA interactions. Science 282:296–298
16. Endo K, Stapleton JA, Hayashi K et al (2013) Quantitative and simultaneous translational control of distinct mammalian mRNAs. Nucleic Acids Res. 41:e135
17. Imai Y, Matsushima Y, Sugimura T et al (1991) A simple and rapid method for generating a deletion by PCR. Nucleic Acids Res 19:2785
18. Hansson MD, Rzeznicka K, Rosenbäck M et al (2008) PCR-mediated deletion of plasmid DNA. Anal Biochem 375:373–375
19. Saito H, Fujita Y, Kashida S et al (2011) Synthetic human cell fate regulation by protein-driven RNA switches. Nat Commun 2:160
20. Smith AM, Fuchs RT, Grundy FJ et al (2010) Riboswitch RNAs: regulation of gene expression by direct monitoring of a physiological signal. RNA Biol 7:104–110
21. Eisenstein RS (2000) Iron regulatory proteins and the molecular control of mammalian iron metabolism. Annu Rev Nutr 20:627–662
22. Stapleton JA, Endo K, Fujita Y et al (2012) Feedback control of protein expression in mammalian cells by tunable synthetic translational inhibition. ACS Synth Biol 1:83–88
23. Suzuki Y, Ishihara D, Sasaki M et al (2000) Statistical analysis of the 5′ untranslated region of human mRNA using 'Oligo-Capped' cDNA libraries. Genomics 64:286–297

Chapter 14

Guanine-Tethered Antisense Oligonucleotides as Synthetic Riboregulators

Masaki Hagihara

Abstract

Regulation of gene expression by short oligonucleotides (antisense oligonucleotides), which can modulate RNA structures and inhibit subsequent associations with the translation machinery, is a potential approach for gene therapy. This chapter describes an alternative antisense strategy using guanine-tethered antisense oligonucleotides (G-ASs) to introduce a DNA–RNA heteroquadruplex structure at a designated sequence on RNA targets. The feasibility of using G-ASs to modulate RNA conformation may allow control of RNA function by inducing biologically important quadruplex structures. This approach to manipulate quadruplex structures using G-ASs may expand the strategies for regulating RNA structures and the functions of short oligonucleotide riboregulators.

Key words Antisense oligonucleotides, Guanine quadruplexes, DNA–RNA hybrids, Reverse transcriptase stop assay, Circular dichroism analysis, UV melting analysis

1 Introduction

Natural riboswitches are folded RNA structures within mRNA that sense metabolites and regulate protein biosynthesis in cells [1–3]. The coupling of ligand binding to structural changes in mRNA is a central regulatory mechanism of riboswitches. Thus, development of small molecules that modulate the RNA structure and function has been a current topic of intensive research [4]. Such designed ligand–RNA pairs would enable one to construct artificial gene regulation systems, which may provide genetic tools for both basic biological study and clinical research.

The function of RNAs depends largely on their tertiary structures, in which RNAs form active ternary structures through specific hydrogen bonds, such as Watson–Crick and Hoogsteen bonds. Among a variety of RNA ternary structures, guanine quadruplexes are a unique structure, in which four guanine bases engage in Hoogsteen hydrogen bonds to build a guanine tetrad and incorporate

Atsushi Ogawa (ed.), *Artificial Riboswitches: Methods and Protocols*, Methods in Molecular Biology, vol. 1111, DOI 10.1007/978-1-62703-755-6_14,

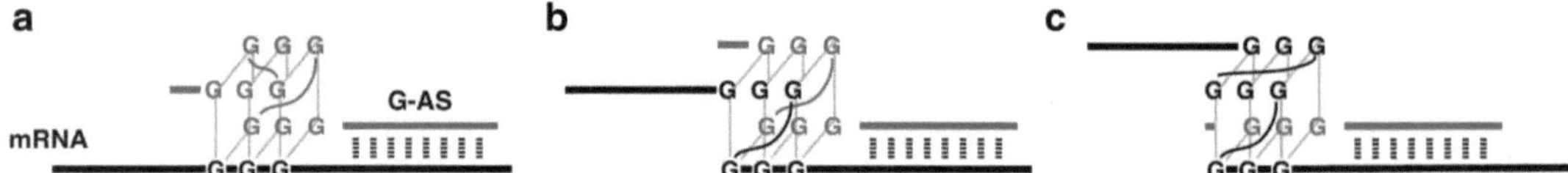

Fig. 1 Illustrations of the induced hybrid guanine quadruplex structure on mRNA templates by guanine-tethered antisense oligonucleotides. A guanine-tethered antisense (G-AS, shown in *gray*) hybridizes to a target RNA, and contiguous guanines in both the RNA and the G-AS then associate to form a quadruplex. The drawing shown in (**a**) and (**c**) show a 1:2 split G-quadruplex mode for the number of GGG sequences, and the drawing shown in (**b**) uses a 2:2 split G-quadruplex mode

a metal ion, such as potassium and sodium, inside the tetrad [5, 6]. Recent reports have shown that quadruplex formation of mRNAs plays a significant role in biological processes, including translation and translocation [7, 8]. Thus, controlled introduction of quadruplex structures into mRNAs may provide opportunities for regulation of biological processes.

Antisense oligonucleotides are relatively short single-stranded DNAs designed to abrogate protein function by hybridizing directly with oligonucleotides to exposed regions of the targeted mRNA [9, 10]. A recently reported novel antisense strategy using guanine-tethered antisense (G-AS) oligonucleotides confers an aberrant DNA–RNA heteroquadruplex structure [11]. G-ASs comprise two functionally independent domains; one is an antisense domain that binds specifically to the target RNA in the same way as conventional antisense oligonucleotides, and the other involves contiguous guanine runs tethered at the 5′-end of the antisense domain that is designed to assemble into heteroquadruplex structures together with a contiguous guanine run in the target RNAs (Fig. 1). The key concept of G-ASs is based on the cooperative formation of DNA–RNA heteroquadruplex structures with the assistance of hybridization of antisense sequences. The combination of antisense recognition and self-assembly of contiguous guanines affords predictable and sequence-specific DNA–RNA heteroquadruplex structures, which effectively induce structural changes in RNA templates. The induced quadruplex structures produced effectively inhibit reverse transcription on the native RNA sequences.

A reverse transcriptase-mediated enzymatic analysis (*RTase* stop assay) [12] together with circular dichroism (CD) spectroscopy [13, 14] and UV thermal melting experiments [15–17] using the model assembly of RNA and G-AS oligonucleotides provides information about the induced guanine quadruplex structures on RNA templates in association with G-ASs. Detailed investigation of DNA–RNA hybrid heteroquadruplex formation reveals that at least one GGG sequence in the target RNA is the minimal sequence for the introduction of a stable heteroquadruplex by G-ASs. This feasible and effective approach to controlling RNA structures by

G-ASs opens the possibility of using G-ASs as synthetic riboregulators for gene regulation. The following steps describe the design guidelines for G-ASs and the characterization of the G-AS-induced hybrid guanine quadruplex structures.

2 Materials

Great care should be taken to avoid inadvertently introducing contaminating RNases into the RNA sample during experiments. Always wear clean gloves while handling RNA samples and use RNase-free reagents to prevent contamination by nucleases from the surface of the skin or from dusty laboratory equipment. Change gloves often and keep tubes closed whenever possible. Prepare all solutions using nuclease-free water and DNase- and RNase-free molecular grade reagents, and store all reagents at room temperature unless otherwise indicated.

2.1 RTase Stop Assay Components

1. Nuclease-free TE buffer: 10 mM Tris–HCl, pH 8.0, 1 mM EDTA, molecular biology grade.
2. 0.1× TE buffer (dilute nuclease-free TE buffer with nuclease-free water).
3. 10 μM antisense oligonucleotides in 0.1× TE buffer (store at –20 °C) (*see* **Notes 1** and **2**).
4. 3 μM RNA stock solution in 0.1× TE buffer (store at –20 °C) (*see* **Notes 3** and **4**).
5. 1 μM Texas Red-labeled cDNA primer in 0.1× TE buffer (store at –20 °C, keep in a dark box to protect from light) (*see* **Note 5**).
6. 100 units/mL ReverTra Ace reverse transcriptase (Toyobo, Tokyo, Japan) (store at –20 °C) (*see* **Note 6**).
7. dNTP (10 mM each of dATP, dGTP, dCTP, and dTTP) (Promega) (store at –20 °C).
8. Blue dextran stock: 30 mg/mL blue dextran (Sigma-Aldrich) dissolved in 50 mM EDTA, pH 8.0 (store at –20 °C).
9. Reaction buffer (5×): 250 mM Tris–HCl, pH 8.0, 50 mM dithiothreitol, 500 mM KCl (*see* **Note** 7).
10. Extension mixture: 24 μL of dNTP mixture, 10 μL of 150 mM $MgCl_2$, 6 μL of reverse transcriptase (add reverse transcriptase just before use and mix well).
11. Precipitation buffer: 10 mL of 95 % ethanol, 1 mL of 3 M sodium acetate, pH 5.3.
12. Loading solution: mix deionized formamide and blue dextran stock in a ratio of 5:1 (store at –20 °C).

13. Standard thermal cycler (e.g., TaKaRa PCR Thermal Cycler Dice).
14. Automated fluorescent DNA sequencer (e.g., Hitachi SQ5500 automated DNA sequencer).

2.2 CD Analysis Components

1. 100 μM oligoribonucleotide stock solutions in TE buffer (store at −20 °C) (*see* **Notes 8** and **9**).
2. 100 μM antisense oligonucleotide stock solution in TE buffer (store at −20 °C).
3. CD measurement buffer (2×): 20 mM lithium cacodylate, pH 7.0 (*see* **Note 10**), 200 mM KCl.
4. Quartz cuvettes (10 mm path length).
5. CD spectrometer (e.g., Jasco 715 spectrometer).
6. Temperature controller (e.g., Eppendorf ThermoStat plus).

2.3 UV Melting Analysis Components

1. 100 μM oligoribonucleotide stock solutions in TE buffer (store at −20 °C) (*see* **Note 9**).
2. 100 μM antisense oligonucleotide stock solution in TE buffer (store at −20 °C).
3. UV melting measuring buffer (2×): 20 mM lithium cacodylate, pH 7.0 (*see* **Note 11**).
4. 1 M KCl, molecular biology grade.
5. Quartz cuvettes and caps.
6. UV–Vis spectrophotometer equipped with a cell changer device (e.g., Shimadzu PharmaSpec UV-1700 spectrophotometer).
7. Temperature-control device (e.g., Shimadzu Peltier element-controlled TMSPC-8 system).

3 Methods

3.1 RTase Stop Assay

1. Mix 6 μL of reaction buffer, 3 μL of cDNA primer, and 3 μL of the RNA sample, and add 11 μL of water to 0.2 mL PCR tubes.
2. Anneal cDNA primer and the RNA sample by heating tubes at 80 °C for 5 min and place tubes at room temperature for 10 min.
3. Add 3 μL of the G-AS to each reaction tube and place tubes at room temperature for 15 min.
4. Immediately dispense 4 μL of extension mixture into each reaction tube containing RNA–primer–G-AS mixture and incubate tubes at 42 °C for 30 min in the thermal cycler.

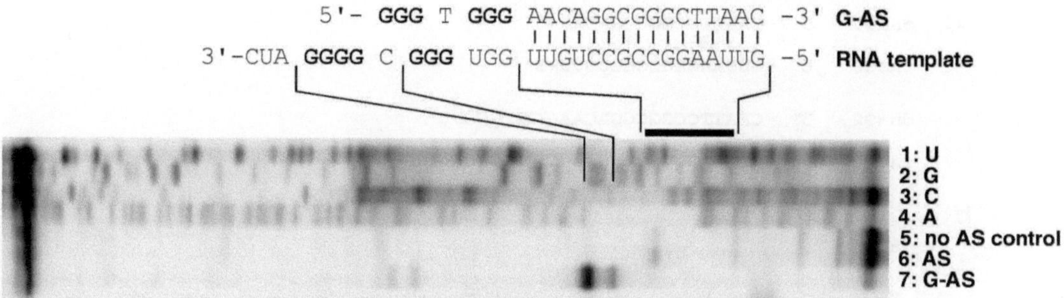

Fig. 2 Guanine-tethered antisense sequences inhibit reverse transcription of an RNA template bearing two runs of the GGG sequence. An RNA–AS heteroduplex sequence is shown in the figure. The *bold line* indicates the AS hybridization site. The lane markers U, G, C, and A indicate the bases on the template RNA. A partial genomic RNA sequence from the central polypurine tract of HIV-1 (nucleotides 4581–4756), which is involved in nuclear import of HIV-1, was chosen as the target sequence

5. Transfer precisely 25 μL of reaction mixture to fresh 1.5 mL tubes and add 75 μL of precipitation buffer to all tubes, vortex, and centrifuge at 14,000 × *g* for 15 min at room temperature.
6. Remove samples from the centrifuge and carefully remove supernatants.
7. Wash pellet with 100 μL of 70 % ethanol and centrifuge at 14,000 × *g* for 3 min.
8. Remove supernatants and allow pellet to air dry for the minimum time required to remove any residual ethanol. Overdrying will make redissolving the pellet difficult.
9. Add 10 μL of loading solution per tube, resuspend by vortexing several times, and centrifuge briefly.
10. Heat samples for 3 min at 90 °C and quickly move to ice.
11. Analyze 1–2 μL of the samples on a denaturing polyacrylamide gel and analyze with a fluorescent DNA sequencer (Fig. 2) (*see* **Note 12**).

3.2 CD Analysis of DNA–RNA Quadruplexes

1. Determine the concentration of oligonucleotides by measuring absorbance at 260 nm in a UV–Vis spectrophotometer and verify concentration of each oligonucleotide using the extinction coefficient (Fig. 3a) (*see* **Note 13**).
2. Dilute RNA oligonucleotide and antisense DNA samples with ultrapure water in a total volume of 175 μL, and then add 175 μL of CD measurement buffer to achieve a final concentration of 5 μM for each strand.
3. Start refolding of secondary structure by heating to 90 °C for 5 min followed by slowly cooling to 25 °C (0.5 °C/min) in a programmable temperature controller.

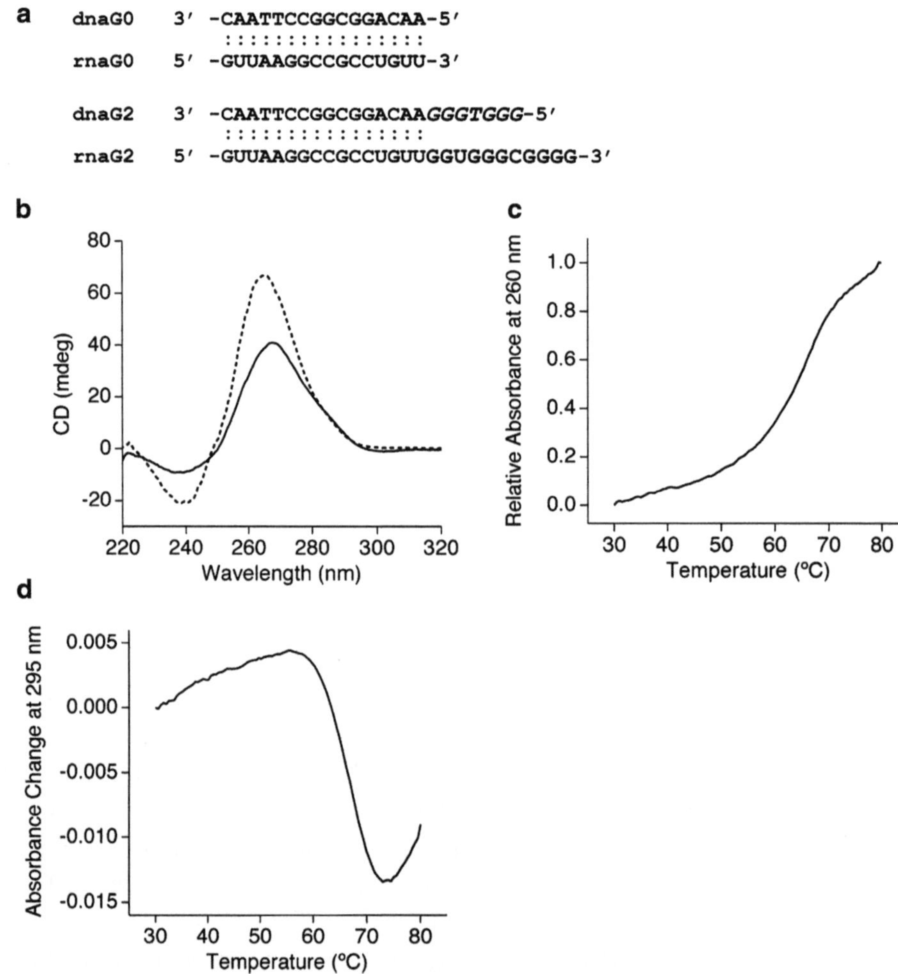

Fig. 3 Biophysical properties of a model assembly of RNA–DNA hybrids modified with runs of contiguous guanines. (**a**) Oligonucleotide sequences used in these analyses. Guanine-tethered sequences are shown in *italic letters*. (**b**) CD spectra of rnaG0 and dnaG0 (*black line*) and rnaG2 and dnaG2 (*dotted line*). (**c, d**) UV-melting profile of DNA–RNA hybrids of rnaG2 and dnaG2 measured at (**c**) 260 nm and (**d**) 295 nm

4. Transfer refolded DNA–RNA samples to a quartz cuvette.
5. Record the CD spectra (Fig. 3b) (*see* **Notes 14** and **15**).

3.3 UV Melting Analysis of DNA–RNA Quadruplexes

1. Determine the concentration of oligonucleotides by measuring absorbance at 260 nm in a UV–Vis spectrophotometer and verify the concentration of oligonucleotides using the extinction coefficient (Fig. 3a) (*see* **Note 13**).
2. Mix 50 μL of Tm-measuring buffer, 5 μL of 1 M KCl, 5 μL of oligonucleotide stock solution, 5 μL of antisense oligonucleotide stock solution, and ultrapure water to a final volume of 100 μL.

3. Mix components in 0.2 mL PCR tubes, heat the tubes for 5 min at 90 °C, and cool the tubes to room temperature slowly in a standard thermal cycler to maintain the proper secondary structures of DNA–RNA hybrids (*see* **Note 16**).
4. Transfer samples to the quartz cuvette and overlay mineral oil to limit evaporation if the cap does not fit perfectly on the cuvette.
5. For a typical experiment, record absorbance changes at 295 nm and 260 nm to examine the stability of the quadruplex and duplex, respectively (*see* **Note 17**); use an integration time of 1 s and a temperature gradient of 0.5 °C/min.
6. Represent the absorbance profiles as a function of the temperature (Fig. 3c, d) and check the quality of the data (*see* **Note 18**).
7. Calculate the melting points of the duplex and the quadruplex structures by determining the melting temperatures at the half-maximum change of the signal (*see* **Note 19**).

4 Notes

1. The general principle for designing G-AS oligonucleotides depends on the simple features of the guanine quadruplex formation [18, 19] described below. (1) The formation of one guanine quadruplex unit requires four consecutive GGG repeats. (2) Guanine quadruplexes can be divided into two parts; one possesses three GGG repeats, and another part possesses one GGG repeat, namely, the 3:1 split mode; or both possess two GGG repeats, namely, the 2:2 split mode. (3) Two single-stranded oligonucleotides having a total of four GGG repeats can associate together to form a hybrid quadruplex structure when brought into close proximity. To design G-ASs for sequence-specific introduction of guanine quadruplex structures, the association between four units of unpaired GGG in close proximity and a covalent connection of the antisense domain and the guanine-rich region are essential for introducing stable hybrid guanine quadruplex structures on the RNA templates. For example, when two adjacent GGG sequences are found in the target RNAs, the design G-AS is accomplished simply by modifying two runs of contiguous GGG units at the 5′-terminal of the antisense region.
2. The stability of DNA–RNA heteroquadruplexes may depend on both the sequence of the guanine tethers in the G-ASs and the length of the antisense region. If generation of stable heteroquadruplexes is desired, the sequences of tethers and the length of the antisense can be optimized.

3. Transcription by T7 RNA polymerase is generally better than solid-phase chemical synthesis for producing RNA longer than 30 nucleotides. The DNA templates in the T7 RNA polymerase reaction can be either a linearized plasmid or PCR products that contain the T7 promoter sequence.
4. The *RTase* stop assay is extremely sensitive to cleavage of the RNA templates by contaminating nucleases. Thus, the use of RNA templates of the highest purity is critical for successful operation of the assay. Polyacrylamide gel purification of RNA templates is strongly recommended.
5. The cDNA primer is designed to hybridize to a specific sequence downstream of the consecutive guanine-rich sequences in the target RNA. The optimum length of the cDNA primer is 17–25 nucleotides.
6. Several commercially available reverse transcriptases have been tested for their appropriateness in evaluating the stability of guanine quadruplexes. Among the enzymes tested, ReverTra Ace, an M-MLV reverse transcriptase genetically engineered by the introduction of two point mutations in the polymerase and RNase H domains, showed superior performance characteristics for detecting guanine quadruplexes in RNA templates.
7. The reaction buffer used here is designed to promote the formation of DNA–RNA heteroquadruplexes. The concentration of potassium chloride will largely affect the stability of quadruplex structures and can be optimized [20].
8. Measurement of the CD spectra of nucleic acids is operationally similar to that of the UV–Vis absorbance spectrum. However, sample degradation can lead to significant variations in the resulting CD spectra, especially in the wavelength range of 190–260 nm. Great care should be taken in the preparation of samples.
9. The oligonucleotides should be ordered as HPLC purified for greater purity. The purity and integrity of the oligonucleotides should be verified by denaturing polyacrylamide gel electrophoresis.
10. To prepare a 200 mL volume of 20 mM sodium cacodylate at pH 7.0, dissolve 0.55 g of cacodylic acid in 150 mL of double-distilled water. Adjust the pH of the solution using a 1 M NaOH solution until the desired pH is reached. Adjust the volume to 200 mL with RNase-free water.
11. For thermal melting analysis, the optimum buffer system should have a temperature-independent pKa. A cacodylate buffer (pKa = 6.3) or a phosphate buffer is preferred to the more commonly used Tris, HEPES, and MOPS buffers [21].

12. (Anticipated results) In the *RTase* stop assay, an *RTase* proceeds along the RNA until the enzyme encounters stable a guanine quadruplex structure. Inhibition of the *RTase* reaction results in the production of truncated DNA transcripts, which can be detected by DNA sequencing gel analyses.
13. To examine the formation of DNA–RNA heteroquadruplex structures, it is critical to mix the DNA and RNA strands in an equimolar concentration.
14. The following settings are used for CD measurements: resolution, 1 nm; bandwidth, 1.0 nm; sensitivity, 50 mdeg; response time, 8 s; accumulation, 4 times; scan rate, 50 nm/min; and temperature, 25 °C. Baseline spectra are recorded with 1× CD measuring buffer in the cuvette and subtracted from the spectra of the sample.
15. (Anticipated results) CD spectra of a DNA–RNA hybrid lacking guanine regions (rnaG0 and dnaG0) exhibit a typical A-form structure [22], showing a broad positive peak around 270 nm and a weak negative peak around 250 nm. By contrast, the spectra of the DNA–RNA hybrid involving two runs of the GGG sequence in each strand (rnaG2 and dnaG2) show a strong positive peak at 270 nm and a weak negative peak at 240 nm; this is assigned as a characteristic signature of a parallel quadruplex formation [23].
16. This step also ensures removal of dissolved gas from the samples and avoidance of bubble formation, which will interfere with UV measurements during melting analysis.
17. The formation of the DNA–RNA heteroguanine quadruplex structure could be monitored by the absorbance changes at 295 nm, because hyperchromism is observed when consecutive GGG sequences associate to form guanine quadruplexes [16, 17]. In parallel, the absorbance profile at 260 nm should be recorded to monitor the denaturation of DNA–RNA duplex structures. Monitoring the thermal denaturation profiles by measuring CD spectra changes at 270 nm, the characteristic signature of parallel quadruplex structures, works equally well [24].
18. If the melting temperatures are too high to be determined because of highly stable DNA–RNA hybrids, decreasing the KCl content of the solution may allow a more precise Tm determination. Decreasing the KCl content will destabilize the DNA–RNA guanine quadruplex structures, leading to decreased stability of DNA–RNA hybrids [20].
19. (Anticipated results) The DNA–RNA hybrid lacking the guanine consecutive sequences (rnaG0 and dnaG0) shows only duplex melting. The melting temperature of the hybrid with quadruplexes (rnaG2 and dnaG2) measured at 260 nm is

almost identical to that of the quadruplex measured at 295 nm, which suggests a cooperative melting of the duplex and the quadruplex in the DNA–RNA hybrid.

20. (Caution) The cacodylate buffer used in these procedures contains arsenic and can produce arsenic gas when mixed with acids. The solution containing a cacodylate buffer should be disposed of as toxic waste according to the institution's guidelines.

Acknowledgment

This work was supported by a grant from the Ichiro Kanehara Foundation.

References

1. Roth A, Breaker RR (2009) The structural and functional diversity of metabolite-binding riboswitches. Annu Rev Biochem 78:305–334
2. Serganov A, Patel DJ (2007) Ribozymes, riboswitches and beyond: regulation of gene expression without proteins. Nat Rev Genet 8:776–790
3. Mandal M, Breaker RR (2004) Gene regulation by riboswitches. Nat Rev Mol Cell Biol 5:451–463
4. Link KH, Breaker RR (2009) Engineering ligand-responsive gene-control elements: lessons learned from natural riboswitches. Gene Ther 16:1189–1201
5. Burge S, Parkinson GN, Hazel P et al (2006) Quadruplex DNA: sequence, topology and structure. Nucleic Acids Res 34:5402–5415
6. Lane AN, Chaires JB, Gray RD et al (2008) Stability and kinetics of G-quadruplex structures. Nucleic Acids Res 36:5482–5515
7. Kumari S, Bugaut A, Huppert JL et al (2007) An RNA G-quadruplex in the 5′ UTR of the NRAS proto-oncogene modulates translation. Nat Chem Biol 3:218–221
8. Arora A, Dutkiewicz M, Scaria V et al (2008) Inhibition of translation in living eukaryotic cells by an RNA G-quadruplex motif. RNA 14:1–7
9. Stein CA, Cheng YC (1993) Antisense oligonucleotides as therapeutic agents—is the bullet really magical? Science 261:1004–1012
10. Eckstein F (2007) The versatility of oligonucleotides as potential therapeutics. Expert Opin Biol Ther 7:1021–1034
11. Hagihara M, Yamauchi L, Seo A et al (2010) Antisense-induced Guanine Quadruplexes inhibit reverse transcription by HIV-1 reverse transcriptase. J Am Chem Soc 132:11171–11178
12. Hagihara M, Yoneda K, Yabuuchi H et al (2010) A reverse transcriptase stop assay revealed diverse quadruplex formations in UTRs in mRNA. Bioorg Med Chem Lett 20:2350–2353
13. Maurizot JC (2000) Circular dichroism of nucleic acids: nonclassical conformations and modified oligonucleotides. In: Berova N, Nakanishi K, Woody RW (eds) Circular dichroism: principles and applications, 2nd edn. Wiley-VCH, New York, pp 719–739
14. Porumb H, Monnot M, Fermandjian S (2002) Circular dichroism signatures of features simultaneously present in structured guanine-rich oligonucleotides: a combined spectroscopic and electrophoretic approach. Electrophoresis 23:1013–1020
15. Sun XG, Cao EH, He YJ et al (1999) Spectroscopic comparison of different DNA structures formed by oligonucleotides. J Biomol Struct Dyn 16:863–872
16. Mergny JL, Phan AT, Lacroix L (1998) Following G-quartet formation by UV-spectroscopy. FEBS Lett 435:74–78
17. Mergny JL, Lacroix L (2009) UV melting of G-Quadruplexes. In: Egli M, Herdewijn P, Matusda A, Yogesh S (eds) Curr Protoc Nucleic Acid Chem, Wiley-VCH, New York, Chapter 17: Unit 17.1
18. Xiao Y, Pavlov V, Niazov T et al (2004) Catalytic beacons for the detection of DNA and telomerase activity. J Am Chem Soc 126:7430–7431
19. Deng M, Zhang D, Zhou Y et al (2008) Highly effective colorimetric and visual detection of

nucleic acids using an asymmetrically split peroxidase DNAzyme. J Am Chem Soc 130:13095–13102

20. De Cian A, Guittat L, Kaiser M et al (2007) Fluorescence-based melting assays for studying quadruplex ligands. Methods 42:183–195
21. Fukada H, Takahashi K (1998) Enthalpy and heat capacity changes for the proton dissociation of various buffer components in 0.1 M potassium chloride. Proteins 33:159–166
22. Steely HT Jr, Gray DM, Ratliff RL (1986) CD of homopolymer DNA-RNA hybrid duplexes and triplexes containing A-T or A-U base pairs. Nucleic Acids Res 14:10071–10090
23. Jin R, Gaffney BL, Wang C et al (1992) Thermodynamics and structure of a DNA tetraplex: a spectroscopic and calorimetric study of the tetramolecular complexes of d(TG3T) and d(TG3T2G3T). Proc Natl Acad Sci U S A 89:8832–8836
24. Benz A, Hartig JS (2008) Redesigned tetrads with altered hydrogen bonding patterns enable programming of quadruplex topologies. Chem Commun (Camb) 34:4010–4012

Chapter 15

In Vitro Selection of Allosteric Ribozymes that Sense the Bacterial Second Messenger c-di-GMP

Kazuhiro Furukawa, Hongzhou Gu, and Ronald R. Breaker

Abstract

Recently, a number of study have shown the ligand-dependent allosteric ribozymes can be harnessed as biosensors, high-throughput screening, and agents for the control of gene expression in vivo, called artificial riboswitches. In this chapter, we describe how in vitro selection can be used to create an allosteric ribozyme that senses bacterial second messenger cyclic-di-GMP (c-di-GMP). A hammerhead ribozyme was joined to a natural c-di-GMP class I riboswitch aptamer via communication modules. Both c-di-GMP-activating and -inhibiting ribozyme can be obtained by this approach.

Key words Allosteric ribozyme, Aptamer, Hammerhead ribozyme, In vitro selection, Cyclic di-GMP

1 Introduction

Allosteric ribozymes are RNA enzymes whose activity is modulated by the binding of an effector molecule to an aptamer domain [1, 2]. Most allosteric ribozymes are engineered by conjugating an aptamer to a ribozyme through a communication module, that is, ligand binding to the aptamer induces a conformational change, which is transmitted to the ribozyme, the activity of which is modified. Our group [3–8] and others [9–13] have been applied directed evolution strategies to engineer allosteric ribozymes or artificial riboswitches. Also, our group recently found that in nature cells use allosteric ribozymes, i.e., cyclic di-GMP class II (c-di-GMP-II) riboswitch, to regulate gene expression [14]. In *Clostridium difficile* strains, a c-di-GMP-II aptamer is found adjacent to a group I intron. Group I introns are ribozymes that catalyze the splicing of the RNA molecule in which they are embedded. In this case, the outcome of the splicing reaction catalyzed by the intron is controlled by the natural allosteric ribozyme in response to c-di-GMP levels.

Atsushi Ogawa (ed.), *Artificial Riboswitches: Methods and Protocols*, Methods in Molecular Biology, vol. 1111, DOI 10.1007/978-1-62703-755-6_15,

c-di-GMP is a bacterial second messenger produced by the GGDEF domain-containing diguanylate cyclase and regulates many biological processes including the planktonic traits (motility) and communal or sessile traits (synthesis of bacterial surface components such as biofilm, exopolysaccharide, pili, and stalks) of several bacterial groups by its intracellular concentration [15]. Due to the widespread functions of c-di-GMP, the method to quantify the level of this second messenger is required to understand a wide range of bacterial phenomena. High-performance liquid chromatography (HPLC) is a conventional method to monitor the concentration of bacterial metabolites; however, it requires multiple semi-purification steps to remove other metabolites that have similar chemical properties to the target molecule. The allosteric ribozyme is an attractive alternative to HPLC due to its high affinity and specificity to c-di-GMP. The natural c-di-GMP-II riboswitch might be useful for quantitative analysis of c-di-GMP. However, it is not a versatile approach because of low rate constant of the group I intron and requirement of guanosine 5′-triphosphate (GTP) for its catalytic activity.

We have used the method described below to create an engineered allosteric ribozyme resulting from the fusion of a c-di-GMP-I aptamer [16] to a hammerhead ribozyme for simple monitoring of c-di-GMP levels (Fig. 1). Several engineered allosteric ribozymes derived from the hammerhead ribozyme show equal rate constant to the minimum ribozyme core, suggesting that they could be used for rapid quantification of ligands in chemical or biological samples. Moreover, allosteric ribozymes could be used for high-throughput screening to facilitate the rapid discovery of compounds that target the aptamer [17–19].

2 Materials

2.1 Construction of the Starting Pool

1. Chemically synthesized oligonucleotides for DNA templates to prepare the RNA pool shown in Fig. 1a. Oligonucleotide 1: 5′-*TAATACGACTCACTATAGG*CGTAGCCTGATGAG**NNNN**-GCACAGGGCAAACCATTCGAAAGAGTGGGACGCAAAGCCTCCGGCCTAAACCAGAAGACATGGTAG-3′. Oligonucleotide 2: 5′-GGCGCAGCTACGTGGCTTTCACCACGTTTC-G**NNNNN**GTAACCCCGCTACCTACCATGTCTTCTGGTTTAG-3′. Underlined nucleotides are complementary each other at their 3′ ends. T7 promoter sequences are shown in italic.
2. Chemically synthesized oligonucleotides for DNA templates to prepare the RNA pool shown in Fig. 1b. Oligonucleotide 3: 5′-*TAATACGACTCACTATAGG*CACCTGATGAG**NNNNNNN**GCACAGG-GCAAACCATTCGAAAGAGTGGGACGCAAAGCCTCCGGCCTAAACCAGAAGACATGGTAG-3′. Oligonucleotide 4: 5′-CTTAGGCACTACGTGGCTTTCACCA

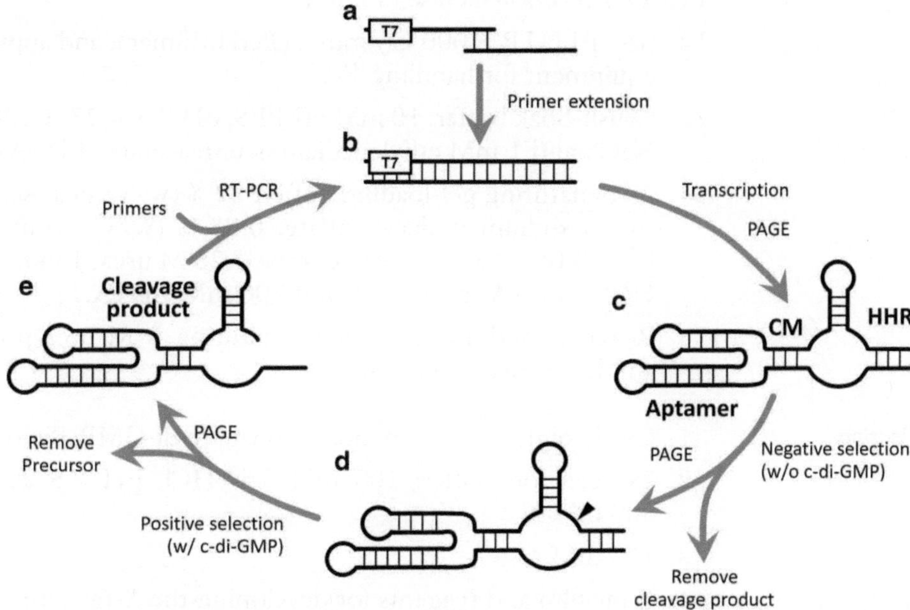

Fig. 1 (**a**) Constructs used for the selection of allosteric ribozymes responding to c-di-GMP. The ribozyme consists of a hammerhead ribozyme (HHR) joined to a class I c-di-GMP aptamer via communication module (CM) consisting of random sequence (N) nucleotides. The three stems that form the HHR are designated I, II, and III and the three stems that form the aptamer are labeled P1, P2, and P3. An *arrowhead* identifies the site of hammerhead-mediated cleavage. (**b**) Modified construct for the selection of c-di-GMP-activated ribozyme

CGTTTCG**NNNNNNNN**G-TAACCCCGCTAC<u>CTACCATG TCTTCTGGTTTAG</u>-3′. Underlined nucleotides are complementary each other at their 3′ ends. T7 promoter sequences are shown in italic.

3. 5× reverse transcription buffer: 250 mM Tris–HCl, pH 8.3 at 23 °C, 375 mM KCl, 15 mM $MgCl_2$.
4. 100 mM dithiothreitol (DTT).
5. 10× deoxynucleoside 5′-triphosphate (dNTP) mix: 2 mM each of deoxyguanosine 5′-triphosphate (dGTP), deoxyadenosine 5′-triphosphate (dATP), deoxythymidine 5′-triphosphate (dTTP), deoxycytidine 5′-triphosphate (dCTP).
6. Reverse transcriptase (e.g., SuperScript II, Invitrogen).
7. 10× transcription buffer: 500 mM HEPES, pH 7.5 at 23 °C, 150 mM $MgCl_2$, 20 mM spermidine and 50 mM DTT.
8. 3 M sodium acetate, pH 5.2.
9. 100 % ethanol.
10. 5× nucleoside 5′-triphosphate (NTP) mix: 12.5 mM each of guanosine 5′-triphosphate (GTP), adenosine 5′-triphosphate (ATP), thymidine 5′-triphosphate (TTP), cytidine 5′-triphosphate (CTP).

11. T7 RNA polymerase (T7 RNAP).
12. [α-^{32}P] UTP, 3,000 Ci/mmol (PerkinElmer), and appropriate equipment for handling ^{32}P.
13. Crush-Soak buffer: 10 mM HEPES, pH 7.5 at 23 °C, 200 mM NaCl, and 1 mM ethylenediaminetetraacetate (EDTA).
14. 2× denaturing gel-loading buffer: 32 % (w/v) sucrose, 0.16 % (w/v) sodium dodecyl sulfate, 0.08 % (w/v) bromophenol blue, 0.08 % (w/v) xylene cyanol, 7.25 M urea, 144 mM Tris–HCl, 144 mM boric acid, and 200 mM EDTA.
15. Reagents and apparatus for denaturing 8 M urea polyacrylamide gel electrophoresis.

2.2 Selection

1. Cyclic-diguanosyl 5′-monophosphate (c-di-GMP, Biolog).
2. 2× selection buffer: 100 mM Tris–HCl, pH 7.5 at 23 °C, 200 mM NaCl.
3. 100 mM $MgCl_2$.
4. X-ray film and reagents for developing the X-ray film.
5. PhosphorImager cassettes and PhosphorImager (e.g., Storm PhosphorImager, GE HealthCare).

2.3 Reverse Transcription and Polymerase Chain Reaction

1. Chemically synthesized oligonucleotides for the amplification of the cleaved products. Primer A: 5′-TAATACGACTCACTATAGGCGTAGCCTGATGAG-3′. Primer B: 5′-GGCGCAGCTACGTGGCTTTCACCACGTTTCG-3′. Primer A’: 5′-TAATACGACTCACTATAGGCACCTGATGAG-3′. Primer B’: 5′-CTTAGGCACTACGTGGCTTTCACCACGTTTCG-3′.
2. Reverse Transcriptase (e.g., SuperScript II, Invitrogen).
3. 10× PCR buffer: 100 mM Tris–HCl, pH 8.3 at 23 °C, 400 mM KCl, 15 mM $MgCl_2$, and 0.1 % gelatin.
4. *Taq* DNA polymerase.
5. Reagent and apparatus for agarose gel electrophoresis.

3 Methods

3.1 General Selection Strategy

The construct used for the selection of allosteric ribozymes that respond to c-di-GMP was created by linking the P1 stem of the c-di-GMP class I aptamer derived from *Vibrio cholera* [16] to stem II of the hammerhead ribozyme via a communication module (CM) (Fig. 1). The P1 stem of the aptamer was chosen for CM since the stability of P1 stem is highly involved in the affinity between the aptamer and the ligand. Also, there is an important tertiary interaction between stems P2 and P3 [20, 21], so that using these stems for CM might disrupt key structural contacts within the aptamer. The CM also replaces a majority of the stem II of the hammerhead ribozyme—a structural element that is a critical determinant of the ribozyme activity [22, 23]. The CM within the resulting tripartite construct will provide a sampling of alternative stem II elements that might respond to c-di-GMP binding in the adjacent aptamer domain and confer either positive or negative allosteric control on the adjoining ribozyme domain (*see* **Note 1**). This strategy has been applied to the selection of various allosteric ribozymes (6–8). We have first used the construct shown in Fig. 1a for both c-di-GMP-inhibited and -induced ribozymes. However, the selection for the induced ribozymes was unsuccessful probably because shorter CM (four and five randomized nucleotides) disfavors allosteric induction due to steric clash between important nucleotides of the aptamer and ribozyme. Therefore, we redesigned the construct by elongating randomized sequences (Fig. 1b). This second construct allowed us to obtain the c-di-GMP-induced ribozymes (*see* **Note 2**).

The outline of the selection scheme for the c-di-GMP-activated allosteric ribozyme is shown in Fig. 2. The RNA population of the initial generation (G0) was generated by primer extension of two synthetic oligonucleotides with reverse transcriptase (Fig. 2a), followed by transcription of about 20 pmol (10–20 copies) of the resulting double-strand DNA (Fig. 2b, *see* **Note 3**). After the purification by denaturing PAGE, the full-length RNA precursors were subjected to a negative selection in the absence of c-di-GMP (Fig. 2c). RNAs that remained uncleaved during this reaction were isolated by PAGE and subsequently subjected to a positive selection for self-cleavage in the presence of c-di-GMP (Fig. 2d). The resulting cleavage products were isolated by PAGE and amplified by reverse transcription-polymerase chain reaction (RT-PCR) (Fig. 2e, *see* **Note 4**). The DNA from this stage were transcribed to generate next generation of RNA population, which were then subjected to additional rounds of selective amplification by repeating stages B to E in Fig. 2. The selection for the c-di-GMP-inhibited allosteric ribozyme was done in a same way except for adding c-di-GMP to the transcription and negative selection steps and removing c-di-GMP from positive selection step.

3.2 Construction of the Starting Pool

1. To prepare the double-stranded DNA template for transcription, incubate 40 pmol of each of the two oligonucleotides in 50 μL of 1× reverse transcription buffer supplemented with 10 mM DTT, 200 μM of the four dNTPs, and 8 U/μL reverse transcriptase.
2. Incubate at 37 °C for 2 h.
3. Run a small aliquot out on a 2 % agarose gel to confirm that double-strand DNA was synthesized and is the correct size.

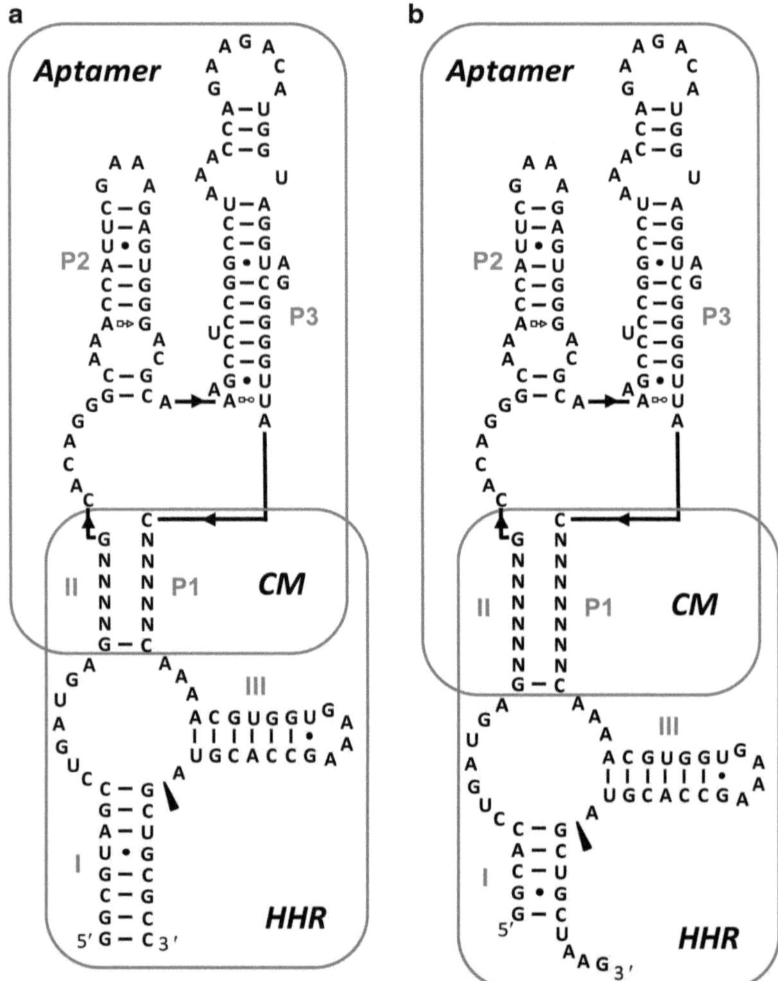

Fig. 2 The selection scheme for the c-di-GMP-activated allosteric ribozyme. (A) Two synthetic oligonucleotides are extended with reverse transcriptase to generate double-strand DNA encoding randomized sequence. (B) Transcription with T7 RNA polymerase. (C) The full-length RNAs (precursor) are purified by PAGE, and subject to negative selection in the absence of c-di-GMP. The cleavage products are removed by PAGE. (D) The precursor are subjected to positive selection in the presence of c-di-GMP. (E) The PAGE-purified cleavage products are reverse transcribed and amplified by PCR

4. Precipitate the double-stranded DNA by adding 0.1 vol of 3 M sodium acetate and 2.5 vol of cold ethanol. Pellet the mixture by centrifugation at 15,000×*g* for 20 min.
5. Resuspend the DNA to 50 μL of 1× transcription buffer supplemented with 2 mM of the four NTPs, 1,000 units of T7 RNA polymerase, and 10 μCi of [α-^{32}P]UTP. For the c-di-GMP-inhibited ribozyme selection, 100 μM c-di-GMP was added to

the mixture to favor isolation of ribozymes that are inactive when bound to c-di-GMP.

6. Incubate at 37 °C for 2 h (*see* **Note 5**).
7. Add 2× denaturing loading buffer to the reaction mixture.
8. Separate full-length precursor RNAs from 3′ cleavage products, premature transcription products, and unincorporated [α-^{32}P]UTP by denaturing 8 % PAGE.
9. Use an autoradiogram of the gel to excise the gel band containing the precursor RNAs.
10. Elute the RNA from the diced gel slices overnight at 4 °C using crush soak buffer.
11. Transfer the solution containing the eluted RNAs into a fresh tube.
12. Concentrate the sample by ethanol precipitation and centrifuge as in **step 4**.
13. Resuspend the RNA in sterile deionized water.

3.3 Selection of Allosterically Inhibited Ribozymes

1. As a negative selection, incubate 300 pmol of an initial pool of RNA precursors (internally ^{32}P-labeled RNA) at 23 °C for 2 h in 1× selection buffer in the presence of 10 mM MgCl2 and 100 μM c-di-GMP (*see* **Notes 6** and 7).
2. To concentrate the RNAs from the reactions, precipitate with ethanol and centrifuge as in Subheading 3.2, **step 4**.
3. Resuspend the RNAs in sterile deionized water and combine with an equal volume of 2 × denaturing gel-loading buffer.
4. Separate the precursor RNAs that resisted cleavage during incubation by denaturing 8 % PAGE. A cleavage marker (full-length precursor) should be run in parallel (*see* **Note 8**).
5. Collect the precursor RNAs as in Subheading 3.2, **steps 9–12**.
6. Resuspend the resulting RNAs in the 1× selection buffer and 10 mM $MgCl_2$ at 23 °C for appropriate time in the absence of c-di-GMP (positive selection) (*see* **Notes 9** and **10**).
7. To concentrate the RNAs from the reactions, precipitate with ethanol and centrifuge as in Subheading 3.2, **step 4**.
8. Resuspend the RNAs in sterile deionized water and combine with an equal volume of 2 × denaturing gel-loading buffer.
9. Separate the cleaved products by denaturing 8 % PAGE. A cleavage marker (e.g., cleaved products in the negative selection) should be run in parallel.
10. Image the resulting gel using a phosphorimager and calculate the fraction of the RNAs that are cleaved (*see* **Notes 11** and **12**).
11. Collect the cleaved products as in Subheading 3.2, **steps 9–13** (*see* **Note 13**).

3.4 Selection of Allosterically Inducted Ribozymes

Selection of ribozymes that undergo c-di-GMP induction was conducted using identical method as allosterically inhibited ribozymes (Subheading 3.3), except that c-di-GMP was included in the positive selection reaction, but was excluded in both the transcription and the negative selection reactions (*see* **Note 14**).

3.5 Reverse Transcription and Polymerase Chain Reaction

1. Add 10 pmol of the 3′ primer (primer A or A′) to the resuspended cleaved products.
2. Reverse transcribe the recovered RNAs in 100 μL of 1× reverse transcribe buffer supplemented with 10 mM DTT, 200 μM of the four dNTPs, and 8 U/μL reverse transcriptase.
3. Incubate at 37 °C for 2 h.
4. Prepare PCR mixture by adding 5–10 μL of the reverse transcription reaction and 20–50 pmol each of primers (primers A and B for allosterically inhibited ribozymes, primers A′ and B′ for allosterically inducted ribozymes) to a 100 μL PCR reaction containing 1× PCR buffer, 200 μM each of the four dNTPs, and 1 U/μL *Taq* DNA polymerase (*see* **Note 15**).
5. Amplify full-length double-stranded DNAs using the following PCR cycling parameters: segment 1: 94 °C for 1 min; segment 2: 94 °C for 30 s, 48 °C for 30 s, 72 °C for 30 s; segment 3: 72 °C for 5 min (*see* **Note 16**).
6. Analyze a small aliquot of the PCR reaction on a 2 % agarose gel to check the amount and size of the PCR products.
7. Use the amplified DNAs as templates to transcribe the RNA pool for the next round of selection, as in Subheading 3.2, **step 5**. Save a portion of the PCR products.
8. Sequence the individual clones after several rounds of selection (*see* **Notes 17** and **18**).

4 Notes

1. A typical communication module ranges from a few to a dozen nucleotides. The two constructs (Fig. 1) used in our study contains 9 and 13 random nucleotides, respectively, in the domain of the communication module. These should give us the anticipated RNA libraries with great diversity (262,144 (4^9) and 67,108,864 (4^{13}) pieces of different RNA sequences, respectively).
2. When the hammerhead self-cleaving motif is fused with the large c-di-GMP-I aptamer to create the c-di-GMP-dependent allosteric ribozymes, there may be some steric hindrance between the two moieties that could possibly affect the isolation of active allosteric ribozymes. To minimize the potential steric clash, a longer communication module and truncated stem I of the hammerhead ribozyme was used as shown in Fig. 1b.

3. Considering that a typical run-off transcription produces about 10–20 copies of each DNA sequence, including 20 pmol of DNA template in the initial transcription step should give us more than 200 pmol of the RNA library with guaranteed sequence diversity.
4. When engineering the RNA constructs for allosteric selection, make sure the anticipated difference in size between the full-length RNA precursor and the cleaved RNA product is sufficient (no less than eight nucleotides) to be resolved by denaturing PAGE gels. Also ensure that there are enough fixed sequences flanking the random domain (no less than 13 nucleotides) to allow the efficient PCR amplification of the cleaved product with the appropriate primers.
5. A white precipitate or cloudy solution may be seen during the course of in vitro transcription, resulting from the formation of an insoluble compound between Mg^{2+} and inorganic pyrophosphate, which indicates that a large amount of RNA has been produced. The insoluble compound can be easily removed by PAGE purification.
6. To select for c-di-GMP-inhibited ribozymes, 100 μM c-di-GMP was included in the mixture to ensure that the active allosteric ribozymes did not cleave in the transcription step. The starting concentration of c-di-GMP in the selection reaction is influenced by its binding affinity to the aptamer. The c-di-GMP-I aptamer used in this study has a reported K_D value in the picomolar range. Hence, it is very likely to recover all possible c-di-GMP-dependent allosteric ribozymes by using micromolar to nanomolar concentrations of c-di-GMP, which is significantly above its K_D value.
7. The order of addition of the components is critical for the selection. In the negative selection for c-di-GMP-inhibited ribozymes, RNA was firstly mixed with the assay buffer that does not contain Mg^{2+}. Then c-di-GMP was added and the mixture was annealed to allow the RNAs to prefold in the presence of 100 mM NaCl in the assay buffer. Last, 10 mM Mg^{2+} (final concentration) was added to the mixture to initiate the negative selection.
8. It is better to collect the cleaved product band in the negative selection step, for use as a size marker in the following positive selection step. However, one must pay attention not to cross-contaminate the samples. To avoid the possible cross-contamination during PAGE purification, it is suggested keeping a one- or two-lane distance between the marker and the selection product as the samples are loaded into the gel.
9. In the positive selection for c-di-GMP-inhibited ribozymes, the same order was performed except that no c-di-GMP was mixed with RNA. Also, the same order was applied to select

for c-di-GMP-induced ribozymes, except that c-di-GMP was removed in the negative selection and included in the positive selection.

10. In order to select for the most active effector-dependent ribozymes, the incubation time for the positive selection reaction was reduced after several rounds of selection. Usually in the early rounds of selection, the RNA population was incubated for relatively long time (15 min) to increase the recovery of c-di-GMP-dependent self-cleaving RNAs. As more and more rounds of selection were conducted, the duration of the incubation time was gradually reduced to 1 min.
11. In the early (typically two to three) rounds of selection, the fraction RNAs cleaved in both the positive and negative selection reactions should be at the similar level, because the effector-independent RNAs still dominate the early populations.
12. Usually the radioactive signal of the cleaved product is sufficient to allow detection on an autoradiogram after an exposure for 10–30 min. Using fresh [α-^{32}P] UTP to internally label RNA can enhance the radioactive signal of the cleaved product for better detection.
13. The efficient recovery of cleaved RNAs from the positive selection step is important in the early rounds of selection because the occurrence of c-di-GMP-dependent sequences is rare. To avoid losing most molecules during the ethanol precipitation step, glycogen (10 μg) was added as a carrier to help improve the recovery yield.
14. In order to select for the effector-dependent ribozymes with the best binding affinity, the concentration of c-di-GMP was decreased as more rounds of selection were conducted. Initially 100 μM c-di-GMP was included in the selection to ensure the maximal recovery of c-di-GMP-dependent self-cleaving RNAs. After several rounds of selection, the concentration of c-di-GMP was reduced to 0.3 μM to favor the isolation of the allosteric ribozymes that bind c-di-GMP tightly. However, drastically decreasing the effector concentration may result in the loss of the most sensitive effector-dependent ribozymes. It is suggested that in the later rounds of selection, a 5- to 10-time decrease in the effector concentration is performed each time.
15. During the first couple rounds of selection, it is recommended to amplify all of the cDNA. This will increase the probability that even the rarest of the active molecules will be amplified and propagated in subsequent generations.
16. The number of cycles required for complete amplification depends on the amount of input template DNA. For each round of selection, the number of PCR cycles is estimated from the quantity of input cDNA, which in turn is estimated

from the fraction of cleaved RNA in the positive selection step. Over-amplifying cDNA with more cycles than needed will generate undesired higher molecular weight products due to mispriming.

17. To clone the PCR products, it is very convenient to work on a TOPO TA cloning kit (Invitrogen) because this method does not require restriction sites at both the two ends of the PCR amplified product.
18. To generate DNA templates for the transcription of individual clones for further analysis, either PCR amplifying the plasmid DNA or chemically synthesizing the oligo-strands can be conducted.

Acknowledgements

We thank members of the Breaker laboratory for helpful discussions. RNA research in the Breaker laboratory is also supported by the Howard Hughes Medical Institute.

References

1. Breaker RR (2004) Natural and engineered nucleic acids as tools to explore biology. Nature 432:838–845
2. Famulok M (2005) Allosteric aptamers and aptazymes as probes for screening approaches. Annu Rev Microbiol 59:487–517
3. Tang J, Breaker RR (1997) Rational design of allosteric ribozymes. Chem Biol 4:453–459
4. Tang J, Breaker RR (1998) Mechanism for allosteric inhibition of an ATP-sensitive ribozyme. Nucleic Acid Res 26:4214–4221
5. Soukup GA, Breaker RR (1999) Design of allosteric hammerhead ribozymes activated by ligand-induced structure stabilization. Structure 7:783–791
6. Soukup GA, Breaker RR (1999) Engineering precision RNA molecular switches. Proc Natl Acad Sci U S A 96:3584–3589
7. Koizumi M, Soukup GA, Kerr JNQ, Breaker RR (1999) Allosteric selection of ribozymes that respond to the second messengers cGMP and cAMP. Nat Struct Biol 6:1062–1071
8. Gu H, Furukawa K, Breaker RR (2012) Engineered allosteric ribozymes that sense the bacterial second messenger cyclic diguanosyl 5′-monophosphate. Anal Chem 84:4935–4941
9. Wieland M, Benz A, Klauser B, Hartig JS (2009) Artificial ribozyme switches containing natural riboswitch aptamer domains. Angew Chem Int Ed 48:2715–2718
10. Win MN, Smolke CD (2007) A modular and extensible RNA-based gene-regulatory platform for engineering cellular function. Proc Natl Acad Sci U S A 104:14283–14288
11. Ogawa A, Maeda M (2008) An artificial aptazyme-based riboswitch and its cascading system in *E. coli.* Chembiochem 9:206–209
12. Robertson MP, Ellington AD (1999) In vitro selection of an allosteric ribozyme that transduces analytes to amplicons. Nat Biotechnol 17:62–66
13. Najafi-Shoushtari SH, Famulok M (2005) Competitive regulation of modular allosteric aptazymes by a small molecule and oligonucleotide effector. RNA 11:1514–1520
14. Lee ER, Baker JL, Weinberg Z, Sudarsan N, Breaker RR (2010) An allosteric self-splicing ribozyme triggered by a bacterial second messenger. Science 329:845–848
15. Hengge R (2009) Principles of c-di-GMP signaling in bacteria. Nat Rev Microbiol 7:263–273
16. Sudarsan N, Lee ER, Weinberg Z, Moy RH, Kim JN, Link KH, Breaker RR (2008) Riboswitches in eubacteria sense the second messenger cyclic di-GMP. Science 321:411–413
17. Mayer G, Famulok M (2006) High-throughput-compatible assay for *glmS* riboswitch metabolite dependence. Chembiochem 7:602–604

18. Lünse CE, Schmidt MS, Wittman V, Mayer G (2011) Carba-sugars activate the glmS-riboswitch of *Staphylococcus aureus*. ACS Chem Biol 6:675–678
19. Furukawa K, Gu H, Sudarsan N, Hayakawa Y, Hyodo M, Breaker RR (2012) Identification of ligand analogues that control c-di-GMP riboswitches. ACS Chem Biol 7:1436–1443
20. Smith KD, Lipchock SV, Ames TD, Wang J, Breaker RR, Strobel SA (2009) Structural basis of ligand binding by a c-di-GMP riboswitch. Nat Struct Mol Biol 16:1218–1223
21. Kulshina N, Baird NJ, Ferré-D'Amaré AR (2009) Recognition of the bacterial second messenger cyclic diguanylate by its cognate riboswitch. Nat Struct Mol Biol 16:1212–1217
22. Forster AC, Symons RH (1987) Self-cleavage of virusoid RNA is performed by the proposed 55-nucleotide active site. Cell 50:9–16
23. Fedor MJ, Uhlenbeck OC (1992) Kinetics of intermolecular cleavage by hammerhead ribozymes. Biochemistry 31:12042–12054

Chapter 16

Dual-Selection for Evolution of In Vivo Functional Aptazymes as Riboswitch Parts

Jonathan A. Goler, James M. Carothers, and Jay D. Keasling

Abstract

Both synthetic biology and metabolic engineering are aided by the development of genetic control parts. One class of riboswitch parts that has great potential for sensing and regulation of protein levels is aptamer-coupled ribozymes (aptazymes). These devices are comprised of an aptamer domain selected to bind a particular ligand, a ribozyme domain, and a communication module that regulates the ribozyme activity based on the state of the aptamer. We describe a broadly applicable method for coupling a novel, newly selected aptamer to a ribozyme to generate functional aptazymes via in vitro and in vivo selection. To illustrate this approach, we describe experimental procedures for selecting aptazymes assembled from aptamers that bind *p*-amino-phenylalanine and a hammerhead ribozyme. Because this method uses selection, it does not rely on sequence-specific design and thus should be generalizable for the generation of in vivo operational aptazymes that respond to any targeted molecules.

Key words In vitro selection, RNA aptamer, Ribozyme, Aptazyme, Riboswitch, Synthetic biology

1 Introduction

Synthetic biology in general, as well as construction of novel pathways in cells, needs control devices that can regulate cellular behavior. Using the circuit metaphor, these devices are thought of as switches or gates, while in the microbial chemical factory metaphor, the can be thought of as sensors and control valves. In both cases, we are concerned with similar parameters: the input signal level and the output range; additionally, response time may be an important factor. Natural RNA aptamers are known to bind small molecules in order to regulate several systems, including thiamine synthesis, B12 transport, and the synthesis of several nucleotides [1–9]. In addition, several examples of synthetic aptamer-based gene regulatory riboswitches have been successfully demonstrated in bacteria, yeast and mammalian cells [10]. Extensive work has shown that ligand-responsive, self-cleaving ribozymes (aptazymes) can be engineered by connecting ligand-binding aptamer domains

Atsushi Ogawa (ed.), *Artificial Riboswitches: Methods and Protocols*, Methods in Molecular Biology, vol. 1111,
DOI 10.1007/978-1-62703-755-6_16, © Springer Science+Business Media New York 2014

to the self-cleaving hammerhead ribozyme core structure through secondary structures utilizing a helix-slipping mechanism, in which three or more tracts of nucleotides interact differently based on the physical presence of another molecule [11–14]. This effect changes the three-dimensional structure of the aptazyme, which can cause it to become either more or less active, in terms of the rate of phosphodiester bond self-cleavage. In this chapter, we describe a generalizable dual-selection method for engineering functional aptazymes that can be used as riboswitch parts from individual aptamers and ribozymes.

To illustrate our method for rapidly generating novel aptazymes, we describe the selection of a composite aptazyme structure as a sensor RNA component that is capable of detecting *p-aminophenylalanine* (*p*AF) in in vivo-like conditions. *p*AF is a non-proteinogenic amino acid (MW = 180 g/mol) that is a precursor to the pristinamycin family of antibiotics and other aromatic compounds with potential industrial-relevance. It is cell-permeable and can be produced via a 12 gene engineered pathway [15]. In the current example, the *p*AF-R1-1 aptamer, previously selected to bind the target ligand with micromolar dissociation constant in physiological buffer conditions [16], is coupled to a *Schistosoma mansoni* hammerhead ribozyme. In principle, the same methods can be applied to engineer functional aptazymes from almost any desired ligand-binding aptamer domain and any one of many different self-cleaving phosphodiester bond cleaving ribozymes [17].

The general outline of the first, in vitro, stage of the method is based on work by Link et al. [18], in which they generated theophylline aptazymes from a combinatorial library comprised of the ubiquitous Jenison theophylline aptamer [19] coupled to an *S. mansoni* ribozyme. The combinatorial sequence library used to generate the composite aptazymes structures had three parts of the sequence bridging the aptamer and ribozyme domains randomized, with two 5-base tracts, and one 4-base tract. Although Link et al. found that their aptazymes demonstrated tremendous (285-fold) differences in cleavage rates in the presence versus the absence of ligand in vitro, they were not able to dynamically regulate gene expression in the in vivo context they employed. Later experiments by Carothers et al. showed that model-driven approaches could be employed to design a genetic control mechanism and sequence context in which the Link aptazymes could be assembled into functional aptazyme-Regulated Expression Devices (aREDS) with quantitatively predictable genetic outputs in *E. coli* [15]. The dual-selection method given here was developed in the course of generating *p*AF-responsive aptazyme parts.

Briefly, a combinatorial sequence library is designed where the aptamer is attached to the ribozyme at the base-stem, via randomized tracts, and in vitro transcribed. Additionally, the size of a library is important: transformation efficiency into *E. coli* is limited

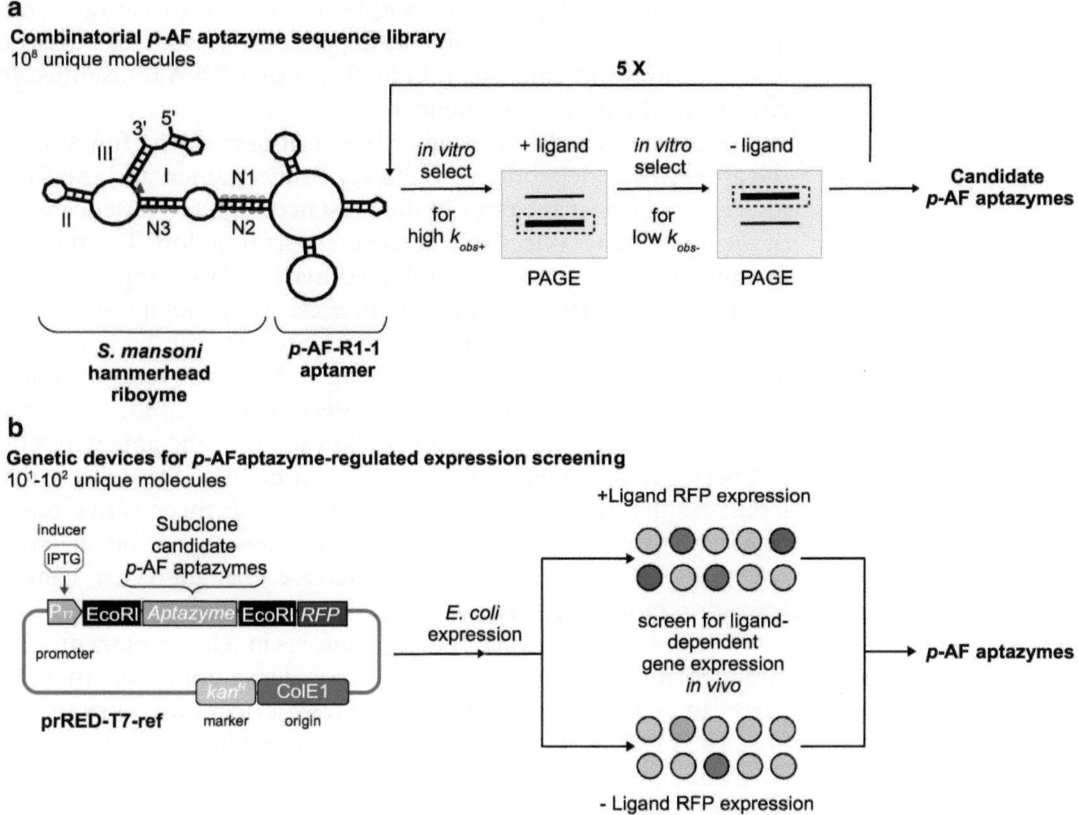

Fig. 1 Dual-selection for ligand-dependent self-cleaving aptazymes. (**a**) To generate aptazymes, a combinatorial sequence library is constructed from individual aptamer and ribozyme domains. In the illustration, a sequence library comprises the *p*AF-R1-1 aptamer ($K_d = 19$ μm) and *S. mansoni* hammerhead ribozyme ($k_{obs} = 1.3$ min^{-1}), shown in secondary structure schematic, was constructed. 14 fully randomized bases within stem I of the hammerhead ribozyme are colored and marked N1, N2, and N3. Aptazyme self-cleavage takes place at the site indicated by the red triangle. PAGE is used to enrich the library pool for aptazymes with fast k_{obs+} by isolating RNA that cleaves rapidly in the presence of ligand (*dashed box at left*) and for aptazymes with slow k_{obs-} by isolating RNA that does not cleave in the absence of ligand (*dashed box at right*). The complete cycle is repeated 4–6 times. (**b**) Candidate aptazymes from the in vitro selection are subcloned into a genetic device for aptazyme-regulated expression screening. A subsequent screen for ligand-dependent regulation of reporter gene expression identifies aptazymes that catalyze self-cleavage reactions inside the cell

to approximately 109, so libraries larger than that cannot be explored solely in vivo [20]. The first stage of (in vitro) selection is then used to enrich for a pool of RNAs that preferentially cleave in the presence of the ligand, but not in the absence of it. In the second stage, the resulting molecules are then cloned into a genetic device (Fig. 1) for aptazyme-regulated expression screening in *E. coli*. Individual bacterial colonies are picked, grown and assayed for differences in reporter gene expression as a function of exogenous ligand added to the growth media. DNA from individual colonies is sequenced to identify individual aptazyme parts exhibiting ligand-dependent phosphodiester bond cleavage activity. Finally, the kinetics of ligand-dependent cleavage are assayed for aptazymes showing function in vivo, using RNA transcribed in vitro from the isolated sequences.

In the context of an aptamer-based ribozyme, the function of the aptazyme is dependent on the k_d of the aptamer, and the folding (k_{fold}) and cleavage rates in the presence (k_{obs+} and absence k_{obs-}) of the ligand that collectively determine part function. The manner in which the selections are conducted makes a large impact on the functionality of the resulting sequences. To generate structures meeting targeted performance criteria, pressure can be applied in terms of time, which affects the folding and catalytic rate constants, and ligand concentrations, which affects the apparent k_d of the aptazyme, to tune the in vitro selection stage. If the aptazymes do not perform as needed, the selections can be biased to achieve different results, for example, the length of time the positive selections are run can be shortened and the length of the negative selections can be lengthened to increase the aptazyme ligand-activation ratios (k_{obs+}/k_{obs-}).

The co-transcriptional folding contexts in which aptazymes are intended to be active are important for determining whether the parts can be used to build genetic control devices functional in vivo [15, 18, 21, 22]. At a minimum, this context includes the cellular environment, pH, temperature, as well as the local flanking sequences and identities of the promoter and RNA polymerase, which affect transcription elongation rates, and thus the kinetics of part folding. In light of these considerations, the first stage, in vitro selection should be performed in physiological buffer conditions [16]. Given local sequence constraints that may prevent aptazyme folding in the context of an elongating transcript in vivo, it is important to realize that the in vivo screen will enable the identification of functional aptazyme parts, but will not tend to exhaustively identify all of the active sequences in a given pool.

To assemble aptazyme component parts into genetic expression devices with quantitatively predictable functions, we previously utilized an underlying control mechanism that targets specific RNA degradation pathways in the cell [23]. The aptazyme self-cleavage reaction produces a 5′ hydroxyl-terminated downstream signal that is degraded by a relatively slow RppH-independent pathway, the effect of which is a predictable increase in mRNA half-life. By developing a biochemical and biophysical model-driven approach, we showed that aptazymes generated and characterized in the ways described here could be assembled into aREDs with the capacity to quantitatively program gene expression levels in vivo. In principle there are many other mechanisms and contexts in which aptazyme parts generated through dual in vitro and in vivo selection could be employed to build sensors and genetic control devices for engineered circuits and pathways [24–26].

Selecting, rather than designing ligand-responsive aptazymes, is a robust method for generating component parts that can be used to engineer aptazyme-regulated expression devices (aREDs) functional in vivo [15]. The use of a two-stage method allows one to take advantage of the high sequence diversity that can be screened with in vitro selection, while utilizing the in vivo selection to rapidly screen the pool of enriched molecules for aptazyme sequences that can undergo self-cleavage in a cellular environment. The generality of the design and procedure should permit its application in a wide range of potential applications where an aptamer to an intermediate, signaling molecule, or protein can be isolated.

2 Materials

Prepare all solutions using RNAase-free water (fresh 18 MΩ pure water from a Milli-Q system for example).

2.1 In Vitro Selection

2.1.1 Solutions and Buffers

1. In vitro Transcription Mix: 40 mM Tris–HCl, pH 7.8, 20 mM each nucleotide triphosphate (NTP), 25 mM $MgCl_2$, 2.5 mM Spermidine, 10 mM dithiothreitol (DTT), 600 μM Top Strand ssDNA, 500 μM Template ssDNA, 50 units/μL T7 Polymerase, 10 units/μL Thermostable inorganic pyrophosphatase (TIPP, New England Biolabs, Inc. USA), 200 units/μL RNAsin (Promega, Inc. USA), 0.1 μCi 32P-ATP (where desired), 0.01 % Triton X-100.
2. 10× TBE Buffer: 108 g of Tris base, 55 g of boric acid, 40 mL of 0.5 mM EDTA in 1 L of RNAse-free water (*see* **Note 1**). Utilize 1× TBE in PAGE gels and the PAGE running buffer, 0.5× TBE in the electro-elution steps.
3. 2× Selection Buffer (2× SB): 130 mM K-Glutamine, pH 7.5, 15 mM NaCl, 10 mM DTT, 1 mM $MgCl_2$.

2.1.2 PAGE Gel

1. APS solution: 25 mg of ammonium per sulfate in 100 mL of RNAse-free water and vortex (*see* **Note 2**).
2. TEMED.
3. Acrylamide (19:1).
4. RNAse-free urea.
5. Clean razor blades.
6. TLC (thin-layer chromatography) plates.
7. UW wand.
8. Glass plates and plastic separators.
9. Binder clips.
10. 2× Loading buffer: 95 % formamide, 5 % 0.5 M EDTA.

2.1.3 Ethanol Precipitation

1. 3 M RNAse-free KCl.
2. RNAse-free pure or 95 % ethanol.
3. Dry ice or a −20 °C freezer.
4. Tabletop centrifuge.

2.1.4 Equipment

1. Whatman Elutrap System.
2. NanoDrop Spectrophotometer (ThermoFisher, Inc. USA).
3. Ready-To-Go RT-PCR beads (GE Healthcare Inc., USA).
4. Plate Reader such as Tecan Safire.
5. Corning Costar plates (or other 96-well plates compatible with your plate reader).
6. Phosphorimager Screen and Reader (such as a Typhoon 8600).

2.2 In Vivo Screen

1. LB media.
2. BL-21 cells (Life Technologies, Inc., USA).
3. Top-100 Chemically competent cells (Invitrogen).
4. Ice.
5. MOPS M-9 media.
6. LB-agar plates with kanamycin (Sigma).
7. 0.5 mM Isopropyl-$_{d}$-1-thiogalactopyranoside (IPTG).

3 Methods

3.1 PAGE Gel

Assemble the glass with spacers, prepare gel by adding acrylamide, urea, and TBE (5 mL of 10× TBE, 12 mL of 20 % acrylamide, 24 g of urea and water to 50 mL); add 50 μL of TEMED and 500 mL of APS last and swirl to initiate the polymerization reaction (*see* **Notes 2** and **3**).

3.2 Running PAGE Gels

1. Pre-run the gels using constant power (18–22 W) for 20–30 min to warm the gel a bit, the gel should be approximately 40–50 °C (*see* **Note 4**).
2. After the pre-run, clean the urea from the wells by utilizing a sterile syringe. Withdraw buffer and expel firmly to flush out the wells, you should see the urea being blown out.
3. Add loading buffer (95 % formamide, 5 % 0.5 M EDTA) to the sample in a 1:1 ratio and gently mix.
4. Load the samples into the gel (*see* **Note 5**), actuate the power and run the gel for 1–1.25 h.
5. Once the gel is complete, separate the gel from the glass plates for UV shadowing using a fluorescent PEI-Cellulose TLC plate (*see* **Note 6**).

6. Briefly illuminate the gel with a UV wand set to a short wavelength (preferably 254 nm), using a marker to mark the band locations.
7. Put aside the TLC plate and place the wrapped gel back on a glass plate, excise the desired band with a clean razor blade.
8. If you are quantifying band intensity, expose the gel to a freshly erased phosphor screen.

3.3 Stage One: In Vitro Aptazyme Selection

In this illustration, we used the *p*AF-binding aptamer *p*AF-R1-1 ($k_d = 30$ μM at 2.5 mM Mg^{2+} [16], which was connected to the *S. mansoni* ribozyme via randomized regions as in Link et al. [18] to construct the starting combinatorial sequence library. Oligos were as follows (Integrated DNA Technologies): LSW-11F 5′-GCT AA TAC GAC TCA CTA TAG GCG AAA GCC GGC GCG TCC TGG ATT CCA CNN NNN CAT GTC CCT ACC ATA CGG GAT TGC CCA GCT TCG GCT GCC ATG CCG GC, LSW-11Rc: 5′-AGC GCG TTT CGT CCT ATT TGG GAC TCA TCA GCN NNN TGT ACC NNN NNC GTA GGC CGG TTA CCG TTT GGC CGG CAT GGC AGC CG and primers LSW-11Fp: 5′-GCT AAT ACG ACT CAC TAT AGG CGA AAG CCG GCG CGT CCT GGA TTC CAC and LSW-11Rp: 5′-AGC GCG TTT CGT CCT ATT TGG G (also *see* **Note 7**).

3.3.1 PAGE Purification of Oligo DNA

The DNA oligos are PAGE-purified, primer-extended, and transcribed in vitro, as detailed below.

1. First, solubilize the oligos in ultrapure water, add 1:1 2× loading buffer and vortex.
2. Purify oligos on a 10–12 % PAGE gel, running gels as above.
3. Illuminate gels with UV light against a PEI-Cellulose TLC plate, mark and excise bands with a razor blade.
4. Elute gel slices (Whatman Elutrap System) at 300 V for 30 min to 2 h into a final volume between 200 and 400 μL (*see* **Note 8**).
5. Ethanol precipitate the eluate with 1/10 volume of 3 M KCl and 2.5 volumes of 95 % ethanol, chill for 10 min on dry ice (solution should become slightly whiter and more viscous), spin for 15 min at 24,000 × *g* in a refrigerated centrifuge. Decant using a pipette and air dry at room temperature or 37 °C.
6. Quantify DNA using a NanoDrop or other spectrophotometer.

3.3.2 In Vitro Selection

1. Prepare transcription mix according to the materials section (*see* **Note 9**).
2. Add the template and top strand oligos (if needed) to the transcription mix.
3. Run transcription at 20–25 °C for 15 min to 2 h.

4. Stop transcription reactions with an equal volume of 2× loading buffer, mix and run out on an 8 % PAGE gel at 22 W for 1–1.25 h.
5. Excise the larger band (full length RNA), elute as above, ethanol-precipitate, and dry the pellet.
6. Resuspend the pellet in water, flicking several times, and lightly vortexing.
7. Add 1× Selection Buffer (SB) with or without the desired ligand and incubate at 37 °C for 5–15 min.
8. In the positive selections (with ligand), the smaller (cleaved) band is excised. For the negative selections, the larger (uncleaved) fraction is excised.
9. Other purification, gel excision, electro-elution, precipitation, are performed as above.
10. The resultant RNA is resuspended in ultrapure water and subjected to 8–12 rounds of RT-PCR using Ready-To-Go beads using the manufacturer's protocol. After 8–12 rounds of PCR, the reactions are visualized on a 2 % agarose gel; if there is no band of the expected length, perform three to four more cycles of basic PCR and recheck (*see* **Note 10**).

3.4 Stage Two: In Vivo Aptazyme Screen

After four to six rounds of in vitro aptazyme selection, the aptazyme parts are screened for ligand-dependent self-cleavage in vivo.

1. The enriched, surviving, sequences are PCR amplified using cloning adapters (For the *p*AF aptazyme selections these are Adapter4A.F: 5′-GAT ATC ATC TCT AGA GGC GAA AGC CGG CGC GTC CTG GAT TCC AC, Adapter4AR: 5′-GAC CTT CTC GAG AGC GCG TTT CGT CCT ATT TGG G), and cloned into prRED-T7-ref (Fig. 1), which harbors the kanamycin (Kan) resistance gene, a T7 promoter, an XbaI-BglII cloning site, and the gene encoding the dsRedExpress Red Fluorescent Protein (RFP) [15] (*see* **Note 11**).
2. The resulting plasmids are then transformed initially into *E. coli* TOP 10 chemically competent cells.
3. For the transformation, add a 2–3 μL of the enriched library DNA to the cells.
4. Incubate on ice for 10 min.
5. Place aliquots in a 42 °C water bath for 15 s, then back on ice for 2 min.
6. Incubate aliquots at 37 °C, shaking, for 75 min.
7. Plate cultures on LB-Kan plates and incubate overnight.
8. The next day, pick individual colonies into test tubes, each with 5 mL of LB medium containing 50 μg/mL kanamycin (LB-Kan).

9. Incubate tubes at 37 °C, shaking, 12 h or overnight.
10. Use a standard plasmid prep to recover plasmids.
11. Re-transform each plasmid into *E. coli* BL-21 chemically competent cells harboring an IPTG-induced T7 expression cassette.
12. Individual colonies are picked and grown for 8 h to overnight in LB-Kan.
13. To assay performance, these cultures are then used to inoculate six tubes containing 5 mL of LB-Kan. Three of these tubes have either none, or saturating concentrations, of the desired ligand (in the current illustration, 0 and 5 mM *p*AF were used). 0.5 mM IPTG is added to all of the tubes to induce transcription from the T7 promoter, and the cultures are incubated with shaking.
14. Samples are then assayed at 12, 24, 36, and 48 h. 100 μL from each culture is taken and added to a 96-well Costar plate and assayed for optical density (OD380) and fluorescence (excitation measured at 557 nm; emission measured at 579 nm) in a plate reader.
15. Samples that exhibit a difference in expression ± ligand are then plasmid-prepped, sequenced and re-transformed into fresh BL-21 cells, and re-assayed to confirm the ligand-dependent cleavage activity, one such resulting structure is shown in Fig. 2.
16. To sequence samples, mini-prep the cultures that exhibit a difference in expression, you can then fold the sequence using an RNA folding program like Vienna RNA, producing an image such as Fig. 2.

3.5 Measuring Aptazyme Kinetics In Vitro

1. To confirm that the sequences exhibiting ligand-dependent differences in gene expression observed in the in vivo screen encode functional aptazymes, you must show that the presence of the ligand significantly increases the cleavage rate constant in vitro.
2. DNA from colonies exhibiting switching behavior is reamplified using the original selection primers with T7 promoter and transcribed with 32 P-ATP. Transcribe sufficient quantities for several experiments (200–500 μL) and purify as above.
3. Resuspend purified RNA in water (100 μL).
4. Add RNA to selection buffer SB with or without the desired target ligand (in triplicate) mix, and hold at the appropriate temperature (e.g., 30, or 37 °C) for 1, 3, 5, 10, and 15 min, stopping the reaction by adding 2× loading buffer and quickly mixing.

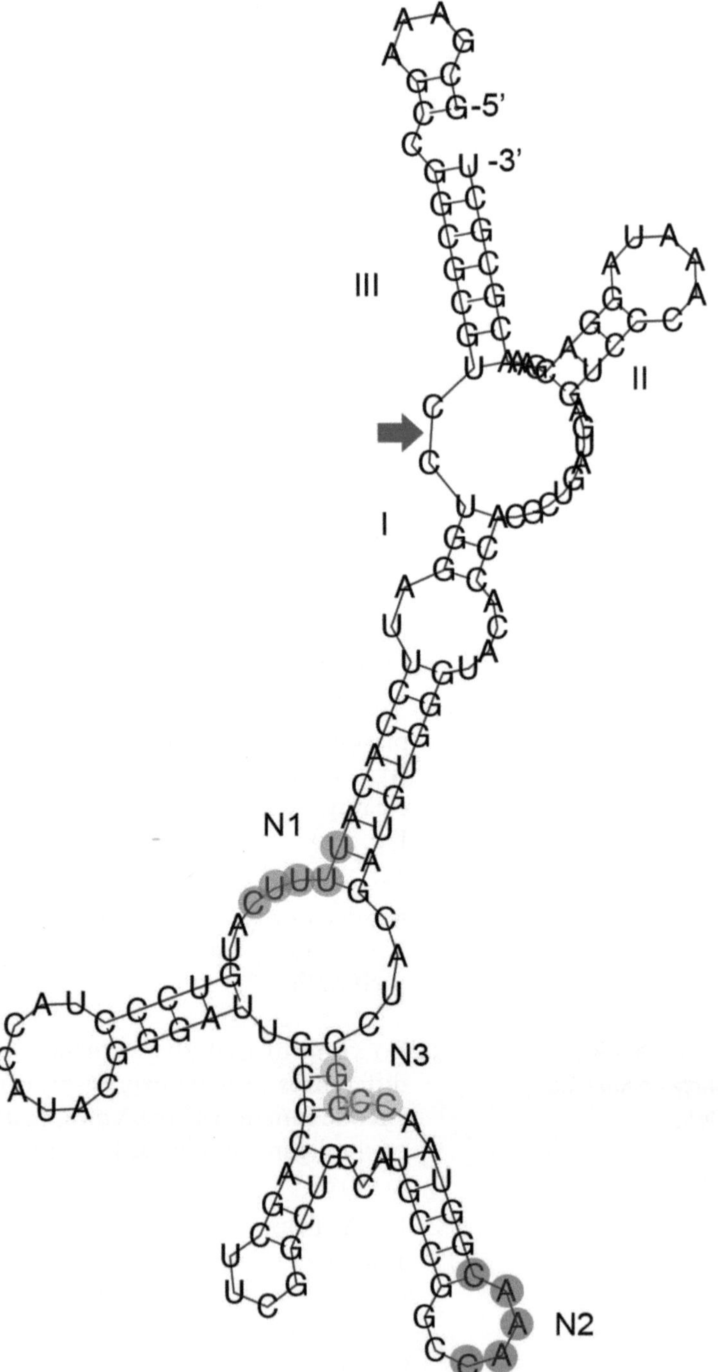

Fig. 2 Winner from the example *p*AF aptazyme selection, W24. Secondary structure of the minimum free energy fold, as computed with Vienna RNAfold with default settings, for the best performing *p*AF aptazyme from the example selection is shown with number and color coding as in Fig. 1

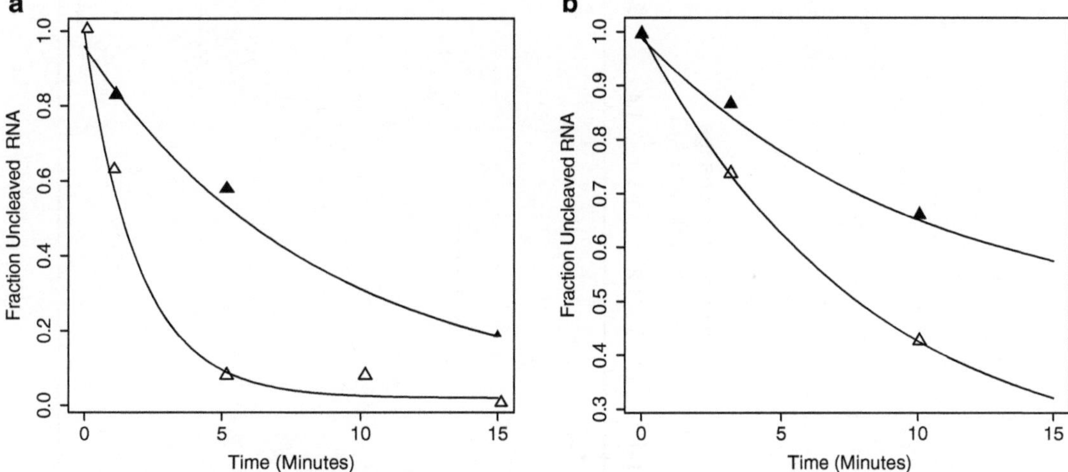

Fig. 3 Measuring the kinetics of aptazyme cleavage in vitro. The kinetics of ligand-dependent cleavage are assayed for aptazymes showing function in vivo, using RNA transcribed in vitro from the isolated sequences. In this figure, results for two isolates from the *p*AF aptazyme selections are shown. The fraction of self-cleaved aptazyme in the presence (*open triangles*) and absence (*solid triangles*) of the target ligand is plotted as a function of time. The data are fitted to a first-order decay curve, overlaid

5. Load samples on a 10–12 % PAGE gel and run for 90 min.
6. Expose the gel to a phosphorimager screen for 20 min, and scan on a Typhoon 8600 (50 μm resolution), or similar equipment.
7. Measure bands and analyze for volume (counts over an area).
8. Plot the fraction of aptazyme RNA cleaved as a function of time and use nonlinear regression to calculate the aptazyme cleavage rate constant [18]. An example is shown in Fig. 3.

To further confirm that functional ligand-responsive aptazymes have been isolated, the expression device plasmids can be re-transformed into *E. coli* BL-21 cells, and re-assayed in triplicate against a range of target ligand concentrations. For instance, Fig. 4 shows that the genetic device encoding the W24 aptazyme exhibits clear, dose-dependent, differences in gene expression upon the addition.

4 Notes

1. For TBE buffer, prepare the solution in a bottle, add 70 % of the water to begin, add a RNAse-free stir bar, and put on a stir plate with heating, get the stirrer rotating and slowly add the dry chemicals, finally adding water to 1 L, and allow to dissolve. If you work in a general molecular biology lab, it is important to keep a separate supply of chemicals to prevent contamination with RNAse.

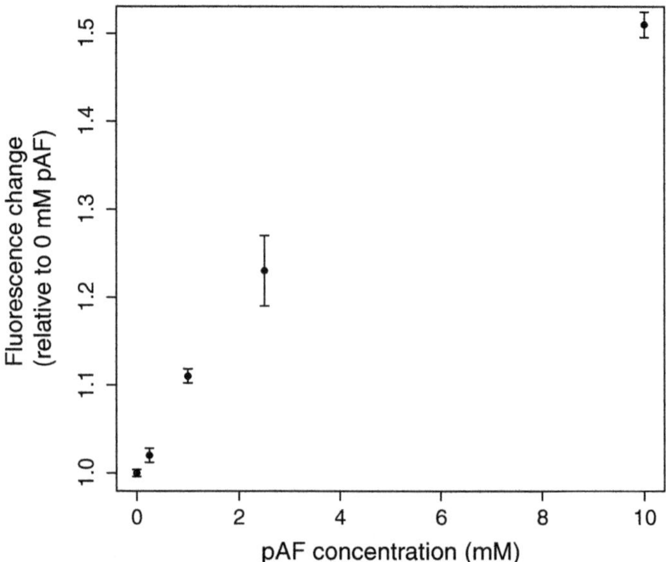

Fig. 4 Aptazyme-regulated genetic expression. The genetic device encoding the W24 aptazyme exhibits clear, dose-dependent, differences in gene expression upon the addition of *p*AF to the media, confirming the isolation of functional ligand-responsive aptazymes. In the example, isolate W24 was assayed in triplicate, with RFP expression normalized to RFU/OD380 and plotted relative to a negative control, as measured 48 h post-induction

2. APS solution should be prepared fresh or stored refrigerated not more than 2 weeks. Prepare by adding 25 mg of ammonium persulfate to 100 mL of ultrapure water, vortex until dissolved.
3. For PAGE, create the gel utilizing glass plates, one 1–2 cm longer than the other, with 1.5 mM spacers and comb (you can utilize either a multichanneled comb to quantify and measure a series of samples, or a single wide comb for purification of a large transcription. Clean the plates with RNAse away or similar reagent, followed by a wash with running Milli-Q water. To pour PAGE gels, assemble the glass with spaces, leave the comb aside for now, the bottoms of the plates should be aligned (The top has the differential height to allow the buffer to flow to the channels but not spill). Secure the plates in place using binder clips, ensuring that they are springy enough to hold the spacers and glass together tightly. Carefully pour the solution into the plate-rig, ensuring that no bubbles form; you can shift the position or lightly tap the rig to allow the bubbles to rise to the surface. Add the comb into the top of the gel. Watch the edges of the rig to ensure that there are no leaks. The gel should be ready in 30–45 min.
4. When running gels, we utilize constant power (as opposed to voltage), from 18–22 W. It is important to ensure that the gel

does not overheat (the gel should be approximately 40–50 °C, if it feels very hot, turn down the power), high temperature could break the glass plates as well as cause bands to smear.

5. Utilize a long-tipped pipette to load the samples into the desired wells of a PAGE gel, putting the solution at the bottom of each well. You may choose to utilize dye (xylene cyanol and bromphenol blue) with the sample or as tracking/ladder in a separate well. The sample's index of refraction is different from the buffer, so you should be able to see it being injected into the well; do not overfill the wells (you may see it flow over the sides of the well) that will prevent a clean separation.

6. Once the gel is complete, remove the rig, remove the edges, and very carefully separate the gel from the glass plates. The first plate should separate, if it is difficult, use a tweezer to get it started but proceed slowly to not damage the gel. Once the first plate is removed, add a piece of saran wrap in its place, ensuring that it is flush; you can then add a fluorescent PEI-Cellulose TLC plate on top of the saran wrap, flip the gel over and slowly remove the second plate. Replace with a piece of saran wrap. If you are removing a band, find a dark place or turn off the lights, briefly illuminate the gel with a UV wand set to a short wavelength (preferably 254 nm), using a marker to mark the band locations. Put aside the TLC plate and place the wrapped gel back on a glass plate, excise the desired band with a clean razor blade. If you are quantifying band intensity, expose the gel to a freshly erased phosphor screen for 15 min to 4 h, depending on the intensity of the 32P used, scan the screen (we utilize a Typhoon 8600 with 50 μm resolution). Utilize the included software package to measure the volume (counts in a given area).

7. To evolve a new custom aptazyme, create oligos that encompass the aptamer, the randomized section and ribozyme core, and PCR primers that regenerate the otherwise-cleaved portion of the ribozyme (Fig. 1). Since the pool is randomized, the final structure cannot be elucidated directly, but it can be biased towards forming the binding/cleavage domains by ensuring that the aptamer and ribozyme base stems are conserved by one to three bases (additional randomization may actually result in increased activity differential ± ligand Order DNA for the oligos from your preferred vendor. You will purify the oligos yourself, so simply have them de-salted. Upon receipt, solubilize the oligos in ultrapure water, add 1:1 2× loading buffer and vortex. Purify your oligos on a 10–12 % PAGE gel, running gels as above. Illuminate gels with UV light against a PEI-Cellulose TLC plate, mark and excise bands with a razor blade. Illuminate the gels in a dark room and minimize the time both you and the gel are exposed to UV, both for safety and to minimize damage to the nucleic acid library.

8. When using an Elu-trap device to electroelute DNA or RNA from gel slices, elution buffer should be prepared as 0.5× TBE buffer, and fresh buffer utilized each time. Assemble the elution modules each time, and place gel slices close to the capture section. Fill the device with buffer such that the slices are just covered by the buffer. Ensure that any unused modules are blocked from the buffer, as they will "short out" the current. Carefully draw the eluted material from the capture section and aliquot into microcentrifuge tubes.
9. For in vitro transcription, pools and the first two rounds are conducted in large volumes (1 mL) while remaining rounds are 100–200 L volume. Running the transcription at 20–25 °C will result in less mass of RNA being produced than at 37 °C, but will yield more full-length transcripts. Longer transcription times will result in a greater number of RNA molecules, but a larger proportion of cleaved RNA molecules.
10. Stage One in vitro selection cycles may need to be repeated approximately 4–6 times until the pool of evolved RNA molecules is dominated by sequences that undergo substantially more cleavage in the presence than in the absence of ligand, as judged by the relative sizes of the corresponding bands on the selection PAGE gel. After each round, template DNA from each step should be marked at stored at −80 °C.
11. We note that common subcloning techniques, such as SLIC and CPEC, may not be successful for owing to the substantial secondary structure present in the cognate single stranded-DNAs [27].
12. Create the gel utilizing glass plates, one 1–2 cm longer than the other, with 1.5 mM spacers and comb (you can utilize either a multichanneled comb to quantify and measure a series of samples, or a single wide comb for purification of a large transcription. Clean the plates with RNAse away or similar reagent, followed by a wash with running Milli-Q water. Assemble the glass with the spacers, leave the comb aside for now, the bottoms of the plates should be aligned (The top has the differential height to allow the buffer to flow to the channels but not spill). Secure the plates in place using binder clips, ensuring that they are springy enough to hold the spacers and glass together tightly. Prepare gel by adding acrylamide, urea, and TBE, add TEMED and APS last and swirl to initiate the polymerization reaction. Carefully pour the solution into the plate-rig, ensuring that no bubbles form; you can shift the position or lightly tap the rig to allow the bubbles to rise to the surface. Add the comb into the top of the gel. Watch the edges of the rig to ensure that there are no leaks. The gel should be ready in 30–45 min.

References

1. Vitreschak AG, Rodionov DA, Mironov AA, Gelfand MS (2004) Riboswitches: the oldest mechanism for the regulation of gene expression. Trends Genet 20:44–50
2. Stormo GD, Ji Y (2001) Do mRNAs act as direct sensors of small molecules to control their expression? Proc Natl Acad Sci U S A 98:9465–9467
3. Patel DJ, Suri AK, Jiang F et al (1997) Structure, recognition and adaptive binding in RNA aptamer complexes. J Mol Biol 272: 645–664
4. Tuerk C (1997) Using the SELEX combinatorial chemistry process to find high affinity nucleic acid ligands to target molecules. Methods Mol Biol 67:219–230
5. Klug SJ, Hüttenhofer A, Famulok M (2000) In vitro selection of RNA aptamers that bind special elongation factor SelB, a protein with multiple RNA-binding sites, reveals one major interaction domain at the carboxyl terminus. RNA 5:1180–1190
6. Lorsch JR, Szostak JW (1994) In vitro selection of RNA aptamers specific for cyanocobalamin. Biochemistry 33:973–982
7. Winkler WC, Breaker RR (2003) Genetic control by metabolite-binding riboswitches. Chembiochem 4:1024–1032
8. Nahvi A, Sudarsan N, Ebert MS et al (2002) Genetic control by a metabolite binding mRNA. Chem Biol 9:1043–1049
9. Nahvi A, Barrick JE, Breaker RR (2004) Coenzyme B 12 riboswitches are widespread genetic control elements in prokaryotes. Nucleic Acids Res 32:143–150
10. Liu CC, Arkin AP (2010) The case for RNA. Science 330:1185–1186
11. Hall B, Hesselberth JR, Ellington AD (2007) Computational selection of nucleic acid biosensors via a slip structure model. Biosens Bioelectron 22:1939–1947
12. Lynch SA, Desai SK, Sajja HK, Gallivan JP (2007) A high-throughput screen for synthetic riboswitches reveals mechanistic insights into their function. Chem Biol 14:173–184
13. Suess B, Fink B, Berens C et al (2004) A theophylline responsive riboswitch based on helix slipping controls gene expression in vivo. Nucleic Acids Res 32:1610–1614
14. Soukup GA, Breaker RR (1999) Engineering precision RNA molecular switches. Proc Natl Acad Sci U S A 96:3584–3589
15. Carothers JM, Goler JA, Juminaga D, Keasling JD (2011) Model-driven engineering of RNA devices to quantitatively program gene expression. Science 334:1716–1719
16. Carothers JM, Goler JA, Kapoor R, Lara LD, Keasling JD (2010) Selecting RNA aptamers for synthetic biology: investigating magnesium dependence and predicting binding affinity. Nucleic Acids Res 38:2736–2747
17. Thompson KM, Syrett HA, Knudsen SM, Ellington AD (2002) Group I aptazymes as genetic regulatory switches. BMC Biotechnol 2:21
18. Link KH, Guo L, Ames TD et al (2007) Engineering high-speed allosteric hammerhead ribozymes. Biol Chem 388:779–786
19. Jenison RD, Gill SC, Pardi A, Polisky B (1994) High-resolution molecular discrimination by RNA. Science 263:1425–1429
20. Dower WJ, Cwirla SE (1992) Guide to electroporation and electrofusion. Academic, San Diego
21. Isambert H, Siggia E (2000) Modeling RNA folding paths with pseudo-knots: application to hepatitis delta virus ribozyme. Proc Natl Acad Sci U S A 97:6515–6520
22. Xayaphoummine A, Viasnoff V, Harlepp S, Isambert H (2007) Encoding folding paths of RNA switches. Nucleic Acids Res 35:614–622
23. Deana A, Celesnik H, Belasco JG (2008) The bacterial enzyme RppH triggers messenger RNA degradation by 5-prime pyrophosphate removal. Nature 451:355–358
24. Bayer TS, Smolke CD (2005) Programmable ligand-controlled riboregulators of eukaryotic gene expression. Nat Biotechnol 23:337–343
25. Wieland M, Hartig JS (2008) Improved aptazyme design and in vivo screening enable riboswitching in bacteria. Angew Chem Int Ed 47:2604–2607
26. Weigand J, Sanchez M, Gunnesch EB et al (2008) Screening for engineered neomycin riboswitches that control translation initiation. RNA 14:89–97
27. Hillson NJ, Rosengarten RD, Keasling JD (2011) j5 DNA assembly design automation software. ACS Synth Biol 1:14–21

Chapter 17

In Vivo Screening for Aptazyme-Based Bacterial Riboswitches

Charlotte Rehm and Jörg S. Hartig

Abstract

In many synthetic biology applications, modular and easily accessible tools for controlling gene expression are required. In addition, in vivo biosensors and diagnostic devices will become more important in the future to allow for noninvasive determination of protein, ion, or small molecule metabolite levels. In recent years synthetic RNA-based switches have been developed to act as signal transducers to convert a binding event of a small molecule (input) into a detectable output. Their modular design allows the development of a variety of molecular switches to be used in biochemical assays or inside living cells. RNA switches developed by our group are based on the *Schistosoma mansoni* hammerhead ribozyme, a self-cleaving RNA sequence that can be inserted into any RNA of interest. Connection to an aptamer sensing a small molecule renders the cleavage reaction ligand-dependent. In the past we have successfully designed and applied such hammerhead aptazymes for the allosteric control of both bacterial and eukaryotic gene expression by affecting transcription elongation, translation initiation, or mRNA stability. In order to yield functional switches optimization of the connecting sequence between the aptamer and the HHR needs to be carried out. We have therefore developed an in vivo screening protocol detailed in this chapter that allows the identification of functional aptazymes in bacteria.

Key words Riboswitch, Allosteric ribozyme, Biosensor, Hammerhead aptazyme, In vivo screening, Aptamer, Gene expression, Regulation

1 Introduction

Aptamers are short nucleic acids motifs which can specifically recognize small molecule interaction partners or even ions with high specificity and affinity [1]. They were originally obtained by systematic evolution of ligands by exponential enrichment (SELEX) [2, 3]. However, Breaker and coworkers later proofed the existence of RNA aptamers in the 5′ untranslated regions (5′ UTR) of many bacterial messenger RNAs (mRNAs). There they form selective binding pockets for a respective metabolite and serve as sensors of intracellular metabolite concentrations in regulatory RNA elements called riboswitches [4–6]. Metabolite binding to the

Atsushi Ogawa (ed.), *Artificial Riboswitches: Methods and Protocols*, Methods in Molecular Biology, vol. 1111, DOI 10.1007/978-1-62703-755-6_17,

aptamer domain induces structural rearrangements in an adjoining expression platform and consequently switches expression of the downstream gene on or off by affecting transcription elongation or translation initiation. The major advantages of the use of riboswitches in vivo are that protein factors are not required to induce changes in gene expression and that the regulatory active device is covalently attached to the encoded message.

Synthetic RNA-based devices containing aptamers have been developed that process the binding of the aptamer ligand (input) and convert it into a variety of detectable physical signals (output), thereby acting as gene regulatory switches or biosensors, both in biochemical assays or inside living cells [1, 7–9]. In our group we are constructing artificial riboswitches using the *Schistosoma mansoni* hammerhead ribozyme (HHR) as central RNA scaffold [10]. The HHR is a short self-cleaving RNA motif formed by three helical arms (stems I–III) surrounding a conserved catalytic core (Fig. 1b).

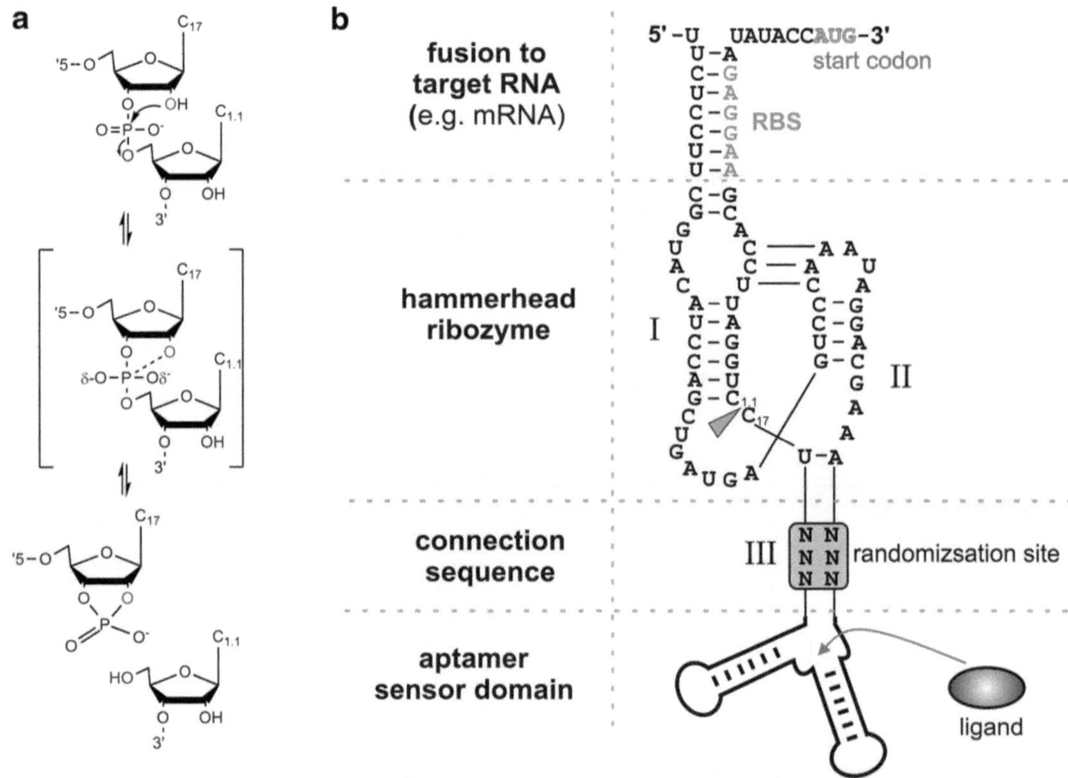

Fig. 1 The hammerhead ribozyme as catalytically active domain. (**a**) Reaction mechanism of the intramolecular strand-cleavage reaction by in-line nucleophilic attack of the 2′ hydroxyl group. (**b**) Example of a hammerhead aptazyme inserted into the 5′ untranslated region of a bacterial mRNA. The RBS is incorporated into stem I and blocked from access by the ribosome, an aptamer is attached at stem III. Stabilizing tertiary interactions between stem I and II are shown. The cleavage site is marked by the *arrowhead*. Variability of the randomization site allows screening for activating as well as inactivating aptazymes

In the catalytically active conformation site-directed RNA strand cleavage (in cis) occurs by 2′-hydroxyl group of C_{17} attacking the phosphodiester, thus generating a 2′-3′-pentacyclic phosphodiester and a free 5′-hydroxl group (Fig. 1a) [11]. The cleavage activity strongly depends on the presence of divalent metals ions, e.g., Mg^{2+}, and is further enhanced by tertiary interactions between stem I and II enabling efficient cleavage at low Mg^{2+} concentrations [12].

In order to achieve allosteric control of the cleavage reaction aptamers can be attached to the HHR. Breaker and coworkers first demonstrated allosteric control of a HHR by connecting an ATP-aptamer with the ribozyme, creating a so-called aptazyme [13]. In the past we have successfully constructed hammerhead aptazymes (HHAz) by connecting theophylline- and thiamine pyrophosphate-aptamers at stem III (Fig. 1b) and thereby rendering the cleavage activity ligand-dependent [14, 15]. Variability of the connecting sequences allows the construction of ON- as well OFF-switching aptazymes. Artificial expression platforms can be attached at stem I to create functional switches of gene expression. In further studies we and others have shown that is possible to attach the aptamer at any of the three stems of the HHR [15–18]. However, preserving stem I and II interactions is advisable as to avoid special growth condition, e.g., high Mg^{2+} concentrations. In particular, we have constructed a very efficient artificial riboswitch in bacteria by expanding the full-length HHR by addition of a helix to stem I while at the same time maintaining stem I and II interactions, and thereby preserving fast cleavage rates [19].

In bacteria sequestration of the ribosomal binding site has been employed to switch gene expression (Fig. 1b) [14, 15, 20]. In case of ON-switches, activation of the HHR cleaves the mRNA. This liberates the ribosomal binding site, which is now accessible by the ribosome, so that translation can occur. Alternatively, mRNA stability is regulated in eukaryotic cells. Ribozyme-mediated detachment of the stabilizing 5′-cap structure leads to degradation of the mRNA. In this case ON-switches are achieved by inhibiting HHR cleavage upon ligand binding, so that translation can proceed [21, 22]. In addition to mRNAs other RNA classes can be regulated: Attachment of a theophylline-dependent HHAz to the 5′ end of an *E. coli* tRNA has enabled us to switch the amino acid identity in proteins post-transcriptionally [20, 23]. Incorporation of a thiamine pyrophosphate-dependent HHAz into the *E. coli* 16S rRNA allowed control of the rRNA integrity; upon thiamine addition to the medium cleavage of the 16S rRNA and subsequent degradation occurs [24]. Yokobayashi and coworkers have employed the design of our theophylline-HHAz to switch gene expression in eukaryotes via regulation of RNAi. In their approach a primary microRNA (pri-miRNA) hybridizes with an inhibitory strand. Activation of the HHAz by theophylline frees the pri-miRNA, which is then further processed by the intracellular RNAi machinery to yield a functional miRNA [25].

Regarding non-small molecule inducers, our group has recently developed HHR based genetic switches reacting to a physical stimulus, namely a change in temperature, by attachment of a RNA thermometer to the HHR scaffold [26]. Furthermore, we have constructed a novel riboregulatory system in which allosteric control of the HHR is achieved by small trans-acting RNAs via formation of a RNA–RNA duplex structure [27].

Taken together, HHAzs have proven to be versatile tools owing to their modular nature; input as well as output domains can be varied and easily remodeled to fit one's needs by standard molecular biology methods. Incorporation of different aptamer domains changes the ligand dependency of the aptazyme, modifications of the artificial expression platform enable the regulation of a variety of RNA classes as outlined above. As the secondary structure of nucleic acids is primarily determined by Watson–Crick base pairing, rational design can be applied to connect any given aptamer with the HHR and thereby reprogram the allosteric regulation of the encoded message or regulatory RNA [7]. However, Breaker and coworkers have shown that the nucleotides in the sequence connecting the aptamer and the ribozyme are crucial for relaying the binding event to the catalytic domain, thereby enabling switching [28]. Optimization of the connecting sequence is necessary to improve the performance of the HHAz for a given aptamer. In our group we apply an in vivo screening protocol to detect suitable switches from a pool of mutants carrying randomized connecting sequences. Optimization of the connecting sequence can also be performed in vitro [28]; however, our approach ensures that all selected clones are already functional in vivo in bacteria.

To generate more complex switching pathways or biosensors it is crucial that a greater number of sensor domains will be explored. Herein, we describe the experimental details that allow reprogramming the ligand-specificity of the HHAz in order to control mRNA translation in *E. coli*. However, the general strategy can as well be used to change ligand-dependency for the control of other RNA classes. ON- as well as OFF-switching riboregulators can be identified using our protocol. First we will explain the construction of mutant libraries, followed by our method for screening and initial characterization of functional riboswitches.

2 Materials

All solutions are prepared using ultrapure water and analytical grade reagents are used. Reagents are stored at room temperature unless otherwise noted.

2.1 PCR

1. 3 M Na-acetate, pH 5.2: Dissolve 20.4 g $CH_3COONa\text{-}3H_2O$ in 50 mL H_2O. Adjust to pH 5.2 with HCl.
2. 50, 70, and 99 % ethanol.

3. 0.5 M EDTA, pH 8.0: Dissolve 73.08 g EDTA and 30 g NaOH pellets in 400 mL H_2O, adjust pH 8.0 using NaOH, adjust volume to 500 mL.
4. 5× TBE buffer: Dissolve 54 g Trizma base, 27.5 g boric acid, 20 mL 0.5 M EDTA (pH 8.0) in 1 L of H_2O. Adjust to pH 8.3, if needed, with concentrated HCl. Dilute to 0.5× in H_2O before use.
5. 6× agarose gel loading buffer: Dissolve 4 mL 50 mM Tris–HCl, pH 7.6, 12 mL 100 % glycerol, 2.4 mL 0.5 M EDTA, 6 mg bromophenol blue, 6 mg xylene cyanol in 20 mL, prepare aliquots of 1 mL. Can be stored at −20 °C for several months.
6. Ready-to-use DNA size standard (e.g., GeneRuler 1 kb DNA Ladder (Thermo Scientific)).
7. Agarose gels. For 0.8 % (w/v) dissolve 0.8 g agarose in 100 mL 0.5× TBE buffer by boiling in the microwave. Dissolved agarose can be stored in an incubator at ≥70 °C for several days (*see* **Note 1**).
8. DNA oligonucleotides as primers to be used in PCR (*see* Subheading 3.1).
9. DpnI restriction enzyme and NEB buffer 4 (NEB).
10. Ethidium bromide gel staining solution. Dilute 200 μg in 400 mL H_2O. Can be kept in the dark at room temperature and reused (*see* **Note 2**).
11. Destaining solution: 400 mL. H_2O. Can be kept at room temperature and reused.
12. PCR cycler.
13. PCR template: e.g., eGFP expression vector (pET16b_eGFP_wtHHR or pQE31-J06-wtHHR2, AG Hartig, University of Konstanz).
14. Phusion Hot Start II High-Fidelity DNA polymerase, 5× HF buffer and 100 % DMSO (NEB). Store at −20 °C.
15. 2 mM dNTP Mix. Store at −20 °C.
16. Quick Ligase, 2× Quick Ligation Reaction Buffer (NEB). Store at −20 °C. Thaw Quick Ligation Reaction Buffer on ice.
17. DNA Gel Extraction Kit (e.g., Gel DNA Recovery Kit (Zymo Research)).
18. DNA Purification Kit (e.g., DNA Clean and Concentrator-5 (Zymo Research)).
19. Tabletop centrifuge.
20. Agarose gel electrophoresis chamber and power supply.
21. Razor blade.
22. UV light table.

2.2 Cell Culture and Screening

1. 96-deep-well plate and 96-well plate incubator (e.g., Heidolph Inkubator 1000 and Titramax 1000) (*see* **Note 3**).
2. Air-permeable adhesive seals for 96-well-plates.
3. 1,000× carbenicillin stock solution (100 mg/mL): Dissolve 1 g carbenicillin in 10 mL 50 % (v/v) ethanol. Can be stored at −20 °C for several weeks.
4. Electro-competent *E. coli* (e.g., *E. coli* BL21 (DE3) gold, Stratagene) (*see* **Note 4**). 80 μL aliquots frozen at −80 °C.
5. Electroporator (e.g., Eppendorf Electroporator 2510) and electroporation cuvettes (e.g., Electroporation cuvettes, 0.1 cm (Bio-Rad)).
6. Fluorescence plate reader (e.g., TECAN M200).
7. LB-Carb-medium: For LB (Lennox) dissolve 10 g tryptone, 5 g yeast extract, 5 g NaCl in 1 L H_2O, adjust to pH 7.0, if necessary, and autoclave. Supplement with 1 mg/L carbenicillin before use. Can be stored at 4 °C for up to 4 weeks.
8. LB-Carb-agar plates: Dissolve 10 g tryptone, 5 g yeast extract, 5 g NaCl, and 10 g agar-agar in 1 L H_2O and autoclave. Let cool until hand-warm, then supplement with 1 mg/L carbenicillin. Pour plates in 15 cm petri dishes. Plates can be stored at 4 °C for up to 6 weeks.
9. Kit for plasmid isolation (e.g., Zyppy Plasmid Miniprep Kit).
10. SOC medium: Dissolve 20 g tryptone, 5 g yeast extract, 0.6 g NaCl, 0.2 g KCl in 900 mL H_2O. Adjust pH to 6.8–7.0 using NaOH. Adjust volume to 960 mL and autoclave. Add 10 mL 1 M $MgCl_2$, 10 mL 1 M $MgSO_4$, and 20 mL 1 M glucose from sterile filtered stocks. Store 1 mL aliquots at −20 °C for several months.
11. Toothpicks sterilized by autoclaving.
12. Flat, transparent 96-well plates.

3 Methods

3.1 Construction of Randomized Riboswitch Libraries

New aptamer sequences are incorporated into the HHR using whole plasmid PCR as done for site-directed mutagenesis (Fig. 2). Randomization of the connecting sequence is achieved by using primers with unbiased random positions generated during solid phase DNA synthesis using a 1:1:1:1 mixtures of nucleoside phosphoramidites. For successful ligation of the PCR product phosphorylation of one primer at the 5′ end is required. We recommend purification of the long primers (≥45 nts) by denaturing PAGE, as described by Sambrook and Russel in Molecular Cloning (*see*: Preparation of Denaturing Polyacrylamide Gels and Purification of Synthetic Oligonucleotides by Polyacrylamide Gel Electrophoresis).

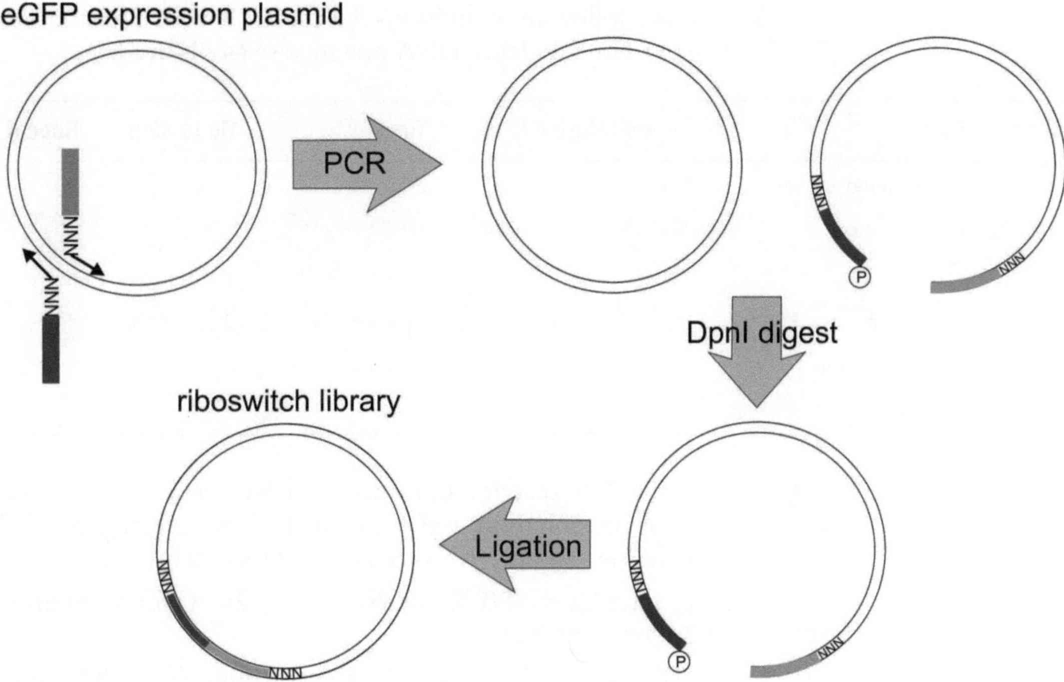

Fig. 2 Library construction. Whole plasmid PCR is used for incorporation of new aptamer sequences using primers carrying randomized sequences that will constitute the connecting sequence. After PCR the template plasmid is degraded by DpnI digest. The purified PCR product is ligated yielding a library of randomized riboswitches

1. Prepare the reaction mixture as follows and split the mixture into 3 PCR tubes. The reaction can be scaled up, if necessary. We recommend using no more than 30 μL per tube:

Starting concentration	Material	Volume	Final concentration
5×	HF buffer	16	1×
2 mM	dNTP mix	8	200 μM
10 μM	Forward primer	4.0	500 nM
10 μM	Reverse primer	4.0	500 nM
10 ng/μl	Template	1.6	16 ng
100 % (v/v)	DMSO	2.4	3 % (v/v)
2 U/μl	Phusion Hot Start DNA polymerase	0.8	1.6 U
	H_2O	43.2	
	Total	80	

2. Use the following conditions for the PCR with Phusion Hot Start II High-Fidelity DNA polymerase (*see* **Note 5**).

Step	Stage	Temperature (°C)	Time	Go to step	Repeat
1	Initial denaturation	98	30 s		
2	Denaturation	98	10 s		
3	Annealing	60	30 s		
4	Extension	72	15–30 s/kb	2	24
5	Final extension	72	7 min		
6	Cooling	4	Pause		

3. Pool the PCR reaction mix into one 1.5 mL reaction tube. For ethanol precipitation add 5 μL of 3 M Na-acetate, pH 5.2, followed by 165 μL 99 % ethanol and mix by inverting.
4. Place samples at −80 °C for 20 min or −20 °C for a minimum of 2 h or overnight.
5. Centrifuge samples in a tabletop centrifuge at 12,000 × *g* for 15 min at room temperature. Discard supernatant and wash once with 70 % ethanol. Resuspend the pellet in 44 μL H_2O.
6. For digestion of the template vector add 5 μL NEB buffer 4 and 1 μL DpnI and mix thoroughly. Incubate at 37 °C for 50 min followed by 10 min incubation at 80 °C for heat inactivation of the enzyme.
7. In the meantime, pour a 0.8 % (w/v) TBE-agarose gel and let it solidify. Once the sample is ready, add 10 μL of 6× agarose gel loading buffer and mix. Load the sample in a long well and load 2.5 μL DNA size standard in a small well to run alongside. Run the gel at 10 V/cm for 75–90 min.
8. After the run, transfer the gel into ethidium bromide staining solution for 10 min, then place the gel in destaining solution for 10 min for improving contrast. As exposure to UV light may introduce mutations to the DNA, cut the gel vertically excising the marker and a small part of the band with your sample. Visualize the DNA bands over a UV light table.
9. Excise the appropriate band in the correct size, piece the gel back together and excise the corresponding band in the gel that has not been exposed to UV light and purify the DNA fragment using a DNA gel extraction kit according to manufacturer's protocol.
10. Determine the DNA concentration by measuring A_{260}.
11. For ligation of the PCR product combine 50 ng of the purified PCR product in 9 μL of water with 10 μL 2× Quick Ligation Reaction buffer and add 1 μL Quick Ligase, mix well, and briefly centrifuge samples. Incubate at 25 °C for 15 min.

12. Purify the DNA sample by using a DNA purification kit according to the manufacturer's instructions to remove the ligation buffer for better transformation efficiency. Elute ligated DNA using 6 μL H_2O.

3.2 Screening for Functional Aptazymes

A pool of mutants carrying randomized connecting sequences is obtained by transformation of the randomized plasmid pool into *E. coli* (Fig. 3). Single clone colonies can be picked and screened for functional riboswitches in the presence and absence of the respective ligand.

1. Thaw 80 μL aliquot of electro-competent *E. coli* on ice for 15 min. In the meantime, thaw 1 mL aliquot of SOC medium and preheat to 37 °C, place LB-Carb-agar plates on 37 °C and chill one electroporation cuvette on ice along with the purified, ligated DNA sample from the previous step.
2. Add 1.0 μL of the purified sample to 80 μL competent cells and carefully mix on ice. Transfer the cells into the pre-chilled electroporation cuvette and transform by electroporation. Immediately resuspend the cells in pre-warmed 1 mL SOC medium and transfer to a fresh 1.5 mL reaction tube. Incubate for 1 h at 37 °C. Plate the transformed cells on pre-warmed LB-agar plates and incubate plates overnight at 37 °C (*see* **Note 6**).
3. Prepare 96 deep-well plates with 1 mL LB-medium supplemented with carbenicillin to be used as master plates. Save wells in the first or last column of the 96 deep-well plate for control cultures (*see* **Note 7**). Using sterilized toothpicks, pick single colonies from the pool of mutants carrying randomized connecting sequences. Seal with air-permeable adhesive cover. Incubate plates at 37 °C overnight in the 96-well plate incubator.
4. For each master plate of the previous day (**step 3**), prepare two additional destination plates to be used in the screening. The medium added to every second column should be supplemented with the appropriate ligand as shown in Fig. 3.
5. Inoculate the destination plates with clones from the master plates, so that each clone is tested with and without ligand (*see* **Note 8**). Seal with air-permeable adhesive cover. Incubate the destination plates in the 96-well plate incubator at 37 °C overnight.
6. Transfer 100 μL of the out-grown culture into a black 96-well plate and measure expression levels by determining eGFP fluorescence (excitation wavelength $\lambda_{ex}=488$ nm, emission wavelength $\lambda_{em}=535$ nm) (*see* **Note 9**). Potential hits are identified by comparing the fluorescence for each screened clone in the presence and absence of the ligand after subtraction of background fluorescence. In addition measure OD_{600} for OD-correction.
7. Clones of interest are isolated by picking them from the master plate and inoculating fresh LB cultures to be grown under

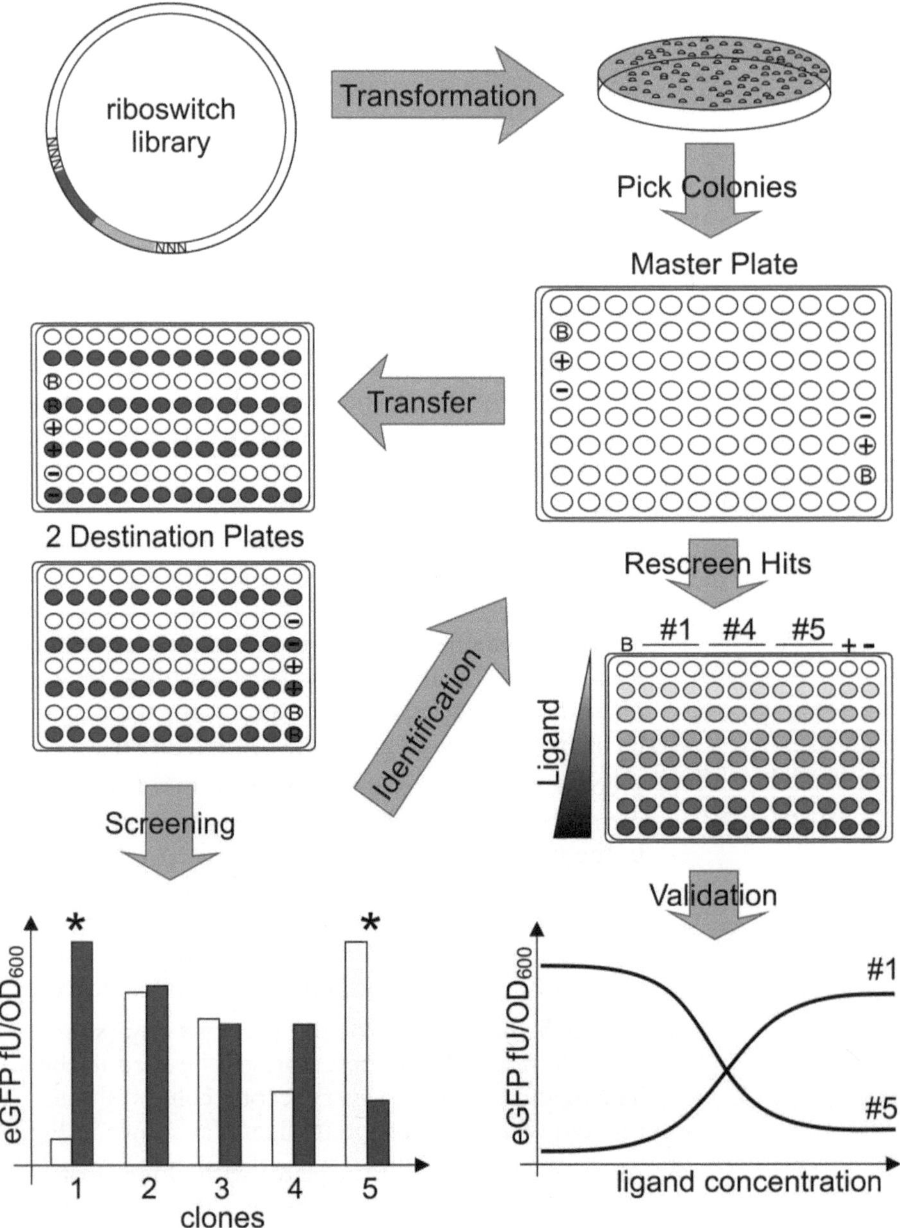

Fig. 3 Screening for functional riboswitches. The plasmid library of randomized riboswitches is transformed into bacteria. Individual clones are picked to construct mutant libraries, which are screened for eGFP expression in the presence (*white*) or absence (*dark grey*) of the aptamer ligand. Controls for background fluorescence correction (B), positive control (+) and negative control (–) are included in the screening. Identified hits are further characterized

vigorous shaking at 37 °C for 8 h for plasmid isolation and preparation of glycerol stocks.

8. Plasmids are isolated using a plasmid isolation kit according to manufacturer's instruction. Integrity of the artificial riboswitch

construct and identity of the randomized connecting sequence are checked by sequencing.

9. The riboswitches are further characterized by growing the clones carrying a potential hit with increasing concentrations of the ligand. Cultures are again prepared in 96-well plates as detailed in **step 4** using a gradient of increasing concentrations of the ligand. Experiments are carried out in triplicate. Plates are incubated at 37 °C overnight and eGFP expression measured to confirm functional, concentration dependent riboswitches.

Pool coverage can be calculated by obtaining the oversampling factor (O_f) which is calculated as $O_f = T/V = -\ln(1 - P_i)$ where T is the number of clones actually screened, V the maximum number of different clones of a randomized sequence (sequence space) and P_i the probability that a particular sequence occurs in the library [29]. Due to the exponential relationship, high pool coverage is only achieved when the theoretical sequence space is oversampled multiple times.

4 Notes

1. For a good separation and resolution of 5–10 kb PCR products, 0.8 % (w/v) agarose gels are sufficient. Higher percentages of agarose (e.g.1.5 % (w/v)) can offer better resolution for smaller products.
2. Ethidium bromide is toxic and mutagenic. Wear nitrile gloves while working with it.
3. Alternatively, for larger screening purposes 384-deep-well plates can be used for cell culture and fluorescence measurements.
4. For the preparation of electro-competent *E. coli* a detailed protocol can be found at http://www.eppendorf.com.
5. Alternative to Phusion Hot Start polymerase, other polymerases can be used. The protocol then needs to be adapted to the appropriate buffer and conditions as required by manufacturer's protocol. Use of a high-fidelity polymerase is advisable. Primers should be designed to exhibit an annealing temperature of 60–63 °C; however, we suggest that a gradient PCR is carried out to determine the optimal annealing temperature.
6. Dilutions of the transformed cells are plated in order to yield single colonies. Dilutions will have to be determined experimentally depending on the transformation efficiency of the electro-competent cells used.
7. As controls one can use cells transformed with a plasmid expressing eGFP without any insert affecting the gene expression (positive control) and a negative control that bears an inactivated hammerhead ribozyme variant by an A to G point

mutant in the catalytic core (Fig. 1b), masking the ribosomal binding site permanently. Cells transformed with an empty vector (e.g., pET16b or pGDR11) can be used to determine background fluorescence. Note that all plasmids should transfer the same antibiotic resistance. Since one master plate will be distributed to two destination plates, we recommend using the top half of the first column and the bottom half of last column for controls, so that each destination plate will receive a set of controls. We also recommend not using the wells in the corners for screening purposes as growth of the cells will suffer from evaporation of the medium.

8. Minimal medium (e.g., M9 or M63 medium) is recommended for the screening process to receive full control over the ligand concentration, especially when screening with naturally occurring ligands. Membrane permeability also has to be considered as not all ligands are readily taken up from the medium, while others may not be stable in medium. In case of TPP switches, we add thiamine to the medium, which is processed by the intracellular thiamine kinase and thiamine phosphate kinase to yield TPP [15, 23]. When using new ligands or media we recommend measuring growth curves for the bacteria, especially to determine potential toxicity of a ligand.

9. It is recommended to wait until eGFP expression has reached a plateau in outgrown cultures (at least 16 h). Use the fluorescence measured from cells transformed with an empty vector to subtract background fluorescence.

References

1. Liu J, Cao Z, Lu Y (2009) Functional nucleic acid sensors. Chem Rev 109:1948–1998
2. Tuerk C, Gold L (1990) Systematic evolution of ligands by exponential enrichment: RNA ligands to bacteriophage T4 DNA polymerase. Science 249:505–510
3. Ellington AD, Szostak JW (1990) In vitro selection of RNA molecules that bind specific ligands. Nature 346:818–822
4. Winkler W, Nahvi A, Breaker RR (2002) Thiamine derivatives bind messenger RNAs directly to regulate bacterial gene expression. Nature 419:952–956
5. Winkler WC, Nahvi A, Sudarsan N, Barrick JE, Breaker RR (2003) An mRNA structure that controls gene expression by binding S-adenosylmethionine. Nat Struct Biol 10:701–707
6. Mandal M, Boese B, Barrick JE, Winkler WC, Breaker RR (2003) Riboswitches control fundamental biochemical pathways in Bacillus subtilis and other bacteria. Cell 113:577–586
7. Vinkenborg JL, Karnowski N, Famulok M (2011) Aptamers for allosteric regulation. Nat Chem Biol 7:519–527
8. Paige JS, Nguyen-Duc T, Song W, Jaffrey SR (2012) Fluorescence imaging of cellular metabolites with RNA. Science 335:1194
9. Breaker RR (2002) Engineered allosteric ribozymes as biosensor components. Curr Opin Biotechnol 13:31–39
10. Blount KF, Uhlenbeck OC (2002) The hammerhead ribozyme. Biochem Soc Trans 30:1119–1122
11. Fedor MJ (2009) Comparative enzymology and structural biology of RNA self-cleavage. Annu Rev Biophys 38:271–299
12. Khvorova A, Lescoute A, Westhof E, Jayasena SD (2003) Sequence elements outside the hammerhead ribozyme catalytic core enable

intracellular activity. Nat Struct Biol 10:708–712
13. Tang J, Breaker RR (1997) Rational design of allosteric ribozymes. Chem Biol 4:453–459
14. Wieland M, Hartig JS (2008) Improved aptazyme design and in vivo screening enable riboswitching in bacteria. Angew Chem 47:2604–2607
15. Wieland M, Benz A, Klauser B, Hartig JS (2009) Artificial ribozyme switches containing natural riboswitch aptamer domains. Angew Chem 48:2715–2718
16. Ogawa A, Maeda M (2008) An artificial aptazyme-based riboswitch and its cascading system in E. coli. Chembiochem 9:206–209
17. Win MN, Smolke CD (2007) A modular and extensible RNA-based gene-regulatory platform for engineering cellular function. Proc Natl Acad Sci U S A 104:14283–14288
18. Win MN, Smolke CD (2008) Higher-order cellular information processing with synthetic RNA devices. Science 322:456–460
19. Wieland M, Gfell M, Hartig JS (2009) Expanded hammerhead ribozymes containing addressable three-way junctions. RNA 15:968–976
20. Klauser B, Saragliadis A, Auslander S, Wieland M, Berthold MR, Hartig JS (2012) Post-transcriptional Boolean computation by combining aptazymes controlling mRNA translation initiation and tRNA activation. Mol BioSyst 8:2242–2248
21. Auslander S, Ketzer P, Hartig JS (2010) A ligand-dependent hammerhead ribozyme switch for controlling mammalian gene expression. Mol BioSyst 6:807–814
22. Ketzer P, Haas SF, Engelhardt S, Hartig JS, Nettelbeck DM (2012) Synthetic riboswitches for external regulation of genes transferred by replication-deficient and oncolytic adenoviruses. Nucleic Acids Res 40:e167
23. Berschneider B, Wieland M, Rubini M, Hartig JS (2009) Small-molecule-dependent regulation of transfer RNA in bacteria. Angew Chem 48:7564–7567
24. Wieland M, Berschneider B, Erlacher MD, Hartig JS (2010) Aptazyme-mediated regulation of 16S ribosomal RNA. Chem Biol 17:236–242
25. Kumar D, Annna CI, Yokobayashi Y (2009) Conditional RNA interference mediated by allosteric ribozyme. J Am Chem Soc 131:13906–13907
26. Saragliadis A, Krajewski SS, Rehm C, Narberhaus F, Hartig JS (2013) Thermozymes: Synthetic RNA thermometers based on ribozyme activity. RNA Biol 10(6):1010–6
27. Klauser B, Hartig JS (2013) An engineered small RNA-mediated genetic switch based on a ribozyme expression platform. Nucleic Acids Res 41(10):5542–52
28. Soukup GA, Breaker RR (1999) Engineering precision RNA molecular switches. Proc Natl Acad Sci U S A 96:3584–3589
29. Reetz MT, Kahakeaw D, Lohmer R (2008) Addressing the numbers problem in directed evolution. Chembiochem 9:1797–1804

Chapter 18

Engineered Riboswitch as a Gene-Regulatory Platform for Reducing Antibiotic Resistance

Libing Liu and Shu Wang

Abstract

Antibiotic resistance (AR), the ability of a microorganism to withstand the effects of antibiotics, is a growing and increasingly serious global public health problem. Enzymatic activation of antibiotics though the production of β-lactamase is one of the main mechanisms causing AR. Synthetic riboswitch containing aptazyme is constructed in *E. coli* to regulate the expression of β-lactamase through small molecule–aptamer interactions, which sharply reduces the antibiotic resistance of the engineered bacteria.

Key words Antibiotic resistance, Riboswitch, Theophylline-specific aptazyme, Plasmid construction, Gene regulatory systems, β-Lactamase activity

1 Introduction

Riboswitches, which are based on small molecule–RNA interaction, have attracted much attention due to their wide applications in controllable gene expression [1–4]. Here, we present a method for construction of a synthetic riboswitch containing theophylline-specific aptazyme, which can regulate the genetically modified expression of β-lactamase in response to theophylline. The theophylline can induce the conformation change of the riboswitch to activate the ribozyme. The cleavage of active site results in the decrease of β-lactamase gene expression via mRNA degradation pathway. Through reducing the amount of β-lactamase, the antibiotic resistance can be ultimately regulated. The small molecule theophylline can be used as a co-agent to preserve the efficacy of traditional antibiotics, which provides a potential strategy to reduce antibiotic resistance instead of discovering new antibiotics.

Resistance to β-lactam antibiotic agents in gram-negative *E. coli* is primarily mediated by β-lactamase that can hydrolyze the β-lactam ring to inactivate the drugs [5]. Our goal is to construct a riboswitch in *E. coli* for regulating the expression of β-lactamase so as to reduce antibiotic resistance. The riboswitch of hammerhead

Atsushi Ogawa (ed.), *Artificial Riboswitches: Methods and Protocols*, Methods in Molecular Biology, vol. 1111,
DOI 10.1007/978-1-62703-755-6_18,

ribozyme is constructed inside the expression vector pBV220 [6]. The theophylline-specific aptazyme sequence is inserted at a downstream location of ampicillin-resistant (Amp^r) gene in the plasmid pBV220. The obtained new plasmid is designated as pBV220-Apt. Precious evidence has shown that integration of the hammerhead ribozyme into the 3′UTR of a reporter gene can suppress protein translation in a theophylline-dependent fashion in *Saccharomyces cerevisiae* [7]. In the absence of theophylline, the hammerhead ribozyme is inactive and the β-lactamase is highly expressed. Upon adding ampicillin the growth of *E. coli* is unaffected even with a large amount. In the presence of theophylline, the ribozyme will be activated via conformation change of the theophylline-specific aptamer and the cleavage of active site results in the decrease of β-lactamase gene expression by its mRNA degradation. In this case, the β-lactamase expression is greatly inhibited and the antibiotic resistance is sharply impaired. Under this circumstance *E. coli* cells are efficiently inactivated by the addition of ampicillin. Therefore, the growth of bacteria cells can be effectively controlled by theophylline in the presence of ampicillin, where the theophylline acts as a co-agent to preserve the efficacy of traditional antibiotics.

2 Materials

Prepare all solutions using ultrapure water (by purifying deionized water with Milli-Q water at 25 °C) and analytical grade reagents. Prepare and store all reagents at room temperature (unless indicated otherwise).

2.1 Construction of Plasmid pBV220-Apt

1. 0.5 mg/mL pBV220 solution in water (*see* **Note 1**).
2. AvrII (NEB) (*see* **Note 2**).
3. XhoI (NEB) (*see* **Note 2**).
4. DNA purification kit (Tiangen).
5. 10 μM oligonucleotides of aptazyme template and primers in ultrapure water (*see* **Note 3**). Aptazyme template: GCTGTCA CCGGATGTGCTTTCCGGTCTGATGAGTCCGTGTTC CT GATACCAGCATCGTCTTGATGCCCTTGGCAGCAG TGGACGAGGACGAAACAGC.

   Primer 1: ATCCTAGGGCTGTCACCGGATGTGC.

   Primer 2: ACCTCGAGGCTGTTTCG TCCTCG. Store at −20 °C.
6. dNTPs (TaKaRa) (*see* **Note 4**).
7. Pfu DNA polymerase (Tiangen) (*see* **Note 5**).
8. T4 DNA ligase (TaKaRa) (*see* **Note 5**).
9. DH5α competent cells (Tiangen) (*see* **Note 6**).

10. TOP10 competent cells (Tiangen) (*see* **Note 6**).
11. LB (Luria–Bertani) liquid medium: 10 g/L Tryptone, 5 g/L Yeast extract, 5 g/L NaCl, pH 7.4 (*see* **Note 7**). Store at 4 °C.
12. LB solid/agar medium: 15 g/L agar in LB liquid medium (*see* **Note 7**). Store at 4 °C.
13. 50 mg/mL ampicillin solution (*see* **Note 8**). Store at –20 °C.

2.2 Growth Curves of E. coli TOP10 Cells

1. 100 mM theophylline solution in ultrapure water (*see* **Note 9**). Store at 4 °C.

2.3 Assay for β-Lactamase Activity

1. 100 mM nitrocefin solution in ultrapure water (*see* **Note 9**). Store at 4 °C.

2.4 Equipment

1. Thermal cycler.
2. Shaking incubator.
3. Plate-reading spectrophotometer.
4. Centrifuge.

3 Methods

3.1 Construction of Plasmid pBV220-Apt Containing Aptazyme Sequence

1. Digest 1 μg of pBV220 by 1 unit of AvrII and XhoI at 37 °C for 1 h in a total reaction volume of 50 μl. And then purify the product by using the DNA purification kit (*see* **Note 10**). Store it at –20 °C until use.
2. Generate the aptazyme sequence by PCR amplification using the following conditions: 100 μM dNTPs, 0.4 μM primer 1, 0.4 μM primer 2, 5 μl of 10× buffer, 1 U of pfu DNA polymerase, 1 μl of 0.1 μM aptazyme template in a total reaction volume of 50 μl; 35 rounds of 95 °C for 30 s, 50 °C for 60 s, and 65 °C for 60 s (*see* **Note 11**). And then purify the PCR product by using the DNA purification kit (*see* **Note 10**).
3. Digest 5 μl of the PCR product by 1 unit of AvrII and XhoI at 37 °C for 1 h in a total reaction volume of 50 μl (*see* **Note 12**). And then purify the product by using the DNA purification kit (*see* **Note 10**).
4. Ligate 3 μl each of digestion product of pBV220 and aptazyme sequence with 1 μl of T4 DNA ligase in 1× ligation buffer in a total volume of 20 μl at 4 °C for 16 h (Fig. 1) (*see* **Note 13**).
5. Mix 3 μl of ligation product and 100 μl of DH5α or TOP10 gently on ice for 5 min. And then put the mixture in 42 °C hot water for 90 s (*see* **Note 14**). After that, add 500 μl of LB liquid medium into the mixture and shake at 37 °C for 1 h. Spread 100 μl of the mixture on the LB solid plate with 50 μg/mL ampicillin and keep them in the incubator at 37 °C for 12 h.

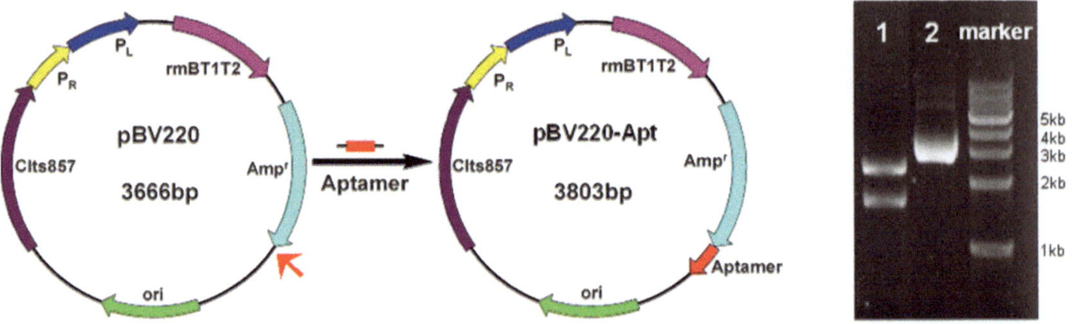

Fig. 1 Construction map of pBV220-Apt containing the aptazyme sequence and its electrophoretic analysis (*Line 2*) and restriction enzymes (AvrII and XhoI) cutting (*Line 1*) (reproduced from ref. 6 with permission from the Royal Society of Chemistry)

3.2 Growth Curves of E. coli TOP10 Cell

1. Transfer a colony of *E. coli* TOP10 cells harboring the plasmid pBV220-Apt from the LB/agar plate containing 50 μg/mL ampicillin into culture tubes containing 5 mL of LB media supplemented with 50 μg/mL ampicillin. Then, incubate them at 37 °C overnight in the shaking incubator (*see* **Note 15**).
2. Add 15 μL of the overnight bacteria liquid into culture tubes containing 5 mL of fresh LB media supplemented with 50 μg/mL ampicillin and appropriate concentrations of theophylline, and shake the culture tubes at 37 °C with a speed of 200 rpm in the shaking incubator (*see* **Note 16**).
3. Record the optical density at 600 nm for each concentration of theophylline by using a plate-reading spectrophotometer at 1 h intervals over 12 h (Fig. 2) (*see* **Note 17**).

3.3 Assay for β-Lactamase Activity

1. Transfer two separate colonies of *E. coli* TOP10 cells harboring pBV220 and pBV220-Apt plasmids into separate culture tubes containing 5 mL of LB media. Then, incubate them at 37 °C overnight in the shaking incubator (*see* **Note 18**).
2. Add 15 μL of the overnight culture into culture tubes containing 5 mL of fresh LB media supplemented with 50 μg/mL ampicillin and appropriate concentrations (e.g., 0, 0.2, 0.4, 0.6, 0.8, 1.0 mM) of theophylline (*see* **Note 19**).
3. After the optical density at 600 nm reaches to the range of 0.5–0.6, collect 1 mL of bacterial cells by centrifugation with 3,000 × *g* for 4 min at 4 °C.
4. Resuspend the precipitation with 100 μL of pre-cooling purified water (*see* **Note 20**).
5. Freeze and thaw the suspension for 10 times by using liquid nitrogen and 37 °C water.
6. Centrifuge it with 10,000 × *g* for 10 min.

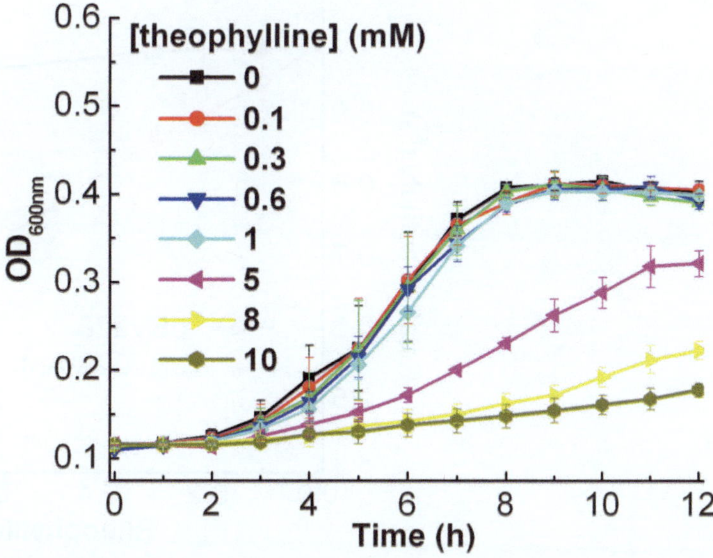

Fig. 2 Growth curves of *E. coli* in the presence of varying concentrations of theophylline. The error bars represent the standard deviation of three measurements (reproduced from ref. 6 with permission from the Royal Society of Chemistry)

7. Collect the supernatant in a new tube and store on ice.
8. Use nitrocefin as the chromogenic cephalosporin substrate to evaluate the β-lactamase activity [8]. Add 5 μL of the supernatant, 2 μL of nitrocefin, and 93 μL of purified water to each well in a 96-well plate. Measure the change of the optical density at 492 nm in 10 min with a plate photometer at room temperature (Fig. 3). All measurements are conducted in triplicate (*see* **Note 21**).

4 Notes

1. pBV220 is a kind of plasmid DNA and easily digested by DNase. Therefore, it is strongly recommended that water is sterilized at 121 °C and 102.9 kPa for 20 min before use. pBV220 solution in water can be stored at -20 °C for a few months. Plasmid pBV220 can be stored in ethanol for a longer time. Avoid repeatedly freezing and thawing. Other expression vector can also be used instead of pBV220.
2. Store at -20 °C before and after use. During experiments, it should be placed on ice. We find that it is better to centrifuge with 10,000 ×*g* for 10 s before opening the lid of the tube.
3. Oligonucleotides of aptazyme template and primers are a kind of DNA and easily digested by DNase. Thus, powder and solution should be stored at -20 °C before and after use.

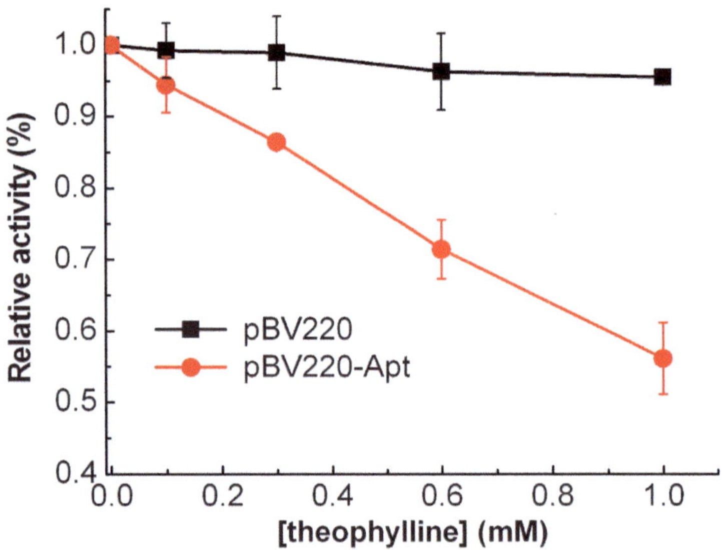

Fig. 3 Riboswitch mediated β-lactamase relative activity in *E. coli* TOP10 cells as a function of varying concentrations of theophylline from 0 to 1.0 mM. The error bars represent the standard deviation of three measurements (reproduced from ref. 6 with permission from the Royal Society of Chemistry)

It is necessary to centrifuge with 10,000 × *g* for 10 s before opening the lid of the tube. During experiments, all DNA samples should be placed on ice.

4. We find it can be stored at 4 °C for 1 week or −20 °C for long time. Avoid repeatedly freezing and thawing.
5. It should be stored at −20 °C. Avoid repeatedly freezing and thawing.
6. It should be stored at −20 °C within 3 months or −80 °C for a longer time. During experiments, it should be placed on ice. DH5α and TOP10 competent cells should be reserved as individual package, for the competent state will be lost after thawing.
7. We find that solution pH is in the range of 7.0–7.4 after tryptone, yeast extract, and NaCl are dissolved. It does not matter about the pH deviation. Otherwise, pH can be adjusted by using 1 N HCl. All media should be sterilized at 121 °C and 102.9 kPa for 20 min.
8. Ampicillin is easy to dissolve and easily loses the effectiveness in water. It should be divided into small aliquots (e.g., 1 mL) and stored at −20 °C after sterilizing with a 0.22 μm Millipore filter.
9. It should be divided into small aliquots (e.g., 1 mL) to avoid contamination after sterilization by using a 0.22 μm Millipore filter.
10. Follow the kit protocol to complete the purification steps.

11. Several reasons can account for PCR failure. First, PCR is very sensitive to DNA contamination, which gives rise to nonspecific DNA amplification. Exogenous contamination DNA can be addressed by optimizing experimental protocols and procedures. It is necessary to set up a separate PCR area for the analysis or purification of PCR products, to use disposable plasticware, and to thoroughly clean the work surface between reaction facilities. Second, primer-design techniques are highly important for improving PCR product yield and avoiding the formation of false positive products. The use of alternative buffer components or polymerase enzymes is beneficial.
12. Constant temperature is very critical for maintaining enzyme activity. Volume of enzyme in the reaction cannot exceed 1/10 of the total. Digestion time can be extended approximately to 2 h if the ligation product cannot be obtained in the **step 5** of this section.
13. T4 DNA ligase easily loses the enzyme activity, so the ligation reaction solution is usually kept at 4 °C. Ligation reaction should be carried out at 4 °C for 16 h to obtain a ligation product. We find that it is optimal to supplement additional 1 μl of T4 DNA ligase after 8 h.
14. It is necessary to incubate the competent *E. coli* cells for heat-shocking at 42 °C for exact 90 s. This procedure can transiently result in the formation of pores on the cell membrane, which facilitates the entry process of exogenous DNA (plasmid).
15. *E. coli* colonies on the LB plate contain both pBV220 and pBV220-Apt. Therefore, we pick several single colonies and send them to BGI Company for sequence to determine which colony carries the contracted plasmid pBV220-Apt.
16. It is necessary to keep a proper shaking speed for *E. coli* growth. We find the range of 150–220 rounds per minute suitable.
17. The optical density at 600 nm represents solution turbidity, which reflects the bacterial growth condition.
18. Plasmid in *E. coli* TOP10 generally loses after cultivation of generations because of unknown reasons. During experiments, we find β-lactamase expresses higher in *E. coli* TOP10 than in *E. coli* DH5α. Therefore, we use DH5α to store and amplify the plasmid pBV220-Apt and use TOP10 to measure data.
19. The overnight culture is regarded as the bacterial seed for the subsequent amplification. Mix it well before seeding.
20. Enzyme is a kind of protein in living cells and is easily digested. Thus, the common operation is to keep the solution on ice or low temperature after this step.
21. One unit of β-lactamase is defined as the amount which produces 1.0 mmol of product per minute.

Acknowledgements

This work was supported by the National Natural Science Foundation of China (Nos. 20725308, 90913014, 20721061 and TRR61) and the Major Research Plan of China (No. 2006CB932102).

References

1. Winkler WC, Nahvi A, Roth A, Collins JA, Breaker RR (2004) Control of gene expression by a natural metabolite-responsive ribozyme. Nature 428:281–286
2. Sharma V, Nomura Y, Yokobayashi Y (2008) Engineering complex riboswitch regulation by dual genetic selection. J Am Chem Soc 130: 16310–16315
3. Bayer TS, Smolke CD (2005) Programmable ligand-controlled riboregulators of eukaryotic gene expression. Nat Biotechnol 23:337–343
4. Desai SK, Gallivan JP (2004) Genetic screens and selections for small molecules based on a synthetic riboswitch that activates protein translation. J Am Chem Soc 126:13247–13254
5. Brinas L, Zarazaga M, Saenz Y, Ruiz-Larrea F, Torres C (2002) β-Lactamases in ampicillin-resistant Escherichia coli isolates from foods, humans, and healthy animals. Antimicrob Agents Chemother 46:3156–3163
6. Feng X, Liu L, Duan X, Wang S (2011) An engineered riboswitch as a potential gene-regulatory platform for reducing antibacterial drug resistance. Chem Commun 47: 173–175
7. Win MN, Smolke CD (2007) A modular and extensible RNA-based gene-regulatory platform for engineering cellular function. Proc Natl Acad Sci U S A 104:14283–14288
8. Hedberg M, Lindqvist L, Bergman T, Nord CE (1995) Purification and characterization of a new β-lactamase from Bacteroides uniformis. Antimicrob Agents Chemother 39: 1458–1461

Chapter 19

Construction of Ligand-Responsive MicroRNAs that Operate Through Inhibition of Drosha Processing

Chase L. Beisel, Ryan J. Bloom, and Christina D. Smolke

Abstract

MicroRNAs (miRNAs) offer powerful tools for targeted gene silencing in almost all eukaryotes. These tools have received considerable attention for their utility in both fundamental genetic studies and as therapeutic agents. Rendering individual microRNAs responsive to endogenous or exogenously applied molecules (or ligands) can improve the stringency of silencing and can mediate autonomous control. This chapter describes the construction of ligand-responsive miRNAs that undergo reduced processing and subsequent gene silencing when bound by the recognized ligand. Following a simple set of rules, the engineered microRNAs can be readily modified to target different sequences and to bind different ligands. Individual miRNAs also can be incorporated into the same transcript for tunable, multi-gene silencing.

Key words Aptamers, Drosha, Gene regulation, Mammalian cells, MicroRNA, Riboswitch, RNA interference

1 Introduction

MiRNAs are small regulatory RNAs that carry out gene silencing in most eukaryotes [1]. In mammalian cells, transcribed miRNAs undergo sequential processing by the endoribonucleases Drosha and Dicer to form a ~22-nucleotide "mature" miRNA [2]. The mature miRNA serves as a guide to recruit the RNA-induced silencing complex (RISC) to complementary mRNAs, which are silenced by cleavage, inhibition of translation, or sequestration in inaccessible intracellular bodies [3]. Many of the known targets of natural miRNAs are involved in development, angiogenesis, and cancer formation, underscoring the widespread importance of these RNA regulators [4].

MiRNAs have also received considerable attention as tools for targeted gene silencing [5, 6]. The mature miRNA sequence can be altered to specifically silence different genes within the host organism or to access regulatory networks of endogenous miRNAs.

Atsushi Ogawa (ed.), *Artificial Riboswitches: Methods and Protocols*, Methods in Molecular Biology, vol. 1111, DOI 10.1007/978-1-62703-755-6_19, © Springer Science+Business Media New York 2014

One challenge impeding the implementation of miRNA-based tools is ensuring temporal and spatial control of miRNA activity. The use of inducible promoter systems offers one solution, although these systems require the expression of heterologous proteins that may trigger an immune response, that require large expression constructs, and that cannot be easily modified to accommodate different inducing molecules.

A separate solution is integrating miRNAs with aptamers, RNAs that specifically and tightly bind a selected molecule. Procedures for the in vitro and in vivo selection of aptamers are well established, resulting in aptamers against a large assortment of ligands, including small molecules, metals, and proteins [7–9]. Furthermore, insights into the design of riboswitches have suggested how to rationally combine aptamers and regulatory RNAs for the conditional control of gene expression.

Here, we describe the rational integration of aptamers and miRNAs for conditional gene silencing in eukaryotes. Ligand binding to the aptamer inhibits Drosha processing, reducing the production of mature miRNAs and subsequent gene silencing. This design allows the aptamer and mature miRNA sequence to be readily replaced, facilitating the broad use of conditional miRNAs in applications ranging from genetic studies to the development of "smart" therapeutics.

2 Materials

1. pCS1815 expression vector.
2. Standard molecular biology equipment and reagents.
3. Restriction enzymes AgeI, AsiSI, ClaI, and PacI.
4. *E. coli* JM110 (*see* **Note 1**).
5. Standard equipment for mammalian cell culture.
6. Fugene HD transfection reagent (Promega).
7. Dulbecco's Modified Eagle Medium supplemented with 10% FBS (Life Technologies).
8. Opti-MEM (Life Technologies).
9. 24-well cell culture plates (BD Biosciences).
10. Trypsin (Life Technologies).
11. HEK293 cell line.
12. HEK293 GFP cell line.
13. Flow cytometer with 488 nm laser with 525 nm and 610 nm long pass filters and a 550 nm dichroic mirror (Beckman Coulter).

3 Methods

We first describe the design of ligand-responsive miRNAs that target a desired gene and bind a specific ligand. As a model, we target the gene encoding green fluorescent protein (GFP) and integrate an aptamer that recognizes the small molecule theophylline [10]. Next, we cover the cloning of designed microRNAs into an expression vector from our laboratory, although other expression vectors can be readily used (*see* **Note 2**). Finally, we introduce assays to evaluate the activity of the designed miRNAs in mammalian cell culture.

The protocol below describes the expression of a miRNA that silences the expression of chromosomally encoded GFP in *trans*. However, there are two other possible configurations to regulate gene expression with the ligand-responsive miRNA: cis regulation and a combination of *cis* and *trans* regulation (*see* **Note 3**). Under *cis* regulation, a ligand-responsive miRNA is placed in the 3′ untranslated region of a non-targeted gene. Ligand-dependent processing of the miRNA deactivates the transcript, silencing the upstream gene. Under combined *cis* and *trans* regulation, a ligand-targeting miRNA is placed in the 3′ untranslated region of a targeted gene. This configuration imparts regulation through miRNA processing and miRNA-mediated silencing, where both steps are inhibited by the ligand.

3.1 Design of Ligand-Responsive MicroRNAs

1. The ligand-responsive miRNA is comprised of two parts, a mature miRNA and an aptamer, grafted onto the base miRNA miR-30a (Fig. 1a). The mature miRNA is 22 nucleotides and is complementary to the target mRNA (*see* **Note 4**). The paired nucleotides must be complementary with the exception of the first nucleotide on the 5′ end to preserve the secondary structure of the miRNA. The example sequence in Fig. 1a shows a potent sequence for silencing the expression of GFP (top) and the expression of La (bottom).
2. The aptamer is placed at the base of the lower stem. Normally, the unstructured base of the stem serves as a ruler for Drosha cleavage [11]. When the base of the stem is structured, such as through base pairing, Drosha does not effectively bind and cleave the miRNA [11–13]. To exploit this feature of processing by Drosha, the aptamer must be inserted such that it is unstructured in the absence of ligand and tightly bound in the presence of ligand. Fortunately, most aptamers contain stems at the top and bottom of the binding region; these regions can be removed in order to insert the aptamer into the miRNA. The only limitation is that the aptamer must exhibit a minimal bulge size in the absence of ligand (*see* **Note 5**). Below this limit, Drosha processing is substantially inhibited [14]. If the bulge of the selected aptamer is too small, the loop of the aptamer can be included, as illustrated for the hypoxanthine aptamer in Fig. 1b.

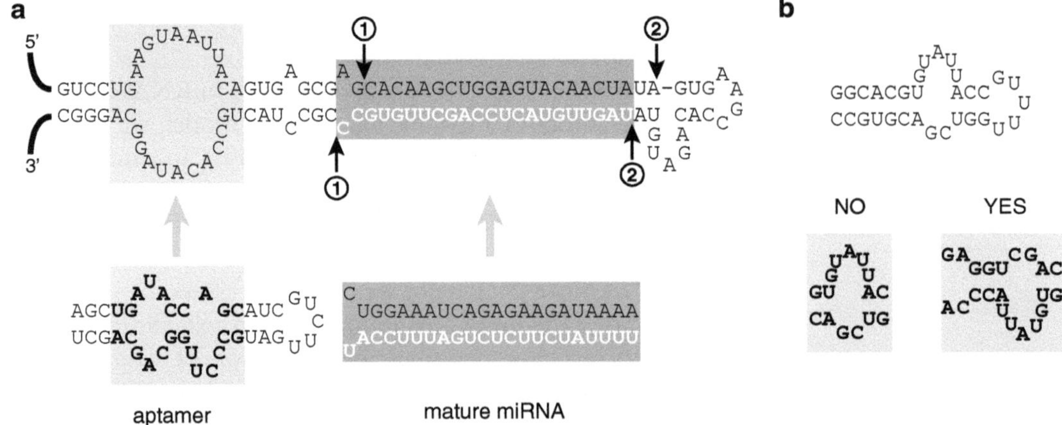

Fig. 1 Design of ligand-responsive miRNAs. (**a**) MiRNAs undergo sequential processing by Drosha (*circle 1*) and Dicer (*circle 2*), where one of the two final strands, the mature miRNA, is retained by RISC for the silencing of complementary mRNAs. The design builds on miRNA-30a through the integration of a mature miRNA sequence (*white letters, dark gray box*) and the binding domain of an aptamer (*bolded letters, light gray box*). The binding domain of the aptamer is inserted below the lower stem of the miRNA. In the absence of ligand, the aptamer will be unstructured, allowing Drosha processing. In the presence of ligand, the bound aptamer will be structured, inhibiting Drosha processing. The displayed design depicts the insertion of a mature miRNA targeting the La gene in place of a mature miRNA targeting the GFP gene, and the insertion of the binding domain of the theophylline aptamer in place of a large bulge. (**b**) The aptamer core must be of sufficient size to allow Drosha processing in the absence of ligand. In cases where a small aptamer, such as the hypoxanthine aptamer as shown, is selected, the aptamer binding domain and part of the loop can be integrated. Note that the binding domain is placed adjacent to the lower stem to ensure that ligand binding inhibits Drosha processing

3.2 Construction of Ligand-Responsive MicroRNAs

1. Once the ligand-responsive miRNA is designed, the miRNA must be cloned into an expression vector. We use a construct (pCS1815, Fig. 2a) that can express the miRNA from the pol II CMV promoter or the pol III U6 promoter. The construct additionally expresses DsRed-Express as a transfection marker. The plasmid can be isolated from *E. coli* cells using standard plasmid preparation techniques. The miRNA is inserted into the restriction sites AgeI and ClaI downstream of the CMV promoter. The intervening CD19 gene serves as a surface marker for transfection efficiency and for miRNA processing.
2. The miRNA is assembled into linear, double-stranded DNA that can be cleaved with AgeI/ClaI restriction enzymes and ligated into the plasmid (Fig. 2b). The linear DNA includes four restriction sites adjacent to the stem below the aptamer: AgeI/AsiSI on the 5′ end, and PacI/ClaI on the 3′ end. The internal restriction sites allow the sequential insertion of multiple miRNAs as described below. The linear DNA encoding the miRNA also contains four base pairs flanking the ends of the restriction sites for efficient cleavage by AgeI and by ClaI.

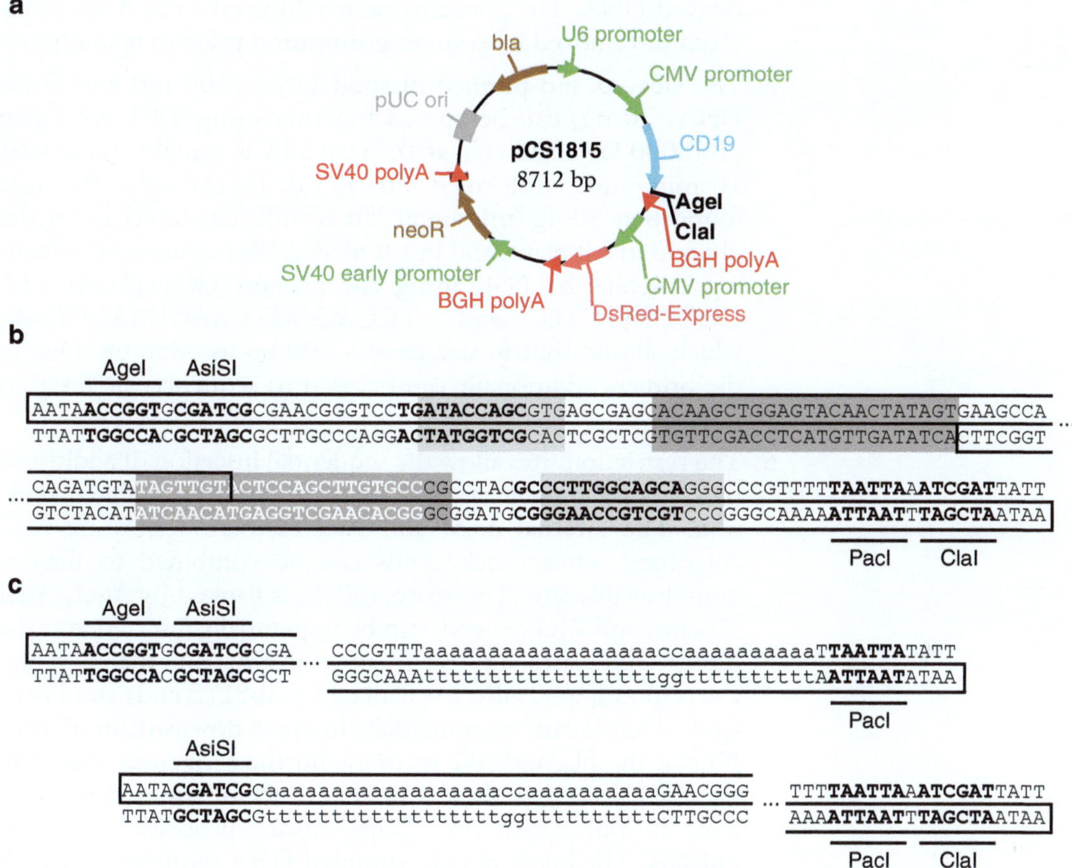

Fig. 2 Cloning designed miRNAs into the pCS1815 expression plasmid. (**a**) Plasmid map of pCS1815. Designed miRNAs are cloned into the AgeI/ClaI restriction sites downstream of the gene encoding the CD19 surface marker. (**b**) Sequence of linear double-stranded DNA encoding the designed miRNA. The DNA includes four restriction sites for the insertion of the double-stranded DNA into the expression plasmid. The *boxed* nucleotides indicate the oligonucleotide sequences needed to PCR amplify the full-length DNA. (**c**) Sequences for the sequential insertion of multiple miRNAs. The A-rich spacer (*lowercase letters*) helps ensure processing of each miRNA. The *top sequence* can be cleaved with AgeI/PacI and inserted into AgeI/AsiSI upstream of a miRNA in pCS1815. The *bottom sequence* can be cleaved with AsiSI/ClaI and inserted downstream of a miRNA in pCS1815

3. The linear DNA is synthesized with two overlapping oligonucleotides, which are boxed in Fig. 2b (*see* **Note 6**). The two oligonucleotides can serve as primers and DNA templates in a standard PCR, where six cycles are sufficient for the synthesis of the linear DNA. The PCR-amplified DNA is then column-purified to remove any remaining oligonucleotides and unincorporated dNTPs.
4. The plasmid DNA and the linear DNA are digested with AgeI and ClaI and column-purified to remove the small cleaved fragments. The plasmid DNA can be further treated with a phosphatase to reduce the frequency of ligation of partially

cleaved DNA. The phosphatase would need to be heat-inactivated or removed by column purification prior to ligation.

5. The cleaved and purified plasmid DNA (200 ng) and linear DNA (20 ng) can be ligated together using T4 DNA ligase (100,000 U), incubated at 16 °C in a PCR machine for at least 10 min, and transformed into *E. coli* JM110 cells. We have found that 50 μg/mL ampicillin is sufficient to select for the plasmid in LB plates and in LB media. We recommend screening colonies by PCR using the primers GCTGTGACTTT GGCTTATCTG and TCCAGGGTCAAGGAAGGCAC, which should shift in size from ~ 200 bp to ~350 bp. One of the primers additionally can be used to sequence the inserted miRNA.
6. The restriction sites allow the sequential insertion of additional miRNAs for tunable or combinatorial control of gene expression. The internal AsiSI and PacI restriction sites are isoschizomers whose sticky ends can be combined to form a non-cleavable site. Therefore, miRNAs flanked by AgeI/AsisI (5′ end) and PacI (3′ end) can be sequentially inserted into the AgeI/AsiSI sites upstream of miRNAs in the plasmid. Correspondingly, miRNAs flanked by AsiSI (5′end) and PacI/ClaI (3′ end) can be sequentially inserted downstream of miRNAs in the plasmid. We recommend the sequences shown in Fig. 2c. Note that these sequences include an A-rich spacer that, in our experience, helps ensure processing of each miRNA. The linear double-stranded DNA sequences for each inserted miRNA can be generated as described above.

3.3 Measurement of Ligand-Responsive MiRNA Activity in Cell Culture

1. The cloned plasmid first is isolated from the JM110 cells. The purity of the DNA sample is a major factor in transfection efficiency, with highly pure plasmid DNA lacking endotoxins resulting in the most reproducible, high efficiency transfections (*see* **Note 7**). The original pCS1815 plasmid or the same plasmid containing a miRNA with a scrambled mature miRNA sequence should be isolated; this plasmid will serve as a control when assessing the silencing of GFP expression.
2. HEK 293 cells stably expressing GFP are seeded in 24 well plates at a density of 100,000 cells/well in Dulbecco's Modified Eagle Medium (DMEM) on Day 1. An exogenous ligand can be added to the medium at this time or during the transfection. HEK 293 cells lacking GFP can be additionally seeded to serve as a "DsRed-only" control for color-compensation.
3. On Day 2, Fugene HD and Opti-MEM are allowed to equilibrate to room temperature. Then purified plasmid (500 μg) is mixed with 25 μL of Opti-MEM and 1.5 μL of Fugene HD. This mixture is incubated at room temperature for 30 min then added to a single well of cells and mixed with the medium

by gentle shaking. If multiple wells of cells are to be transfected with the same plasmid (for experimental replicates or to test different biological conditions) the mixture of transfection reagents can be scaled accordingly. We recommend at least two biological replicates for each condition tested as well as single-color controls for both green and red and a no-color sample. The transfected cells are then incubated for 48 h at 37 °C and 5 % CO_2.

4. On Day 4, the cells are harvested by aspirating away the media and adding 100 μL of trypsin to each well. Once the cells have become detached (~15 min), the trypsin is neutralized by adding 200 μL of DMEM and the medium is gently pipetted up and down to break up clumps. The cells are then added to a 96-shallow-well, round bottom plate for flow cytometry (*see* **Note 8**).
5. A "GFP-only" control (untransfected HEK293 GFP cells) and a "DsRed-only" control (HEK293 cells transfected with the plasmid) are first run through the flow cytometer to set any required compensation for the green and red signals. The "GFP-only" control additionally is used to set a gate separating DsRed-positive (transfected) and DsRed-negative (untransfected) cells; all of these cells should be contained within the DsRed-negative gate. When running the experimental samples, the analysis will be restricted to DsRed-positive cells that have received the plasmid containing the miRNAs.
6. Once all of the samples have been run on the flow cytometer, the data are analyzed using appropriate analysis software, such as FlowJo. The knockdown efficiency of a microRNA is determined by dividing the median fluorescence of GFP in DsRed-positive cells transfected with the tested plasmid by the GFP fluorescence in DsRed-positive cells transfected with a control plasmid harboring a microRNA with a scrambled targeting sequence.

$$\%\ Knockdown = 100\%\left(1-\frac{Median\ GFP\ fluorescence\ (DsRed-positive\ cells\ with\ tested\ plasmid\ )}{Median\ GFP\ fluorescence\ (DsRed-negative\ cells\ with\ control\ plasmid\ )}\right)$$

The responsiveness of a miRNA to the ligand is measured by the fold change. The fold change is calculated by dividing the percent knockdown of a miRNA in cells cultured in medium without or with the ligand:

$$Fold\ change = \frac{1-\%\ Knockdown\,(+ligand)}{1-\%\ Knockdown\,(-ligand)}$$

The fold change is expected to be greater than one because of reduced silencing in the presence of ligand.

4 Notes

1. The cloning strain must be *dam*⁻/*dcm*⁻ for efficient cleavage of the ClaI restriction site in pCS1815.
2. The designed miRNAs can be expressed from any expression vector as long as the miRNA is encoded downstream of a polII promoter (e.g., CMV, EF1α) or a polIII promoter (e.g., U6). The miRNA also can be encoded as an independent transcript or in the 3′ UTR of a gene. Note that the latter configuration reduces the expression of the upstream gene [14].
3. The observed fold-change in gene expression in response to ligand varies with the configuration. In our experience, the combined *cis* and *trans* configuration exhibits the largest fold-change, whereas the *cis* configuration exhibits the smallest fold-change.
4. The selected mature miRNA sequence has a major impact on the potency of gene silencing. While there are excellent tools available for the design of these sequences [15], we recommend testing at least four different sequences.
5. In our experience, the bulge size should be a minimum of 13 nucleotides. The bulge can include intervening base pairs of limited length.
6. In our experience, no additional purification steps are required beyond the standard purification performed by the DNA synthesis company. Alternatively, the designed miRNA can be ordered from different gene synthesis companies as a single piece of double-stranded DNA.
7. We recommend that plasmid DNA can be harvested using the PureYield™ Plasmid Miniprep System, which yields highly pure, endotoxin-free plasmid DNA.
8. Other fluorescence measurement techniques, such as fluorescence microscopy or a fluorescence microplate reader, can be employed. In the case of a fluorescence microplate reader or any other bulk measurement technique, plasmids expressing GFP and DsRed-Express should be co-transfected so both proteins will be expressed in the same cell. Then bulk GFP fluorescence can be normalized to bulk DsRed-Express fluorescence similar to previous work [16].

Acknowledgments

This work was supported by North Carolina State University (start-up funds to CLB), the National Science Foundation (fellowship to RJB), the National Institutes of Health (RC1GM091298), and the Defense Advanced Research Projects Agency (HR0011-11-2-0002).

References

1. Bartel DP (2009) MicroRNAs: target recognition and regulatory functions. Cell 136:215–233
2. Bartel DP (2004) MicroRNAs: genomics, biogenesis, mechanism, and function. Cell 116:281–297
3. Gu S, Kay MA (2010) How do miRNAs mediate translational repression? Silence 1:11
4. Alvarez-Garcia I, Miska EA (2005) MicroRNA functions in animal development and human disease. Development 132:4653–4662
5. Leisner M, Bleris L, Lohmueller J, Xie Z, Benenson Y (2010) Rationally designed logic integration of regulatory signals in mammalian cells. Nat Nanotechnol 5:666–670
6. Dickins RA, Hemann MT, Zilfou JT, Simpson DR, Ibarra I, Hannon GJ, Lowe SW (2005) Probing tumor phenotypes using stable and regulated synthetic microRNA precursors. Nat Genet 37:1289–1295
7. Tuerk C, Gold L (1990) Systematic evolution of ligands by exponential enrichment: RNA ligands to bacteriophage T4 DNA polymerase. Science 249:505–510
8. Ellington AD, Szostak JW (1990) In vitro selection of RNA molecules that bind specific ligands. Nature 346:818–822
9. Hermann T, Patel DJ (2000) Adaptive recognition by nucleic acid aptamers. Science 287:820–825
10. Jenison RD, Gill SC, Pardi A, Polisky B (1994) High-resolution molecular discrimination by RNA. Science 263:1425–1429
11. Han J, Lee Y, Yeom KH, Nam JW, Heo I, Rhee JK, Sohn SY, Cho Y, Zhang BT, Kim VN (2006) Molecular basis for the recognition of primary microRNAs by the Drosha-DGCR8 complex. Cell 125:887–901
12. Zeng Y, Cullen BR (2005) Efficient processing of primary microRNA hairpins by Drosha requires flanking nonstructured RNA sequences. J Biol Chem 280:27595–27603
13. Zeng Y, Yi R, Cullen BR (2005) Recognition and cleavage of primary microRNA precursors by the nuclear processing enzyme Drosha. EMBO J 24:138–148
14. Beisel CL, Chen YY, Culler SJ, Hoff KG, Smolke CD (2011) Design of small molecule-responsive microRNAs based on structural requirements for Drosha processing. Nucleic Acids Res 39:2981–2994
15. Petri S, Meister G (2013) siRNA design principles and off-target effects. Methods Mol Biol 986:59–71
16. An CI, Trinh VB, Yokobayashi Y (2006) Artificial control of gene expression in mammalian cells by modulating RNA interference through aptamer-small molecule interaction. RNA 12:710–716

Chapter 20

A Three-Dimensional Design Strategy for a Protein-Responsive shRNA Switch

Shunnichi Kashida and Hirohide Saito

Abstract

We have recently developed synthetic short hairpin RNA (shRNA) switches that respond to intracellular proteins and control the expression of target genes in mammalian cells (Kashida et al. Nucleic Acids Res 40:9369–9378, 2012; Saito et al. Nat Commun 2:160, 2011). Here, we describe a method for the three-dimensional (3D) design of a protein-responsive shRNA switch that employs modeling software and known 3D structures of RNA–protein complexes. We were able to predict the effect of steric hindrance between the Dicer enzyme and shRNA-binding protein in silico by superimposing the 3D model of the shRNA switch on Dicer. The function of the designed switch can be evaluated in vitro and in living cells. Our expertise will help utilize the 3D structure of biomacromolecular complexes for the design of functional genetic switches.

Key words 3D design, Riboswitch, RNP, shRNA, Dicer, Steric hindrance, Prediction of switch efficiency, Synthetic biology, RNAi

1 Introduction

One of the main goals in synthetic biology is to control and rewire gene regulatory networks to improve human health and to generate new drugs [1]. Synthetic RNA switches that control target gene expression, depending on the intracellular environment, play a crucial role in achieving these objectives [2–5]. In this chapter, we focus on engineering an shRNA that responds to intracellular proteins and control knockdown of target genes. In the mammalian shRNA processing pathway, an RNase III-class endonuclease, Dicer, recognizes shRNA at the end of its stem and cleaves the shRNA at approximately the 22nd nucleotide from the 3′- and/or 5′-end to produce a functional double-stranded RNA (dsRNA) for performing RNAi [6, 7]. This endonuclease does not discriminate the loop region of the shRNA during its processing, and therefore, the desired sequences can be incorporated into the loop region in general [8].

Atsushi Ogawa (ed.), *Artificial Riboswitches: Methods and Protocols*, Methods in Molecular Biology, vol. 1111,
DOI 10.1007/978-1-62703-755-6_20, © Springer Science+Business Media New York 2014

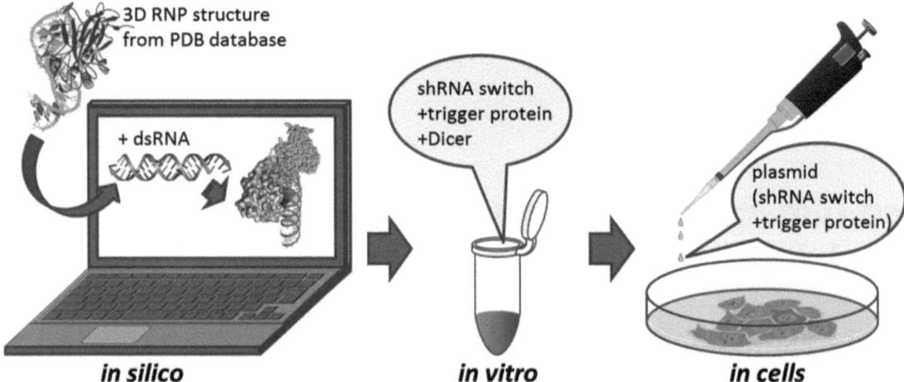

Fig. 1 The 3D design and construction of protein-responsive shRNA switches. The function of the in silico-designed RNA switches can be evaluated using biochemical assays in vitro and gene reporter assays in mammalian cultured cells

It has been previously reported that a small molecule (e.g., theophylline) responsive shRNA switch was constructed by introducing the theophylline-binding RNA motif (aptamer) into the loop region of shRNA. This modification inhibits Dicer activity when the drug is bound to the loop [9, 10]. In addition, naturally occurring protein-responsive shRNA has been discovered in human systems. For example, the Lin28 protein interacts with pre-let-7 microRNA (miRNA) and inhibits the processing of pre-let-7 microRNA into mature let-7 miRNA by Dicer, which leads to the de-repression of let-7 target genes, including oncogenes and cell-cycle genes [11, 12].

To design and construct a variety of synthetic RNA switches, synthetic biologists have utilized the primary and secondary structures of the RNA molecules [10, 13–15]. Alternatively, functional switches have been screened from RNA library sets containing random sequences [16, 17]. However, it is difficult to predict the function of the generated RNA switches prior to experiments in most cases. Thus, we aimed to develop a method for the 3D molecular design of protein-responsive RNA switches with a predictable function. This method will contribute to the investigation of the design principles of functional genetic switches composed of biomolecular complexes [18].

Here, we describe the 3D design of protein-responsive shRNAs in silico based on high-resolution structures of known RNA–protein (RNP) complexes. We evaluated the performance of these shRNAs in vitro and in mammalian cells (Fig. 1). The designed shRNA senses the intracellular RNA-binding protein (trigger protein) and inhibits the activity of Dicer via steric hindrance between Dicer and the RNP complex (The model of a protein-responsive shRNA switch is shown in Fig. 2). Thus, these synthetic shRNA devices can control target gene expression in a predictable manner.

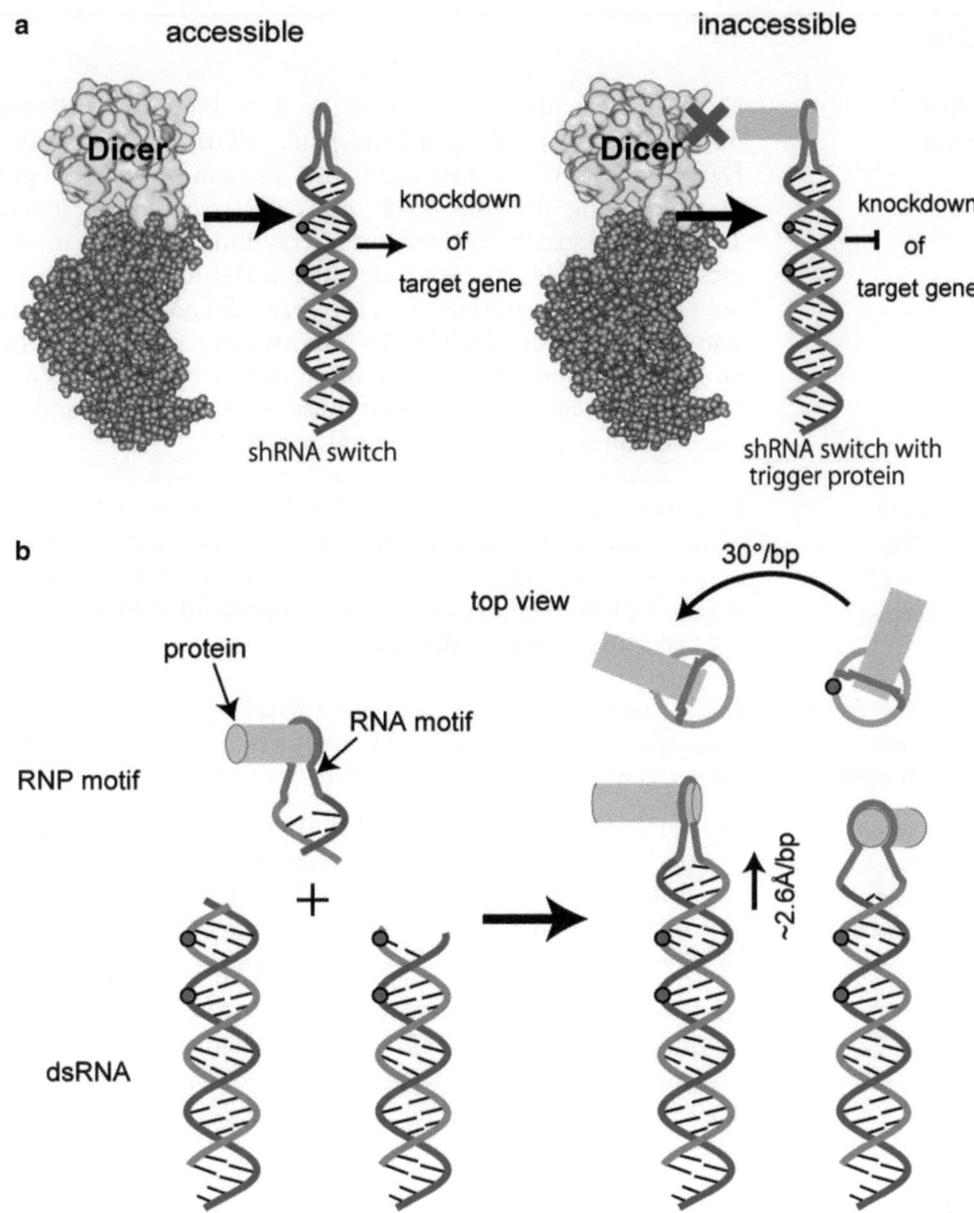

Fig. 2 Schematic representation of the protein-responsive shRNA device mechanism and its bound protein orientation. (**a**) In the absence of the trigger protein, Dicer can access and process the designed shRNA, and the processed shRNA induces the knockdown of its target gene through RNAi (*left*). Dicer is inaccessible to shRNA in the presence of the trigger protein that binds to the loop region of the shRNA because the RNP interaction faces Dicer and inhibits its access (*right*). The prevention of Dicer's function causes the de-repression of the gene knockdown. (**b**) Protein-responsive shRNA switches are designed in silico by connecting an RNP motif to the corresponding dsRNA and Dicer. By specifying the position of the devices with reference to the Dicer cleavage sites and by comparing the bound proteins' locations, we observed that with a 1-bp (base pair) insertion, the bound trigger protein on the shRNA switch is located approximately 2.6 Å more distant from the Dicer cleavage sites and rotates approximately 30° in an anticlockwise direction around the axis of the dsRNA

2 Materials

2.1 3D RNP Structures

The 3D molecular structures of the RNP-binding motifs can be obtained from the Protein Data Bank (PDB, Worldwide Protein Data Bank). We constructed the protein-responsive shRNA switches using the four RNP molecular structures described in Table 1. The structures from both X-ray and NMR structural analyses are available. To generate effective shRNA switches, tight interaction between RNP (e.g., less than a 50 nM dissociation constant; Kd) is preferable. The RNA motif extracted from the parental RNP structure should have more than 1 bp at the 5′/3′-ends because the region of the motif must be connected to and superimposed on the dsRNA region of the shRNA.

2.2 Software and Operating Environment for Macromolecular Design

Discovery Studio (Accelrys) ver. 2.5-3.1 can be used for the modeling. The operation of the software should be conducted in accordance with the instructions of the software. In addition, more than 6 GB of cache memory capacity is recommended to avoid stalling the program during modeling.

2.3 Inhibition Assay of Human Dicer Cleavage for shRNA In Vitro

2.3.1 In Vitro Transcription and Purification of shRNA Switches

For in vitro transcription, we used T7 RNA polymerase and buffer prepared in-house or used a MEGAshortscript™ T7 Kit (AM1354, Invitrogen) with slight modifications (*see* **Note 1**).

1. Single-stranded template DNA (purchase from an oligonucleotide synthesis company): 5′-(complementary strand of the shRNA switch) TATAGTGAGTCGTATTAGC-3′ (*see* also Subheading 3.3 for the sequence design of the shRNA switch).
2. T7 annealing primer: 5′-GCTAATACGACTCACTATA-3′.
3. Transcription reaction mixture: 0.75 μM single-stranded template DNA, 0.75 μM T7 annealing primer, 7.1 μg/mL Pyro phosphatase (Roche), 50 μg/mL bovine serum albumin, 40 mM HEPES–KOH, 20 mM $MgCl_2$, 5 mM DTT, 2 mM

Table 1
Features of the RNP structures used for shRNA switches

PDB ID	Protein	Kd	protein organism	RNA origin	structural analysis method
1RLG	L7Ae	0.46 nM	*Archaeoglobus fulgidus*	Ribosomal RNA	X-ray structural analysis
1URN	U1A	<50 pM	*Homo sapiens*	U1snRNA stem loop II	X-ray structural analysis
1AUD	U1A	<50 pM	*Homo sapiens*	3'UTR of U1A mRNA	NMR
1OOA	NFκB p50	5.4 nM	*Mus musculus*	from *in vitro* selection	X-ray structural analysis

PDB ID	RNA sequence
1RLG	*GCUC*UGACCGAAAGGCGUGAU*GAGC*
1URN	*AA* UCCAUUGCACUCCGGAU *UU*
1AUD	*GGCAG* AGUCCUUCGGGACAUUGCAC *CUGCC*
1OOA	*CAUAC* UUGAAACUGUAAGGUUGGC *GUAUG*

Spermidine, 4.8 mM ATP, 4.8 mM UTP, 4.8 mM CTP, 1.28 mM GTP, 12.8 mM GMP (*see* **Note 2**).

4. T7 RNA polymerase (prepared in-house).
5. 2 U/μL TURBO DNase (Ambion).
6. PD-10 column (GE Healthcare).
7. PD-10 buffer: 0.3 M potassium acetate, pH 6.0, 15 % ethanol.
8. 15 % native gel: 15 % acrylamide–bisacrylamide (29:1), 0.1 % ammonium persulfate, 0.05 % N,N,N′,N′-tetramethylethylenediamine, 0.5× TBE buffer.
9. 5× loading buffer: 0.25 % BPB, 30 % glycerol.
10. 22-μm PES microfilter (Millex-GP).
11. Elution buffer: 0.5 M NaCl, 0.1 % SDS, 1 mM EDTA.

2.3.2 Expression and Purification of the Trigger RNA Binding Proteins Using FPLC

1. Competent cell: BL21 (DE3) pLysS (Promega) or chemically competent *E. coli* cells suitable for transformation and protein expression.
2. Phosphate buffer A: 20 mM NaH_2PO_4, 20 mM Na_2HPO_4, 500 mM NaCl.
3. Phosphate buffer B: 20 mM NaH_2PO_4, 20 mM Na_2HPO_4, 500 mM NaCl, 1 M imidazole.
4. FPLC: AKTApurifier (GE Healthcare).
5. Prepacked column: HisTrap HP column (5 mL, GE Healthcare).
6. 6× Sample buffer: Sample buffer solution with reducing reagent (6×) for SDS-PAGE (Nacalai).
7. 2× dialysis buffer: 40 mM HEPES–KOH, pH 7.8, 230 mM NaCl, 10.8 mM KCl, 1.6 mM $MgCl_2$.
8. Dialysis cassette: Slide-A-Lyzer Dialysis Cassettes (Thermo Scientific).
9. Sample concentrator: Vivaspin 6 concentrator (GE Healthcare).

2.3.3 Inhibition Assay of Human Dicer Cleavage for shRNA In Vitro

1. Recombinant human Dicer: Recombinant human Dicer can be purchased from Genlantis.inc for the experiments. Human Dicer purified in the laboratory can also be used for the experiments.
2. Dicer reaction mixture: 1 μL of 10 mM ATP, 0.5 μL of 50 mM $MgCl_2$, 4 μL of Dicer Reaction Buffer (Genlantis), 1 μL of 0.5 unit/μL Recombinant Human Dicer Enzyme (Genlantis).
3. Binding buffer: Opti-MEM I (Life Technologies).
4. Gel staining buffer: 1× SYBR Green I and SYBR Green II in 0.5× TBE buffer.

2.4 Construction of Plasmids and RNAi Experiments in Mammalian Cells

2.4.1 shRNA Expressing Plasmid

1. ShRNA cloning kit: BLOCK-iT™ Inducible H1 RNAi Entry Vector Kit (Invitrogen).
2. ShRNA top strand (purchase from an oligonucleotide synthesis company): 5′-CACC (strand of the shRNA switch without additional two nucleotides (Fig. 4, 3)) -3′ and ShRNA bottom strand: 5′-AAAA (complementary strand of the shRNA switch without additional two nucleotides (Fig. 4, 3)) -3′.

2.4.2 Trigger Protein Expression Plasmid

pcDNA3.1, 4 and 5 (Life Technologies) or other expression plasmids suitable for expression in mammalian cells can be used.

2.4.3 EGFP Expression Control of the Designed shRNA Switches in Cultured Mammalian Cells

1. Human cultured cells: Maintain 293FT cells (Life Technologies) using Dulbecco's modified Eagle's medium (Nacalai) containing 10 % fetal bovine serum (Cambrex), 2 mM L-glutamine (Life Technologies), 0.1 mM MEM Nonessential Amino Acids (Life Technologies), 1 mM sodium pyruvate (Sigma), 0.5 % Penicillin-Streptomycin (Life Technologies) and 500 μg/mL G418 sulfate at 37 °C under a 5 % CO_2 concentration. HeLa cells, which were also used, and other human cultured cells successfully transfected with plasmids can also be used for the experiment.
2. Lipofection reagent: Lipofectamine 2000 (Life Technologies).
3. Fluorescent microscopy: To observe EGFP expression in cells, we used an IX-81 inverted microscope system (Olympus), a 470- to 495-nm filter for excitation and a 510- to 550-nm filter for emission.
4. Flow cytometry: To detect and quantify EGFP fluorescence, we used a flow cytometer (FACS Aria, BD Bioscience), a 488-nm semiconductor laser beam, and a 530/30-nm emission detector.

3 Methods

The methods can be divided into three steps (Fig. 3) as follows.
Step 1: The three-dimensional design of protein-responsive shRNA in silico based on the 3D molecular structures of the RNP motif using modeling software. The steric hindrance between Dicer and the trigger protein on the shRNA can be predicted from the model, which determines the inhibition efficiency of Dicer cleavage.
Step 2: Evaluate the inhibition of Dicer cleavage using recombinant human Dicer in vitro.
Step 3: Assess the function of the designed shRNA switches in human cultured cells.

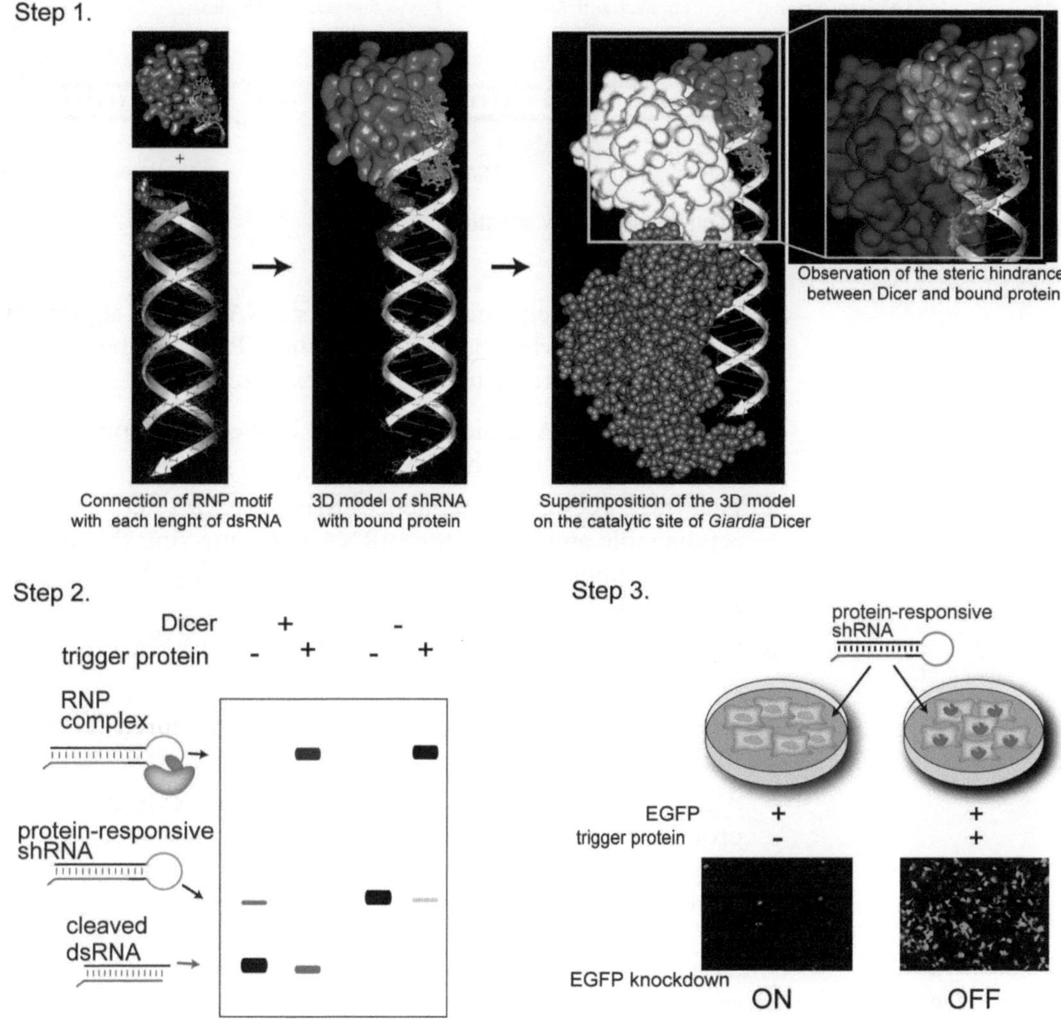

Fig. 3 Overview: 3D molecular design and assessment of the protein-responsive shRNA switches

3.1 Obtaining RNP Motifs

Construct protein-responsive shRNA switches using the specific RNP-binding pairs between a trigger protein and the corresponding RNA sequence. Choose the RNP motifs from known RNP complexes of the trigger proteins and their binding RNA.

3.2 dsRNA Sequence Targeting Specific mRNA

Obtain the dsRNA sequence that induces knockdown of the target gene through RNAi from literature reports or databases. Sequences that start with "G" are preferable because the first "C" in the DNA template is better suited for the efficient transcription of T7 RNA polymerase in vitro and RNA polymerase III in human cells. If there are no known dsRNA sequences, construct and assess the

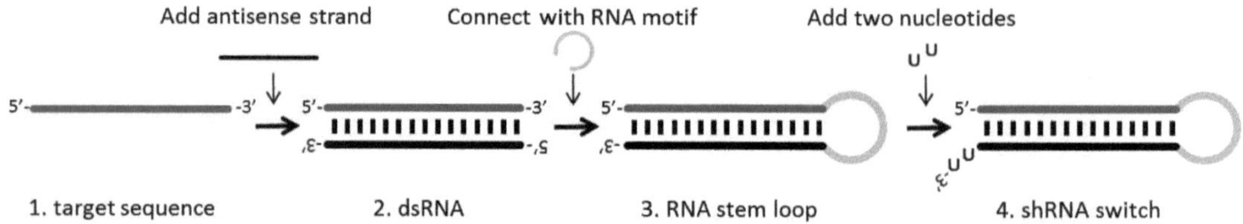

Fig. 4 Design steps of shRNA switch on the secondary structure

candidate dsRNA sequences using an siRNA design algorithm, such as siExplorer, which is available at the URL given below:

http://rna.chem.t.u-tokyo.ac.jp/siexplorer.htm.

1. Input the mRNA sequence or NCBI RefSeq accession number (NM_xxxx) and run the algorithm.
2. Pick the three or four sequences that start with "G" from the results table and design the shRNA by connecting the dsRNA with the appropriate loop (we used a 5′-GAAA-3′ loop or 5′-AGCAUAG-3′ loop).
3. Construct in vitro-transcribed shRNA or shRNA-expressing plasmids.
4. Assess the RNAi efficiency in human cultured cells by transfection.
5. After 24–48 h of transfection, collect the cells, extract the total protein, and quantify the selected protein using western blotting. Alternatively, analyze the knockdown efficiencies of the target fluorescent protein by flow cytometry.

3.3 shRNA Switch Design on the Secondary Structure

Obtain the 3D molecular structure of the RNP motif composed of the trigger protein and its binding RNA motif from the PDB. If the 3D structure is not available, the 3D molecular design step (Fig. 3, Step 1) can be skipped. In this case, the shRNA switches can be constructed based on the secondary structure as follows (Fig. 4).

1. Choose the appropriate RNA motifs. Omit the unnecessary sequences (sequences not having a positive effect on binding affinity) and extract minimal RNA motifs.
2. Design defective RNA motifs that do not bind to the trigger proteins by mutating or deleting the corresponding nucleotides of the motif.
3. Connect the RNA motif with the corresponding length of dsRNA (21–31 bp, also *see* **Note 3**) designed to code the target mRNA sequence to form the stem-loop structure of the RNA (Fig. 4, 1–3).
4. Add two nucleotides ("AG" or "UU") at the 3′ end of the RNA (Fig. 4, 4) to help Dicer recognize the target dsRNA.

3.4 3D Molecular Design of Protein-Responsive shRNA Switches Using Discovery Studio

We describe the 3D molecular design method for protein-responsive shRNA switches using two examples: U1A-responsive shRNA switches (Fig. 3, Step 1) and NFκB (p50)-responsive shRNA switches. U1A is one of the most abundantly expressed mRNA-processing and housekeeping proteins in mammalian cells [19]. NFκB is one of the major transcription factors in eukaryotic cells, and its abnormal expression has been linked to the proliferation of several types of tumor cells [20, 21]. For 3D modeling, the Discovery Studio ver. 2.5-3.1 can be used.

1. Startup Discovery Studio and download the PDB file (.txt) of the selected RNP motifs. The database stores the structure from X-ray structural or NMR analyses (*see* **Note 4** about NMR structure). We obtained the N-terminal domain of U1A (human) and its binding RNA motif in U1snRNA ($U1A_1$-RNA motif) from the PDB (ID: 1URN) and p50 (mouse NFκB) bound to its aptamer RNA (p50-RNA motif) from the PDB (ID: 1OOA) (*see* **Note 5** for using the structure of heterologous RNP). The PDB files from the X-ray structural analysis contained atomic coordinates of the RNP crystal lattice structure, which consists of the one, two, three, four or six identical groups of the RNP (U1A, three groups; p50, two groups), water, ions and other solvent molecules. The groups are symmetrically arranged on the rotation axes.
2. Focus on one RNP group of the crystal lattice and exclude the other groups (uncheck the boxes of the groups in the "Hierarchy window"), and clear the water and solvent molecules from the RNP group.
3. (Optional) Trim the a.a. sequence of the protein (or the protein domain) to be consistent with the sequence of the proteins expressed from the plasmid.
4. Create 22–28 base pairs (bp) of A-form dsRNA. Use "create/grow" at the 3′ function of the "Build and Edit Nucleic Acid" tool and input the sequence from the start of the shRNA switch sequence. The dsRNA consists of a sense strand (the sequence list is indicated as "S" in the RNA duplex group of the hierarchy window) and an antisense strand (indicated as "A"), and the sense strand should correspond with the 5′-end sequence of the stem in the shRNA switch.
5. Choose the RNA structures and change the display style of the atoms from "None" to "Scaled ball and stick."
6. Align the RNA motif and dsRNA three-dimensionally in reference to the nucleotides of the stem end of the RNA motif ($U1A_1$-, 2 bp; p50-, 5 bp; italic RNA sequences in Table 1) and those of the 3′-end of the sense strand.

7. Tether the corresponding atoms of the sugars, bases and phosphate backbones in the reference nucleotides (>2 bp) of the RNA motif and dsRNA using the "Add tether" function in the superimpose protocol.
8. Superimpose the RNA motif on the dsRNA using the "Superimpose by Tether" function in the superimpose protocol. The distance between the RNA motif and dsRNA are minimized using the least squares approximation polynomial and the distance of the corresponding tethered atomic coordinate pairs.
9. Delete the reference nucleotides of dsRNA that overlap with those of the RNA motif.
10. To concatenate the sense strand, RNA motif and antisense strand, connect the oxygen atom of the phosphate and the 5′ carbon atom (C5) of the ribose or connect the oxygen atom of the C5 and the phosphorous atom at the ends of the sequences with a single bond. If the oxygen atom of C5 and that of the phosphate overlap with each other, delete either one of the two and connect the C5 and phosphorus atoms with one oxygen atom.
11. Drag and drop the list of the RNA motif nucleotides in the "Hierarchy window" onto that of the sense strand "S." Then, drag and drop the list of the antisense strand "A" onto "S" to concatenate the list of the sequences.
12. Add two nucleotides ("AG" or "UU") at the 3′-end of the concatenated RNA sequence using the function "Create/Grow at 3′-"
13. Change the display style of the 22nd nucleotide from both ends from "Scaled Ball and Stick" to "CPK" to highlight the Dicer cleavage sites. Human Dicer recognizes the shRNA at the end of its stem and cleaves it at the 22nd nucleotide from the 3′- and/or 5′-end [6].
14. Choose the bound trigger protein, change the display style of the atoms from "None" to "CPK" and add a surface to the protein to visualize the occupied volume of the molecules.
15. Repeat **steps 4–14** to generate the bp series of the shRNA switch (U1A$_1$-, 21–31 bp; p50-, 22–28 bp).
16. To overlay the 22-bp long switch and 23-bp variant, tether the atoms of the base in the 3′ terminal nucleotide of the two. Then, overlay the two switch variants using the "Superimpose by Tether" function. Drag and drop the list of the 23-bp variant onto that of the 22-bp long switch. Then, hide the variant by unchecking the group box.
17. Repeat **step 16** to overlay 22 bp with the next bp variant to overlay all of the bp variants individually.

3.5 Prediction of Steric Hindrance Between Dicer and the Trigger Protein Bound on the shRNA Switch

The entire structure of human Dicer has not been solved. For our 3D model of Dicer, we used the RNase III catalytic domain and peripheral structure of *Giardia* Dicer (PDB ID: 2QVW, 326–385, and 401 C-terminal a.a.; *see* **Note 6**). In this section, we demonstrate how to display the catalytic domain of Dicer, highlight the catalytic sites and superimpose the RNP on Dicer.

3.5.1 Displaying and Modifying the Catalytic Domain and Peripheral Structure of Giardia Dicer

1. Download the PDB file of the crystal structure of *Giardia* Dicer (PDB ID: 2QVW). The crystal structure has four symmetric groups in the crystal lattice.
2. Exclude three groups and focus on one group.
3. Trim the a.a. sequence of the protein without 326–385 and 401 of the C-terminal a.a by unchecking the checked boxes of the nucleotides.
4. Change the display style of the Dicer protein and the four metal ions from "None" to "CPK," and highlight the metal ions and 9 a.a. (E336, E340, D404, D407, E649, D653, E684, D720, and E723) by changing the color of the atoms.

3.5.2 Superimpose the Protein-Responsive shRNA Switch and Bound Trigger Protein on Dicer

1. Copy and paste the 3D molecular model of the protein-responsive shRNA switch variants and bound trigger protein (designed in Subheading 3.4) on the modified Dicer structure displayed in the window.
2. Arrange the positions and orientations of the switch and Dicer three-dimensionally in reference to the two phosphates of the 22nd nucleotides, the highlighted metal ions and the 9 a.a. of Dicer.
3. Choose the modified Dicer and add surfaces to the proteins.
4. Display the modified Dicer and each variant of the shRNA switch with the trigger protein and observe the degree of steric hindrance between them. You can predict which trigger protein of the variants will most strongly collide against Dicer (*see* **Note 7**).

3.6 Inhibition Assay of Human Dicer Cleavage for shRNA In Vitro

3.6.1 In Vitro Transcription and Purification of shRNA Switches

1. Add 30 μL of T7 RNA polymerase to 670 μL of the transcription reaction mixture and mix gently by tapping. Do not spin the mixture down because that may cause precipitation. Incubate for 3–4 h at 37 °C.
2. For DNA template degradation, add 10 μL of TURBO DNase to the reaction and incubate for 20 min at 37 °C.
3. Equilibrate a PD-10 column with 50 mL of PD-10 buffer.
4. To remove any proteins from the mixture via phenol-chloroform extraction, add 700 μL (1× volume) of TE-saturated phenol to the reaction and thoroughly vortex. Then, centrifuge the mixture at >20,000 × *g* at 4 °C for 5 min. Transfer the upper layer to a new microtube.

5. Add 700 μL (1× volume) of chloroform to the tube and thoroughly vortex. Then, centrifuge the mixture at >20,000 × *g* at 4 °C for 5 min. Transfer the upper layer to a new microtube.
6. To remove nucleotide monomers, add the reaction to the previously equilibrated PD-10 column and wash the column with 3 mL of PD-10 buffer (for 55–80 nucleotides of RNA; the buffer volume can be optimized for each RNA size).
7. Elute the shRNA switches with 1 mL of PD-10 buffer.
8. Add 1 mL of (1× volume) ethanol (−20 °C) to the elution, mix thoroughly and incubate for 30 min at −20 °C.
9. Centrifuge the mixture at >20,000 × *g* at 4 °C for 20 min, remove the supernatant and allow the pellet to dry.
10. Pre-run a 15 % native gel for polyacrylamide gel electrophoresis (PAGE) in 0.5× TBE buffer for 20 min at room temperature.
11. Dissolve the pellet with 5× loading buffer and load the sample into the polyacrylamide gel. Perform electrophoresis at room temperature until the blue band of BPB travels to a distance of one-third of the glass plate from the bottom.
12. Place the gel on a silica plate and expose it to UV (265 nm) light to confirm the shRNA switch band, and then excise the band.
13. To elute the shRNA switch, add the excised gel piece and 500 μL of elution buffer (0.5 M NaCl, 0.1 % SDS, 1 mM EDTA) to a new tube and incubate at 37 °C overnight.
14. Filter the elution with a microfilter, add a 2.5 elution volume of ethanol (−20 °C) to the elution, mix thoroughly and incubate for 30 min at −20 °C.
15. Centrifuge the mixture at >20,000 × *g* at 4 °C for 20 min and remove the supernatant. Then, add a 2.5 elution volume of 80 % ethanol to rinse the pellet.
16. Centrifuge the mixture at >20,000 × *g* at 4 °C for 5 min and remove the supernatant. Allow the pellet to dry.
17. Dissolve the pellet with double-distilled water (D2W) and adjust the concentration to approximately 10 μM.

3.6.2 Expression and Purification of the Trigger RNA Binding Proteins Using FPLC

The use of an affinity column and FPLC, such as the AKTA (GE Healthcare) system, is recommended for purification of the trigger RNA binding proteins. For U1A protein purification, we modified a protein expression and purification protocol that has been reported previously [22].

1. Pre-culture competent cells transformed with the His-tagged trigger protein expression plasmid with 10 mL of LB (+antibiotic for selection) at 37 °C and 200 rpm (linear shake) overnight.

2. Transfer 7 mL of the pre-culture into 700 mL of LB (+antibiotic for selection) in flasks and culture at 37 °C and 180 rpm (rotary shake) until the OD_{660} reading of the culture reaches approximately 0.65 (requiring approximately 2.5 h).
3. Add 700 μL of 1 M IPTG (isopropylthio-β-galactoside) or the suitable ligand for induction of T7 RNA polymerase expression in the culture (f.c. 1 mM) and culture for another 4 h.
4. Split the culture in two and collect the cells with centrifugation at 4 °C and 4,170 × *g* for 10 min. The centrifuged pellets can be stored at −80 °C.
5. Suspend one of the pellets with a mixture of 5 mL of phosphate buffer A and 25 μL of phosphate buffer B.
6. Sonicate the suspension for 15 s (interval of 30 s) 12 times on ice.
7. Centrifuge the suspension at >20,000 × *g* at 4 °C for 30 min and collect the supernatant.
8. Filtrate the supernatant through a microfilter.
9. Perform FPLC purification with a prepacked column: Pomp wash a 5 column volume (cv) of 5 mM imidazole buffer (mixing phosphate buffer A:B = 199:1). Inject the supernatant from the sample loop. Wash a 10 cv of 20 mM imidazole buffer. Gradient elute a 10 cv of 20–200 mM imidazole buffer. Elute a 10 cv of 500 mM imidazole buffer.
10. Add 2 μL of 6× Sample buffer to 10 μL of flow-through and fractions whose absorbance at 280 nm are highly detectable by FPLC and incubate at 95 °C for 5 min. After SDS-PAGE, stain the gel with CBB and bleach with water. Check the band for the desired protein.
11. Equilibrate a dialysis cassette with 2× dialysis buffer and add the fraction containing the protein into the dialysis cassette using a 10-mL syringe.
12. Dialyze with 600 mL of 2× dialysis buffer for 2 h at 4 °C, change to fresh buffer and incubate for another 6 h.
13. Measure the absorbance of the protein at 280 nm and roughly estimate the concentration. If the solution is sufficiently concentrated (>40 μM), skip to **step 16**. If not, calculate how many folds the volume must be reduced by to provide sufficient concentration.
14. Equilibrate a sample concentrator with 2× dialysis buffer and add the dialyzed protein solution into the sample concentrator (up to 4 mL).
15. Concentrate the solution with centrifugation at 10,000 × *g* and 4 °C for 15 min. Check the volume and spin again until the volume in the concentrator is reduced to the set volume.
16. Add the protein solution into the same volume of glycerol and store at −30 °C.

3.6.3 The Binding Affinities of the shRNA Switches and the Purified Trigger Proteins

1. Mix 50 nM in vitro transcribed shRNA switch and 0 or 0.1–2 μM purified trigger proteins (the concentration can be adjusted and at least four concentration points are needed to analyze the affinity) in the binding buffer to form the RNP complex. The total volume of the mixture can be adjusted to the loading volume for PAGE (we use 30 μL).
2. Incubate on ice for 30 min.
3. Pre-run a 15 % native gel (use an approximately 10-cm size gel) in 0.5× TBE buffer for 10 min at 4 °C.
4. Add 5× loading buffer to the RNP solution and mix gently by tapping.
5. Load the samples into the 15 % native gel and run PAGE for 40 min at 400 V and 4 °C.
6. Stain the gel with the gel staining buffer for 15 min at room temperature.
7. Wash the gel with D2W a few times.
8. Image the gel with a fluorescent image analyzer to detect the stained RNA.
9. Confirm that the shifted band can be observed from the cognate RNP binding pair but not from the sample that contains the defective RNA motif.
10. Determine the protein concentration using the Dicer cleavage assay in Subheading 3.6.5 to a twofold higher concentration than the concentration that totally shifted the band.

3.6.4 The Efficiency of In Vitro Human Dicer Cleavage for the shRNA Switches

1. Mix 0.5 μL of in vitro-transcribed shRNA switch with the Dicer reaction mixture and water (up to 10 μL). Do not chill the mixture during assembly because this may cause precipitation.
2. Incubate the mixture at 37 °C for 15 h and stop the incubation by adding 1 μL of Dicer Stop Solution to the reaction.
3. Conduct the same procedure detailed in **steps 3–8** in Subheading 3.6.3 on the incubated reaction.
4. Quantify the total fluorescent intensity of the band of cleaved dsRNA and confirm that the cleavage efficiency by human Dicer is the same among the base-pair variants and their negative controls (defective RNA motif) for the shRNA switches.

3.6.5 Inhibition Assay of Human Dicer Cleavage for shRNA In Vitro

1. Mix 0.5 μM in vitro-transcribed shRNA switch with the corresponding trigger protein (The concentration of the protein was determined at **step 10** in Subheading 3.6.3 to form the RNP complex). The total volume of the mixture is up to 3.5 μL.
2. (Optional) Incubate at room temperature for approximately 30 min. The incubation time will depend on the affinity between the RNP complexes.

3. Mix the RNP complex with 1 μL of 10 mM ATP, 0.5 μL of 50 mM $MgCl_2$, 4 μL of Dicer Reaction Buffer, 1 μL of 0.5 unit/μL Recombinant Human Dicer Enzyme, and water (up to 10 μL).
4. Incubate the mixture at 37 °C for 15 h and stop the incubation by adding 1 μL of Dicer Stop Solution to the reaction.
5. Conduct the same procedure as detailed in **steps 3–8** in Subheading 3.6.3 on the incubated reaction.
6. Quantify the total fluorescent intensity of the band of cleaved dsRNA (Fig. 3, Step 2).
7. Calculate the relative inhibition efficiency values of Dicer cleavage using the ratio of the intensity in the presence of the trigger protein to that in the absence of the trigger protein.

3.7 EGFP Expression Control of the Designed shRNA Switches in Cultured Mammalian Cells

The ShRNA cloning Kit can be used to construct the shRNA-switch expressing plasmids. The plasmids should be constructed using the kit, and required DNA sequences should be synthesized in accordance with the kit manual. The annealing steps in the kit manual are modified as bellow:

3.7.1 Construction of shRNA Expressing Plasmids

1. Mix 200 μM of ShRNA top strand and 200 μM of ShRNA bottom strand in Oligo Annealing Buffer of the cloning kit.
2. Incubate the mixture using a thermal cycler: 95 °C for 4 min and incubate at 80 °C for 1 min. Decreased 0.1 °C/s to the melting temperature of the duplex of the top strand and bottom strand. And incubate the mixture at the melting temperature for 5 min, followed by room temperature incubation for 10 min.

Prepare the negative control plasmid that contains a defective RNA motif using the same procedure as indicated.

3.7.2 Trigger Protein Expression Plasmid

The trigger protein expression plasmid is prepared as follows.

1. Design the set of PCR primers for amplifying the region of the gene that codes the trigger protein, whose amino acid (a.a.) sequence corresponds with that of the protein in the RNP complex derived from the PDB. Add each sequence of PCR primer to the cutting site of a selected different restriction enzyme.
2. Amplify the gene by PCR from the total cDNA or plasmid using the designed primers.
3. Clone the gene into destination plasmids between the two different restriction enzyme sites to construct the trigger protein expression plasmid.
4. Verify the sequence.
5. Prepare the negative control plasmid that contains a noncognate protein sequence using the same procedure as indicated.

3.7.3 EGFP Expression Control of the Designed shRNA Switches in Cultured Mammalian Cells

1. Seed 293FT cells (1.5×10^5 per well) or other chosen cultured cells on a 24-well plate and incubate at 37 °C in a CO_2 incubator for 1 day.
2. Co-transfect cells with the following three types of plasmids using 2 μL of the Lipofection reagent: 0.3 μg of the bp variant of protein-shRNA switches expressing plasmids or that of the binding-defective control, 0.3 μg of the trigger protein expression plasmid or that of the binding-defective control and 0.3 μg of the EGFP expression plasmid.
3. Twenty-four hours after transfection, observe and image the transfected cells using a fluorescent microscope (Fig. 3, Step 3).
4. Collect the medium of the 24-well plate and transfer to new tubes.
5. Add 200 μL of 0.25 % trypsin-EDTA (1×) phenol red to each well and incubate at 37 °C in a CO_2 incubator for 2–5 min.
6. Add the collected medium to each well, resuspend in the medium and filter the cell suspension with a 12×75 mm tube and a cell strainer cap.
7. Quantify the mean intensity of the EGFP fluorescence of 30,000 cells per sample using a flow cytometer.
8. To exclude abnormal cells, select the cells whose FCS and SSC signal scopes are identical to those of untreated cells to calculate the mean intensities.

4 Notes

1. The MEGAshortscript™ T7 Kit can be used for the transcription reaction by adding GMP and adjusting the ratio of NTP concentrations as described below:
   (a) Add 4 μL of D2W, 0.5 μL of 100 μM single-stranded template DNA, 0.5 μL of 100 μM T7 annealing primer, 2 μL of T7 10× reaction buffer, 1 μL of 75 mM GTP solution, 4 μL of 75 mM GMP solution, 2 μL of 75 mM ATP solution, 2 μL of 75 mM CTP solution, and 2 μL of 75 mM UTP solution to the tube and mix thoroughly with vortexing.
   (b) Add 2 μL of T7 RNA polymerase and mix gently by tapping. Do not spin the mixture down because this may cause precipitation. Set the total reaction in a 20-μL volume.
   (c) Follow **steps 3–17** in Subheading 3.6.1.
2. The GTP at the 5′-end of RNA causes interferon activation, followed by nonspecific translational repression [23, 24]. To avoid the interferon activation and increase the efficiency of Dicer processing [6], we introduced GMP at the 5′-end

of the RNA by adjusting the concentration ratio of GMP–GTP from 4:1 to 10:1 during the transcription reaction.

3. We confirmed that the shRNA sequences that contain shorter dsRNA regions (<22 bp) are not competent for Dicer processing. We also observed that the longer dsRNA regions (>27 bp) of the shRNA might cause lower transcription efficiency from the shRNA-coding plasmids. To avoid the possible interferon response of the longer dsRNA in the cellular assay, we think that a range of 22–27-bp dsRNA is likely sufficient to generate optimal shRNA switches. Changing the length of the dsRNA region from 22 to 27 bp can rotate the RNA loop by approximately 150°, so the steric hindrance between Dicer and the RNP complex can be optimized within this range in most cases [3].

4. The PDB files from the NMR analysis contain the RNP structure ensemble. Use the representative structure, which is displayed as the first alternative [25].

5. If the PDB does not offer the RNP structure of human, other organisms may be used after verifying the a.a. sequence differences between the proteins of the two organisms (e.g., mouse and human) and confirming that the differences are not critical for the structure and binding affinity. We verified the a.a. sequence differences between mouse and human p50 and confirmed that the six point mutations between these two sequences were either on the terminal region or on the loop far from the RNA-binding domain [3].

6. We used the RNase III catalytic domain and peripheral structure of *Giardia* Dicer because structural studies have indicated that the crystal structure of *Giardia* Dicer is geometrically similar to the nuclease core of human Dicer [26]. Two metal ions are coordinated by 9 a.a. (E336, E340, D404, D407, E649, D653, E684, D720, and E723) in each of the catalytic sites of Dicer [27, 28]. The phosphodiester bonds of the 22nd nucleotides from both the 5′-/3′-ends of the dsRNA are cleaved between the two metal ions. We trimmed the Dicer structure of 2QVW to use the 326–385 and 401 C-terminal a.a. not only to use the domains whose structures are similar to those of human but also to highlight the region at which the steric hindrance between Dicer and the trigger protein bound on the shRNA could occur.

7. Steric hindrance may occur not only between Dicer and the trigger protein but also on the protein-stabilized RNA structural motif because a particular RNA–protein interaction is known to stabilize the RNA structural motif containing the protein-binding site. We have previously shown that both an RNA-binding protein and the protein-stabilized RNA structural motif that has the predicted steric hindrance with Dicer can inhibit RNAi [3].

References

1. Endy D (2005) Foundations for engineering biology. Nature 438:449–453
2. Kashida S, Inoue T, Saito H (2012) Three-dimensionally designed protein-responsive RNA devices for cell signaling regulation. Nucleic Acids Res 40:9369–9378
3. Saito H, Fujita Y, Kashida S et al (2011) Synthetic human cell fate regulation by protein-driven RNA switches. Nat Commun 2:160
4. Xie Z, Wroblewska L, Prochazka L et al (2011) Multi-input RNAi-based logic circuit for identification of specific cancer cells. Science 333: 1307–1311
5. Culler SJ, Hoff KG, Smolke CD (2010) Reprogramming cellular behavior with RNA controllers responsive to endogenous proteins. Science 330:1251–1255
6. Park J-E, Heo I, Tian Y et al (2011) Dicer recognizes the 5′ end of RNA for efficient and accurate processing. Nature 475:201–205
7. Vermeulen A, Behlen L, Reynolds A et al (2005) The contributions of dsRNA structure to Dicer specificity and efficiency. RNA 11: 674–682
8. Tsutsumi A, Kawamata T, Izumi N et al (2011) Recognition of the pre-miRNA structure by Drosophila Dicer-1. Nat Struct Mol Biol 18:1153–1158
9. An C-I, Trinh VB, Yokobayashi Y (2006) Artificial control of gene expression in mammalian cells by modulating RNA interference through aptamer–small molecule interaction. RNA 12:710–716
10. Beisel CL, Bayer TS, Hoff KG et al (2008) Model-guided design of ligand-regulated RNAi for programmable control of gene expression. Mol Syst Biol 4:224
11. Rybak A, Fuchs H, Smirnova L et al (2008) A feedback loop comprising lin-28 and let-7 controls pre-let-7 maturation during neural stem-cell commitment. Nat Cell Biol 10:987–993
12. Nam Y, Chen C, Gregory RI et al (2011) Molecular basis for interaction of let-7 microRNAs with Lin28. Cell 147:1080–1091
13. Beisel CL, Smolke CD (2009) Design principles for riboswitch function. PLoS Comput Biol 5:e1000363
14. Salis HM, Mirsky EA, Voigt CA (2009) Automated design of synthetic ribosome binding sites to control protein expression. Nat Biotechnol 27:946–950
15. Ogawa A (2011) Rational design of artificial riboswitches based on ligand-dependent modulation of internal ribosome entry in wheat germ extract and their applications as label-free biosensors. RNA 17:478–488
16. Sinha J, Reyes SJ, Gallivan JP (2010) Reprogramming bacteria to seek and destroy an herbicide. Nat Chem Biol 6:464–470
17. Muranaka N, Sharma V, Nomura Y et al (2009) An efficient platform for genetic selection and screening of gene switches in Escherichia coli. Nucleic Acids Res 37:e39
18. Liang JC, Bloom RJ, Smolke CD (2011) Engineering biological systems with synthetic RNA molecules. Mol Cell 43:915–926
19. Oubridge C, Ito N, Evans PR et al (1994) Crystal structure at 1.92 A resolution of the RNA-binding domain of the U1A spliceosomal protein complexed with an RNA hairpin. Nature 372:432–438
20. Galardi S, Mercatelli N, Farace MG et al (2011) NF-kB and c-Jun induce the expression of the oncogenic miR-221 and miR-222 in prostate carcinoma and glioblastoma cells. Nucleic Acids Res 39:3892–3902
21. Ben-Neriah Y, Karin M (2011) Inflammation meets cancer, with NF-κB as the matchmaker. Nat Immunol 12:715–723
22. Endoh T, Sisido M, Ohtsuki T (2008) Cellular siRNA delivery mediated by a cell-permeant RNA-binding protein and photoinduced RNA interference. Bioconjug Chem 19:1017–1024
23. Poeck H, Besch R, Maihoefer C et al (2008) 5′-Triphosphate-siRNA: turning gene silencing and Rig-I activation against melanoma. Nat Med 14:1256–1263
24. Kim D-H, Longo M, Han Y et al (2004) Interferon induction by siRNAs and ssRNAs synthesized by phage polymerase. Nat Biotechnol 22:321–325
25. Sutcliffe MJ (1993) Representing an ensemble of NMR-derived protein structures by a single structure. Protein Sci 2:936–944
26. Takeshita D, Zenno S, Lee WC et al (2007) Homodimeric structure and double-stranded RNA cleavage activity of the C-terminal RNase III domain of human dicer. J Mol Biol 374: 106–120
27. Macrae IJ, Li F, Zhou K et al (2006) Structure of Dicer and mechanistic implications for RNAi. Cold Spring Harb Symp Quant Biol 71: 73–80
28. Macrae IJ, Zhou K, Li F et al (2006) Structural basis for double-stranded RNA processing by Dicer. Science 311:195–198

INDEX

Atsushi Ogawa (ed.), *Artificial Riboswitches: Methods and Protocols*, Methods in Molecular Biology, vol. 1111, DOI 10.1007/978-1-62703-755-6, © Springer Science+Business Media New York 2014

MIX
Papier aus verantwortungsvollen Quellen
Paper from responsible sources
FSC® C105338

If you have any concerns about our products,
you can contact us on
ProductSafety@springernature.com

In case Publisher is established outside the EU,
the EU authorized representative is:
Springer Nature Customer Service Center GmbH
Europaplatz 3, 69115 Heidelberg, Germany

Printed by Libri Plureos GmbH
in Hamburg, Germany